Post-Disaster Reconstruction of the Built Environment

Rebuilding for Resilience

Edited by

Dilanthi Amaratunga
Professor of Disaster Management at the Centre for Disaster Resilience
School of the Built Environment
University of Salford, UK

Richard Haigh
Senior Lecturer at the Centre for Disaster Resilience
School of the Built Environment
University of Salford, UK

WILEY-BLACKWELL

A John Wiley & Sons, Ltd., Publication

Library of Congress Cataloging-in-Publication Data

Post-disaster reconstruction of the built environment : rebuilding for resilience / edited by Dilanthi Amaratunga and Richard Haigh.
      p.   cm.
   Includes bibliographical references and index.
    ISBN 978-1-4443-3356-5 (hardcover : alk. paper)   1. Buildings–Repair and reconstruction–Standards.   2. Disasters–Social aspects.   3. Architectural design–Technological innovations.   4. Architecture and society.   I. Amaratunga, Dilanthi. II. Haigh, Richard.
    TH3401.P75 2011
    690′.24–dc22   2011002981

A catalogue record for this book is available from the British Library.

This book is published in the following electronic formats: ePDF 9781444344912; Wiley Online Library 9781444344943; ePub 9781444344929; Mobi 9781444344936

Set in 10/12.5pt Minion by Aptara® Inc., New Delhi, India
Printed and bound in Malaysia by Vivar Printing Sdn Bhd

1   2011

This book is dedicated to disaster-affected communities
around the world

Contents

About the Editors

Professor Dilanthi Amaratunga (r.d.g.amaratunga@salford.ac.uk) is the Professor of Disaster Management at the University of Salford, UK where she leads the University's Centre for Disaster Resilience, responsible for supporting research on disaster management portfolios. She is also the Associate Head of International Development for the School of the Built Environment at the University of Salford. Her research interests include: post-disaster reconstruction including conflict mitigation, gender and projection; capability and capacity building in managing disasters; socio-economic measures for conflict-affected re-construction and women in construction. She is the Co-Editor of the *International Journal of Disaster Resilience in the Built Environment*. She has presented widely at international conferences, has led international disaster management workshops and seminars and is working actively with the United Nations. As a member of the Royal Institution of Chartered Surveyors (RICS), she leads several of their disaster management initiatives. She has supervised and supported a wide range of postgraduate research students and to date she has produced over 200 publications, refereed papers and reports, and has made a large number of presentations in over 25 countries. Dilanthi co-chairs the International Conference on Building Resilience: Interdisciplinary approaches to disaster risk reduction, and the development of sustainable communities and cities' to be held in July 2011 (www.buildresilience.org) which encourages debate on individual, institutional and societal coping strategies to address the challenges associated with disaster risk. It will also explore inter-disciplinary strategies that develop the capacity of a system, community or society potentially exposed to disaster-related hazards, to adapt, by resisting or changing, in order to reach and maintain an acceptable level of functioning and structure. Dilanthi's profile can be found at www.dilanthiamaratunga.net

Dr Richard Haigh (r.p.haigh@salford.ac.uk) is a Senior Lecturer at the Centre for Disaster Resilience, which is based in the School of the Built Environment, University of Salford, UK. He is Programme Director of the School's Disaster Mitigation and Reconstruction Masters programme and also Joint Editor of the *International Journal of Disaster Resilience in the Built Environment*.

Richard undertakes research related to the application of disaster risk reduction in the built environment. His main research interests include the reintegration and rehabilitation of conflict-affected communities in Sri Lanka, and corporate social responsibility in disaster risk reduction. He is working closely with local government and major stakeholders to reduce the level of disaster risk in the Sri Lankan District of Batticaloa as part of the UNISDRs Resilient Cities campaign.

He was previously Joint Coordinator of CIBs Task Group 63 on Disasters and the Built Environment (2006–2010) and Co-Chair of the 2008 Building Education and Research Conference, held in Kandalama, Sri Lanka. The conference focused on the built environment field's role in developing a society's resilience to disasters.

A detailed list of Richard's publications and activities can be found at www.richardhaigh.info.

List of Contributors

Yan Chang is an active researcher on project management requirements for post-disaster reconstruction. She is currently studying for a PhD with the Department of Civil and Environmental Engineering, The University of Auckland, and is also part of the national research team 'Resilient Organisations' where she is working on resourcing for post-disaster reconstruction. Yan holds a BSc degree with First Class Honours in Civil Engineering and a Masters degree in Management Science and Engineering from Central South University in China. Apart from the 2008 Wenchuan earthquake in China, she has been involved in the reconstruction studies of the 2004 Indian Ocean tsunami in Indonesia and the 2009 Victorian 'Black Saturday' bushfires in Australia. Her current research interests are cross-culture disaster reconstruction studies, longitudinal post-disaster recovery, reconstruction housing project management, sustainability planning for disaster reduction, and resource availability in post-disaster situations.

Kanchana Ginige is a full time doctoral researcher at the School of the Built Environment, University of Salford, UK. Her doctoral research focuses on mainstreaming women into disaster risk reduction in the built environment. She earned her BSc degree with First Class Honours in Quantity Surveying from the University of Moratuwa, Sri Lanka in 2005. Kanchana worked as a research assistant in the Department of Building Economics, University of Moratuwa, Sri Lanka for about one year. She then worked as a researcher attached to a project titled 'Constructing Women Leaders' at the Research Institute for the Built and Human Environment, School of the Built Environment, University of Salford, UK in 2007. Currently, she contributes to a CIB/UN-HABITAT initiative on 'Capacity Development for Disaster Mitigation and Reconstruction in the Built Environment' at the School of the Built Environment, University of Salford as the project research assistant.

Gayani Karunasena is a senior lecturer in Building Economics, attached to the Department of Building Economics, University of Moratuwa, Sri Lanka. She obtained her BSc(Hons) degree in Quantity Surveying and MPhil in Construction Information Technology from the same university. Currently, she is reading for her split site PhD in the area of disaster waste management at the University of Salford, UK. She has won awards for outstanding research performance by academic staff for 2008 and 2009, awarded by the University of Moratuwa, Sri Lanka. Her current research interests are on disaster management, construction information technology and value management.

Kaushal Keraminiyage is a lecturer at the University of Salford, UK teaching on both undergraduate and postgraduate courses. He completed his PhD on construction process improvements in 2009. Kaushal is also the co-programme director for the BSc (Quantity Surveying) programme. He is an active member of the research centre for disaster resilience at the University of Salford. His research interests are decision-making in post-disaster reconstruction, ICT for the Built Environment in disaster management contexts, collaborative environments for construction education and research, building capacities of construction in Higher Education Institutions through ICT-enabled collaborations, energy conscious construction through process/ICT co-maturation and virtual learning and research environments for the built environment. Kaushal's publication profile includes edited books, book chapters, journal papers, various reports and international conference papers and presentations. He served as an editor for the proceedings of the CIB International World Congress 2010. He is also a member of the Editorial Board of the *International Journal of Disaster Resilience in the Built Environment* and he has facilitated a number of international research workshops and served as an organising committee member for a number of international conferences.

Udayangani Kulatunga has over 6 years' experience in teaching and research in the UK and Sri Lanka. She is a member of the Centre for Disaster Resilience at the University of Salford. She completed her PhD in performance measurement in construction research and development in 2008 at the University of Salford. She is currently attached to the same university as a lecturer in Quantity Surveying and teaches on both undergraduate and postgraduate courses. Udayangani leads the research themes 'disaster risk reduction' and 'culture and disaster risk reduction' at the Centre for Disaster Resilience at the University of Salford. In 2010, she won an INSPIRE exploratory grant funded by the British Council to explore disaster risk reduction activities in Bangladesh. Her research capabilities have also been rewarded with the New Researcher Scheme funded by the Faculty of Business, Law and the Built Environment, University of Salford to explore the cultural impact on disaster risk reduction. Udayangani was the guest editor for a special issue of *Facilities* journal on performance measurement and management and for a special issue of the *International Journal of Strategic Property Management* on disaster management. Her research output is demonstrated by the number of publications in both journals and international conferences. More details on her experiences can be found at http://www.seek.salford.ac.uk/profiles/U.KULATUNGA.jsp

Dean Myburgh has held senior leadership roles in public and private sector organisations, both in New Zealand and abroad. He is currently a director of two consultancies focused on the facilitation of strategic and operational decision-making related to risk, business continuity and emergency management, organisational change management and process improvement. From 1996 to 2005 he was a member of the Executive Steering Group of the

Auckland Civil Defence Emergency Management Group. Dean is an Industry Researcher on the 'Resilient Organisations' research programme (publications at http://www.resorgs.org.nz/pubs.shtml). His main interest is in the leadership and decision-making aspects related to recovery and resilience and he has recently addressed whole-brained thinking approaches to emergency management (in association with Auckland University of Technology).

Taufika Ophiyandri is a lecturer at the Department of Civil Engineering, University of Andalas, Padang, Indonesia. He completed his BSc in Civil Engineering from the same university and was awarded an MSc in Construction Management from the University of Birmingham, UK. He has been working for the University of Andalas since 1998 and taught engineering economics and construction management. Currently he is a full time doctoral student at the School of the Built Environment, University of Salford, UK. His research interest is on risk management for community-based post-disaster housing reconstruction and is fully funded by the Ministry of National Education, Republic of Indonesia.

Roshani Palliyaguru successfully completed her PhD in Infrastructure Development and Environmental Concerns in Disaster Management in November 2010 at the School of the Built Environment, University of Salford, UK. Having completed her PhD, she is involved in lecturing at the School of the Built Environment, University of Salford and works as a research assistant at the Centre for Disaster Resilience in the same university on the research initiative called CEREBELLA (Community Engagement for Risk Erosion in Bangladesh to Enhance LifeLong Advantage). In 2005, Roshani was recruited at the Department of Building Economics, University of Moratuwa on completion of her BSc (Hons) in Quantity Surveying degree from the same department. In 2006, Roshani was awarded the Graduate Teaching Assistantship at the University of Salford. Since then she has made contributions to the academic discipline in the domain of construction management, including lecturing undergraduates, conducting and publishing research work. From 2006, Roshani has been lecturing on construction economics, construction measurement and economics & management at the University of Salford. Roshani is a researcher in disaster management discipline whereas her research interests are ranged in a widespread area including disaster risk reduction, post-disaster reconstruction, vulnerability reduction for natural hazards, and socio-economic development in post-disaster contexts. With over 5 years of research and teaching experience, Roshani has published over 19 research papers, of which five are in international journals.

Chaminda Pathirage completed his PhD in Knowledge Management from the University of Salford, UK. He is currently working as a lecturer and a programme director at the School of the Built Environment, University of Salford, delivering lectures on both undergraduate and postgraduate courses in the fields of construction management and financial management. Chaminda has worked on several RICS (Royal Institution of Chartered Surveyors)-funded research projects on knowledge management in disaster management and facilities management, and has developed his specific research interest in exploring the role of knowledge management in the disaster management cycle. He is also an Editorial Advisory Board member of the *International Journal of Disaster Resilience in the Built Environment* and an Editorial Review Board Member of *International Journal of Knowledge-Based Organizations*. With over 8 years of research and teaching experience, Chaminda has published extensively in both journals and international conferences, and he is leading the research theme 'Knowledge Management for Disaster Resilience' within the Research Centre for Disaster Resilience, University of Salford.

Regan Potangaroa has been an Associate Professor at the Department of Architecture, UNITEC, New Zealand for the last 7 years. However, his professional background is as a structural engineer with over 25 years experience in 17 different countries. In the last 13 years, he has completed over 50 humanitarian assignments and consequently has seen the 'good, the bad and the ugly' of post-disaster reconstruction. He has worked through the concepts of 'durable solutions' and 'aid dependency', then 'participatory design' and more recently 'build back better' and 'up scaling'. The thread through these has been 'resilience' which he believes holds promise for more effective and efficient assistance for those in disasters.

James Olabode Rotimi's PhD thesis examined the improvements that could be made to disaster-related legislation so that it can facilitate post-disaster reconstruction. His study makes particular reference to the Civil Defence Emergency Management, Resource Management and Building Acts in New Zealand. He has written several papers around this subject area to show how subsisting legislation could become a source of vulnerability after significant disasters. His view is that proper consideration should be given to the peculiarities of post-disaster reconstruction within building and environmental development legislation. A 'business as usual' approach should not be expected to be operable during significant reconstruction programmes. James has over 15 years of teaching and research experience in universities in Nigeria and New Zealand. His background is in construction management and he has industry experience including an associate role in a quantity surveying consultancy before becoming an academic. James holds professional membership of the Nigerian Institute of Building, Chartered Institute of Building and the New Zealand Institute of Building. He currently lectures at the Auckland University of Technology, New Zealand.

Krisanthi Seneviratne is a postgraduate research student in the School of the Built Environment, University of Salford, UK. She is currently reading for her PhD in post-conflict reconstruction, focusing especially on the area of housing.

Krisanthi is working on the ISLAND-II (Inspiring Sri-Lankan reNewal and Development-Phase II) research project which is jointly sponsored by the Royal Institution of Chartered Surveyors (RICS) and the School of the Built Environment. This research aims at increasing the effectiveness of disaster management by facilitating the sharing of appropriate knowledge and good practices relating to the key phases of knowledge capturing within the disaster management cycle. Following completion of her first degree at the University of Moratuwa, Sri Lanka, she worked as an assistant quantity surveyor in UAE, Dubai. Krisanthi joined the Department of Building Economics, University of Moratuwa, Sri Lanka in 2006 as a research assistant and was promoted to lecturer. She has won an Overseas Research Studentships Awards Scheme (ORSAS) for her outstanding academic ability and research potential.

Erica Seville leads the 'Resilient Organisations' research programme (www.resorgs.org.nz) which involves 17 researchers from Canterbury and Auckland Universities as well as key industry players, working to making organisations more resilient in the face of major hazards in the natural, built and economic environments. In addition to leading Resilient Organisations, Erica is also a director of Risk Strategies Research and Consulting where she works with clients from a variety of sectors including major health providers, the mining industry, construction contractors, and critical infrastructure providers to offer strategic risk management advice and proactively build their resilience capabilities. Erica has a Bachelor of Civil Engineering degree and a PhD in risk assessment.

Nuwani Siriwardena is a PhD researcher attached to the School of the Built Environment, University of Salford, UK. She obtained her first degree and Masters degree from the University of Colombo, Sri Lanka. Nuwani has been working at the Department of Commerce, University of Colombo since 2001. She has published and presented several papers within and outside the UK. Her current focus is on stakeholder expectations of post-disaster housing reconstruction in Sri Lanka. Corporate social responsibility, corporate governance, stakeholder theories and budgetary controls are other inspirational areas of research.

Richard Sutton is a Project Manager with a major non-governmental organisation involved in humanitarian and reconstruction work across the globe. Richard graduated with a BSc (Hons) in Construction Management and initially worked in the UK construction industry for a major management contractor before moving to the humanitarian sector. His research interests include the development of socially responsible ways to engage private sector construction firms in post-disaster environments.

Nirooja Thurairajah is a full time PhD researcher at the School of the Built Environment, University of Salford. Her PhD research is on women's empowerment in post-disaster reconstruction. Nirooja holds a BSc degree with First Class Honours in Quantity Surveying from the University of Moratuwa, Sri Lanka. Currently she is working as a researcher for a community engagement project 'Resilient Homes'. She has also worked as a researcher for an ESF-funded project 'Constructing women leaders'. Her research interests are on women's issues within the built environment, organisational policies related to gender equality, community engagement, empowerment of women and post-disaster reconstruction.

Suzanne Wilkinson is an Associate Professor in the Department of Civil and Environmental Engineering at The University of Auckland, New Zealand. Suzanne is part of the national research team 'Resilient Organisations' where she leads a team of researchers working in the area of post-disaster recovery and reconstruction. Recently the team have been examining the post-disaster reconstruction of Samoa following the tsunami which destroyed parts of the coastline and coastline villages and Australia following the Australian bush fires north of Melbourne, which destroyed whole communities. Suzanne also teaches construction management, construction administration and construction law and is leader of the construction management research team.

Foreword

Disasters, both natural and man-made, seem to beset communities across the world with increasingly regularity. It may be, of course, that in such a connected world we are simply more immediately aware of these horrendous events, but this very global access to information presents the opportunity to endeavour to accelerate learning about how best we can predict, mitigate and recover from disasters. This is the context for this book, which puts the spotlight on the pervasive contribution that the construction industry could and should make on a very broad front.

Thus, Professor Amaratunga and Dr Haigh, their dedicated and passionate team of researchers and several other invited authors have taken a whole life cycle approach to the subject, but with a strong emphasis on how to break out of strong negative spirals, by focusing on sustainable human and capital development. Strategic perspectives are connected to practical experiences drawn from Sri Lanka, China, Indonesia and New Zealand. Chapters in the book cover a range of economic, social and technical issues.

This extensive treatment of the subject should meet two connected aims. First it will provide a benchmark of the various ways in which construction can make a positive contribution. Secondly, it is to be hoped that it will raise awareness and motivation within the built environment community to actively grasp these opportunities and to promote and deliver rational, balanced responses to these typically unbounded threats to humankind.

As mankind pushes the boundaries of the Earth's carrying capacity to the limit and injects social disasters, such as wars and famine, into the mix, it seems clear that the issues are both technical and social in character. Thus, I am confident that this initiative will link to a positive stream of activities engaging the many stakeholders necessary to build an effective response to disasters of whatever form.

Professor Peter Barrett
University of Salford, UK

Acknowledgements

The Editors of the book express their gratitude to all the contributing Authors for sharing their knowledge pertaining to various aspects of post-disaster reconstruction, in particular, towards rebuilding for resilience. Without this input detailing various examples in the international context, this book wouldn't have been produced.

Further, the Editors and the Publisher acknowledge the help of those who granted their permission to reproduce the material (as indicated within individual chapters) for the benefit of the readers. Authors have taken every step to secure prior permission in this process, but if there are any omissions and errors, we as Editors apologise on behalf of all Authors and will take steps to rectify any mistakes.

Professor Dilanthi Amaratunga
Dr Richard Haigh

1 Introduction

Richard Haigh and Dilanthi Amaratunga

With growing population and infrastructures, the world's exposure to hazards – of both natural and man-made origin – is predictably increasing. This unfortunate reality will inevitably require frequent reconstruction of communities, both physically and socially. At the same time, it will be vital that any attempt to reconstruct after a disaster actively considers how to protect people and their environment to ensure those communities are less vulnerable in the future.

For the remainder of this book and in common with the Centre for Research on the Epidemiology of Disasters (CRED), which maintains the International Disasters Database (EM-DAT), a disaster is a 'situation or event, which overwhelms local capacity, necessitating a request to national or international level for external assistance; an unforeseen and often sudden event that causes great damage, destruction and human suffering'. For a disaster to be entered into the database at least one of the following criteria must be fulfilled: 10 or more people reported killed; 100 people reported affected; there is declaration of a state of emergency; or, a call for international assistance.

There are wide-ranging origins and causes to the many disasters that have affected communities across the world with ever greater frequency. The term disaster is frequently associated with geo- and hydro-meteorological hazards, such as hurricanes, earthquakes and flooding. Three main categories of natural disasters account for 90% of the world's direct losses: floods, earthquakes and tropical cyclones (Swiss Reinsurance Company, 2010).

The degree to which such disasters can be considered 'natural' has long been challenged. In their seminal paper entitled 'Taking the "naturalness" out of natural disasters', O'Keefe *et al.* (1976) identified the cause of the observed increase in disasters as, 'the growing vulnerability of the population to extreme physical events', not as changes in nature. However, as Kelman (2009) observes, even as early as 1756, Rousseau, in a letter to Voltaire about the earthquake and tsunami that hit Portugal a year earlier, commented that nature did not build the houses which collapsed, and suggested that Lisbon's high population density contributed to the toll.

Post-Disaster Reconstruction of the Built Environment: Rebuilding for Resilience, First Edition.
Edited by Dilanthi Amaratunga and Richard Haigh.
© 2011 Blackwell Publishing Ltd. Published 2011 by Blackwell Publishing Ltd.

More recently, the links between disasters and climate change have increasingly been recognised. There are growing concerns over the threats posed by climatological hazards such as extreme temperatures, drought and wild fires, and the multi-faceted threats associated with sea level change. The scale of human contribution to climate change may still be open to debate, but there is widespread, although many would argue, insufficient concern from politicians, commentators, researchers and the public alike, over its ability to increase the number and scale of hazards, and the potential for resultant impact on communities world-wide. The World Meteorological Organisation (WMO) figures showed that 2008 was the 10th warmest year since reliable records began, meaning that the 10 warmest years on record all occurred in the past 12 years.

Alongside disasters of so called 'natural origin', many other disasters to affect populations in recent times are unquestionably of human origin. Conflict sometimes results in wars and terrorist acts that match or exceed the losses from any 'natural' disaster. Other types of disaster, often referred to as 'technical', result from equipment malfunction or human error. Although less frequent they still have the potential to cause widespread damage to people and property.

Regardless of the origins and causes, as previously noted by the authors (Haigh and Amaratunga, 2010), the consequences to human society are frequently similar: extensive loss of life, particularly among vulnerable members of a community; economic losses, hindering development goals; destruction of the built and natural environment, further increasing vulnerability; and, widespread disruption to local institutions and livelihoods, disempowering the local community.

1.1   A global challenge

In 2008, more than 220 000 people died in events like cyclones, earthquakes and flooding, the most since 2004, the year of the Asian tsunami (Swiss Reinsurance Company, 2010). Meanwhile, overall global losses totalled about US$200 billion, with uninsured losses totalling US$45 billion, about 50% more than in 2007. This makes 2008 the third most expensive year on record, after 1995 when the Kobe earthquake struck Japan, and 2005, the year of Hurricane Katrina in the US. The frequency, scale and distribution of disasters in recent years is further evidence, if any is needed, that hazards – of both natural and man-made origins – are a global problem, threatening to disrupt communities in developed, newly industrialised and developing countries. The developed world cannot afford to be complacent.

But recent disasters also highlight that developing and newly industrialised countries are most at risk: the losses to life and the economy – as a percentage of gross domestic product (GDP) – are far greater. During the last decade of the 20th century, direct losses from natural disasters in the developing world averaged US$35 billion annually (Swiss Reinsurance Company, 2000).

Although a disturbingly high figure by itself, perhaps more worryingly, these losses are more than eight times greater than the losses suffered over the decade of the 1960s.

In part, this high risk felt by developing and newly industrialised countries can be attributed to hazard frequency, severity and exposure. The three main categories of natural disasters that account for the greatest direct losses – as identified earlier, these are floods, earthquakes and tropical cyclones – periodically revisit the same geographic zones. Earthquake risk lies along well-defined seismic zones that incorporate a large number of developing countries. High risk areas include the West Coast of North, Central and South America, Turkey, Pakistan, Afghanistan, India, China and Indonesia. Similarly, the pattern of hurricanes in the Caribbean and typhoons in South Asia, Southeast Asia and the South Pacific is well established. These typically affect Algeria, Egypt, Mozambique, China, India, Bangladesh, Taiwan, Indonesia, Philippines, Korea, Afghanistan, Armenia, Georgia, Iran, Mongolia, Thailand, Argentina, Brazil, Chile, Colombia, Cuba, Ecuador, El Salvador, Guatemala, Honduras, Mexico, Nicaragua and Venezuela. These examples illustrate that to a significant degree, developing countries are unfortunate in being located in regions that are particularly prone to natural hazards. Of course, this correlation is not entirely accidental. The large number of disasters resulting from this high level of exposure has seriously hindered the ability of these countries to emerge from poverty.

Aside from hazard frequency, severity and exposure, the other contributory factor to disaster risk is capacity. Unsurprisingly, newly industrialised and developing countries both tend to lack the capacity to deal with the threats posed by hazards. This capacity needs to be deployed before the hazard visits a community in the form of pre-disaster planning. Effective mitigation and preparedness can greatly reduce the threat posed by hazards of all types. Likewise, capacity can also be deployed following a major disruptive event. The post-disaster response can impact the loss of life, while timely reconstruction can minimise the broader economic and social damage that may otherwise result.

Although frequently represented as discrete stages, there is also recognition that the same are inter-connected, overlapping and multidimensional (see for example McEntire *et al.*, 2002). In particular: the level and quality of pre-disaster planning will largely determine – positively or negatively – the post-disaster response; and, the effectiveness of post-disaster reconstruction will determine to what extent the community remains vulnerable to the threats posed by hazards in the future. This link between sustainable development and mitigation has been referred to by Mileti (1999) as 'sustainable hazard mitigation'.

With this in mind, although this book is focused on post-disaster reconstruction, much of what is discussed within its chapters is intent on ensuring that communities are less vulnerable in the future. The emphasis on reconstruction also recognises that, unfortunately, many communities are left in a perpetual cycle of disasters, as failures in reconstruction efforts prevent them from addressing underlying risk factors.

1.2   Why focus upon the built environment?

As noted in the book's title, the emphasis of its chapters is concentrated on reconstruction of the 'built environment'. As will be explained later, this focus does not mean that the authors are suggesting that reconstruction of the built – or physical – environment should be carried out in a vacuum. Instead, many of the chapters will highlight the importance of linking the physical requirements with broader social, natural, institutional and economic needs. However, this emphasis does recognise the growing recognition that the construction industry and built environment professions have a significant role to play in contributing to a society's improved resilience to disasters (Haigh *et al.*, 2006; Lloyd Jones, 2006). In order to understand this role, it is necessary to understand what constitutes the 'built environment' and the nature of the stakeholders involved in its creation and maintenance.

The environments with which people interact most directly are often products of human initiated processes. In the 1980s the term built environment emerged as a way of collectively describing these products and processes of human creation. The built environment is traditionally associated with the fields of architecture, building science and building engineering, construction, landscape, surveying and urbanism. In Higher Education, Griffiths (2004) describes 'a range of practice-oriented subjects concerned with the design, development and management of buildings, spaces and places'.

The importance of the built environment to the society it serves is best demonstrated by its characteristics, of which Bartuska (2007) identifies four that are inter-related. First, it is extensive and provides the context for all human endeavours. More specifically, it is everything humanly created, modified, or constructed, humanly made, arranged or maintained. Second, it is the creation of human minds and the result of human purposes; it is intended to serve human needs, wants and values. Third, much of it is created to help us deal with, and to protect us from, the overall environment, to mediate or change this environment for our comfort and well-being. Last, is that every component of the built environment is defined and shaped by context; each and all of the individual elements contribute either positively or negatively to the overall quality of environments.

As previously noted by the Editors (Haigh and Amaratunga, 2010), several important consequences for disaster risk can be identified from these characteristics. The vital role of the built environment in serving human endeavours means that when elements of it are damaged or destroyed, the ability of society to function – economically and socially – is severely disrupted. Disasters have the ability to severely interrupt economic growth and hinder a person's ability to emerge from poverty. The protective characteristics of the built environment offer an important means by which humanity can reduce the risk posed by hazards, thereby preventing a disaster. Conversely, post-disaster, the loss of critical buildings and infrastructure can greatly increase a community's vulnerability to hazards in the future. Finally, the individual and local nature of the built environment, shaped by context, restricts our ability to apply generic solutions.

1.3   Resilience in the built environment

The consequences outlined above serve to underline and support the growing recognition that those responsible for the built environment have a vital role to play in developing societal resilience to disasters. The notion of resilience is becoming a core concept in the social and physical sciences, and also in matters of public policy. But what does resilience mean? What are the attributes of resilience? What is needed to create a disaster resilient built environment?

The term resilience was introduced into the English language in the early 17th century from the Latin verb resilire, meaning to rebound or recoil. However, there is little evidence of its use until Thomas Tredgold introduced the term in the early 18th century to describe a property of timber, and to explain why some types of wood were able to accommodate sudden and severe loads without breaking. In 1973, Holling presented the word resilience into the ecological literature as a way of helping to understand the non-linear dynamics observed in ecosystems. Ecological resilience was defined as the amount of disturbance that an ecosystem could withstand without changing self-organised processes and structures.

In subsequent decades, the term resilience has evolved from the disciplines of materials science, ecology and environmental studies to become a concept used by policy makers, practitioners and academics. During this period, there have been a range of interpretations as to its meaning.

For some, resilience refers to a return to a stable state following a perturbation. This view advocates a single stable state of constancy, efficiency and predictability, or, as the ability to absorb strain or change with a minimum of disruption (Horne and Orr, 1998; Sutcliffe and Vogus, 2003). For others, resilience recognises the presence of multiple stable states, and hence resilience is the property that mediates transition among these states. This requires very different attributes, as for example advocated by Douglas and Wildavsky (1982), who define resilience from the perspective of risk as, 'the capacity to use change to better cope with the unknown: it is learning to bounce back' and emphasise that, 'resilience stresses variability'. More recently but in a similar vein, Dynes (2003) associates resilience with a sense of emergent behaviour that is improvised and adaptive, while Kendra and Wachtendorf (2003) argue that creativity is vital.

Further discrepancy can be found in the degree to which resilience should be defined in merely passive terms. Douglas and Wildavsky (1982) focus on the ability to simply 'bounce back' from a 'distinctive, discontinuous event that creates vulnerability and requires an unusual response'. Wildavsky (1988) further characterises resilience as the, 'capacity to cope with unanticipated dangers after they have become manifest' and notes that resilience is usually demonstrated after an event or crisis has occurred. Lettieri *et al.* (2009) suggest a 'contraposition' in the literature between two concepts: resilience and resistance. Resilience they argue focuses on after-crisis activities, while resistance focuses on before-crisis activities. These all suggest a reactive approach whereby resilience is considered a 'pattern rather than a prescribed series of steps or activities' (Lengnick-Hall and Beck, 2003). Others stress a positive approach that suggests resilience is more than mere survival; it involves identifying potential risks and taking proactive steps (Longstaff, 2005). The objective is to build resilience by maximising the capacity to adapt to complex situations (Lengnick-Hall and Beck, 2005). Similarly, Paton *et al.* (2001) write of a paradigm shift that accommodates the analysis and facilitation of growth, whereby resilience 'describes an active process of self-righting, learned resourcefulness and growth'.

Resilience is evidently complex and open to a variety of interpretations but how can it be applied to the built environment? The relationship between disaster risk, resilience and the built environment suggests that a resilient built environment will occur when we *design, develop and manage context sensitive buildings, spaces and places that have the capacity to resist or change in order to reduce hazard vulnerability, and enable society to continue functioning, economically and socially, when subjected to a hazard event*. It is possible to elaborate on this definition by exploring specific characteristics of resilience and how they may be present in the built environment.

Firstly, resilience is seen as the ability to accommodate abnormal or periodic threats and disruptive events, be they terrorist actions, the results of climatic change, earthquakes and floods, or an industrial accident. Identifying, assessing and communicating the risk from such threats and events are therefore vital components. Individuals, communities, organisations and, indeed, nations that are prepared and ready for an abnormal event, tend to be more resilient. Consequently, those responsible for the planning, design and management of the built environment need to understand the diverse hazard threats to buildings, spaces and places and the performance of the same if a disruptive event materialises.

The next characteristic is the ability to absorb or withstand the disturbance while still retaining essentially the same function. This may mean returning to the state or condition that existed before the disturbance occurred, or returning to an improved state or condition. This absorption might be realised through the specification and use of hazard resistant methods, materials and technologies. It might also result from the construction of protective infrastructure, or the protection of critical infrastructure. Such measures may resist the threat, or at least reduce the losses experienced.

As outlined in the opening of this chapter, we live in a world which is constantly evolving, in some cases through natural processes and in other cases through the intervention of mankind. There is common agreement in the literature that systems, organisations and people who are able and willing to adapt tend to be more resilient. Creative solutions, the ability to improvise and the capacity to adapt will be essential in order to address the challenges posed by what is often seen as an unbounded threat.

The ability and willingness to learn is often linked to adaptability and being prepared. The learning may come from studying the lessons of others in a formal manner: by gathering and evaluating data, by conducting research in an objective, independent and balanced manner, and by communicating the findings, conclusions and recommendations.

The ability to absorb or withstand also requires economic and human capacity. A resilient built environment will need to be supported by a strong domestic industry and appropriately skilled professions and trades. A well-developed construction sector and supply chain, which largely comprise of micro-, small- and medium-sized enterprises, provides a strong means to counter the economic shocks that frequently accompany other disasters, while also offering an economic stimulus and livelihood opportunity in the recovery period.

As society becomes more complex, resilient communities tend to be those which are well coordinated and share common values and beliefs. This sense of interconnectedness can be undermined by self-interest and personal gain, resulting in vulnerable societies that are less able and willing to plan for, and react to, disruptive events. Understanding the link between the physical and social environment will be vital in developing connectedness. Culturally sensitive, sustainable and socially responsible planning, design and management of the built environment, have the potential to help develop community cohesion and thus contribute to wider societal resilience.

From this discussion of its characteristics, it is evident that the concept of resilience provides a useful framework of analysis and understanding on how we can plan, design and maintain a built environment that copes in a changing world, facing many uncertainties and challenges. Sometimes change is gradual and things move forward in continuous and predictable ways; but sometimes change is sudden, disorganising and turbulent. Resilience provides better understanding on how society should respond to disruptive events and accommodate change.

1.4   Disasters as a window of opportunity

If this idea of a resilient built environment is appealing, how can it be achieved? A further reason for the book's emphasis on reconstruction is that the post-disaster period provides a window of opportunity to address many of the vulnerabilities usually encountered in a community's built environment. There are several features of this post-disaster period that can be capitalised upon. Firstly, the disaster has destroyed much of the built environment that was improperly designed and vulnerable, creating a fresh start from which to address disaster risk. Furthermore, the experience gained during the disaster typically generates new knowledge, which brings various stakeholders together around a shared awareness of the nature of risk. The mistakes of previous development policies and strategies are exposed and can be addressed. Next and perhaps even more significantly, the political will and desire to act is almost certainly stronger than usual. Any interest in disaster risk reduction that had been forgotten or side-lined before the disaster, will suddenly gain renewed prominence in the recovery period. In a similar vein, the lack of resourcing for risk reduction, any presence of corruption and otherwise weak institutional structures that allowed a vulnerable built environment to be constructed will have been highlighted. Finally, but perhaps most importantly, the post-disaster period often provides a level of resourcing, including considerable external funding, that would be otherwise unattainable. If properly utilised – something that is by no means certain – this additional resource does afford a major opportunity to reduce vulnerability.

The fact that this window of opportunity exists does not mean that the various actors involved in reconstruction will take advantage of it. Although many, if not all, of these features are usually present following a major disaster, even a cursory glance at the countless studies and evaluations of programming after disasters, provides evidence that it is frequently a missed opportunity.

There are a myriad of reasons as to why these failures occur. Humanitarian principles are primarily concerned with addressing acute human suffering. By necessity, a timely response is essential. Anything that slows this response is likely to be a problem. Unfortunately, the well-planned reconstruction of a more resilient built environment will take time. Likewise, humanitarian principles also tend to dictate maintaining independence, neutrality and impartiality. This can dissuade actors from highlighting previous failings, which would otherwise create the necessary political will for change.

Effective reconstruction of the built environment is also competing with many other priorities. Poverty alleviation, improved health and good governance are a few of the many goals usually mainstreamed in the post-disaster recovery period. A more resilient built environment can certainly contribute to these goals, but there will inevitably be a time-lag; other recovery programmes can sometimes appear more appealing due to their ability to deliver short-term results. If the window of opportunity is to be taken advantage of, then advocates of a more resilient built environment will need to demonstrate the vital role it plays in helping society achieve much broader development goals.

A further complication is the natural tension between the need for timely reconstruction and a desire to utilise and where necessary develop local capacity. Institutions and local enterprise to plan and construct the built environment may matter, but they are often simply not there. Government, both national and local, is usually called upon to make critical long-term planning decisions, and to develop and enforce appropriate building regulations. This expectation is made of institutions that have usually failed to achieve this in far less challenging periods. The reality is that large scale reconstruction may have to be undertaken during a period soon after a major part of the civil service has perished, or at least been severely disrupted. At a time when even greater demands are being made of the civil service, its employees are sometimes being laid off, with the damage to the local tax base reducing available funding. At the same time, the local construction industry is suddenly called upon to increase its output to meet the needs of an unprecedented programme of reconstruction activity, while simultaneously familiarising itself with less vulnerable methods and materials. Building human resources and local capacity to address these shortfalls and support reconstruction may take years.

The alternative, to make use of international agencies and private enterprises, understandably raises other concerns. International actors are often accused of poaching the most talented local civil servants and encroaching on a country's independence, while the private sector is accused of disaster profiteering and leaves local industry unable to 'benefit' from the economic opportunities afforded by the disaster.

In summary, there is a window of opportunity, but it is beset with challenges. A pragmatic approach to the development of a resilient built environment needs to include an understanding of these difficulties and their implications for what can actually be done, at least in the short term. While the humanitarian efforts are frequently a rushed process, effective rebuilding for resilience will require reflection, discussion and consensus building. This should not undermine the importance of starting this process early in the recovery phase; indeed, a failure to consider long-term reconstruction goals early in the recovery can lead to wasted or misguided effort, as well as undermine efforts for future resilience. Instead, it recognises the importance of a judicious approach that addresses the complexity of creating resilience.

1.5   Structure of the book

It is evident from the discussion thus far in this opening chapter that the challenges associated with creating a disaster resilient built environment are considerable. The following chapters of this book elaborate on these many challenges while also offering ways to rebuild the built environment for resilience. In doing so, they explore individual, institutional and societal coping strategies for dealing with a range of hazards of both natural and man-made origin.

As noted earlier in this chapter, the frequency, scale and distribution of disasters in recent years demonstrate that the threat posed by these many hazards are a global problem, promising to disrupt communities in developed, newly industrialised and developing countries. Consequently, this book explores strategies in different contexts; many chapters include examples of how the built environment and the community it serves has coped when facing a hazard, and also how that community has responded in the aftermath of a disaster. The examples cover a range of communities and are thus taken from developing and developed countries.

The early chapters focus on the need for capacity in reconstruction. Ginige and Amaratunga begin the discussion in Chapter 2, exploring the challenge of capacity development for post-disaster reconstruction of the built environment. They consider the need to address capacity gaps across a range of built environment-related stakeholders. Within Chapter 3, Seneviratne builds on this opening to consider the human, physical, economic and social impact of the Indian Ocean Tsunami in Sri Lanka. The capacity of the Sri Lankan construction industry to respond to this challenge is examined. Using a very different context, within Chapter 4, Chang, Wilkinson, Potangaroa and Seville look at resourcing for post-disaster reconstruction using a longitudinal case study following an earthquake in China.

Maintaining the capacity theme but with a change of emphasis, Chapter 5 sees Thurairajah explore the opportunity for empowerment in disaster response and reconstruction, and in particular the role of women. In a similar vein, Ophiyandri examines the effectiveness of community-based post-disaster reconstruction in Indonesia in Chapter 6.

The second part of the book moves on to explore organisational perspectives of reconstruction. In Chapter 7, Siriwardena and Haigh highlight the importance of stakeholder engagement, and consider how they can be identified, classified and subsequently engaged in the reconstruction process. In Chapter 8, Kulatunga addresses the challenges associated with managing post-disaster reconstruction projects, including issues such as procurement, contract management, the sourcing of labour, materials and equipment, and financial accountability. Offering a different perspective, in Chapter 9 Rotimi, Wilkinson and Myburgh use examples from New Zealand to examine the problems that can be encountered with legislative provision for construction in a post-disaster environment, while Seneviratne and Amaratunga use Chapter 10 to illustrate the difficult context in which reconstruction often takes place, in this instance with specific reference to post-conflict environments. Sutton and Haigh complete the organisational perspective in Chapter 11. They consider whether there is a role for multi-national construction firms in helping to address the resource gap that usually exists in post-disaster environments: the gap between the volume of construction work to be undertaken and the capacity of the local construction industry.

The final part of the book focuses upon sustainable reconstruction. In Chapter 12, Pathirage explains the importance of effectively managing knowledge among complex sets of stakeholders, with a specific focus on practices and systems integration. Keraminiyage uses Chapter 13 to examine the importance of restoring a community's major infrastructure as a means for rehabilitating communities, while in Chapter 14 Karunasena tackles the challenge of managing the large volumes of construction and demolition waste that invariably arise following a disaster. In Chapter 15, Palliyaguru and Amaratunga emphasise the links between reconstruction and sustainable economic development, while in Chapter 16, Ginige focuses on the link between disaster risk reduction and sustainable development.

Chapter 17 completes the book, with Haigh and Amaratunga reflecting upon the points raised in the earlier chapters and considering the overall contribution to building resilience.

References

Bartuska, T. (2007) The built environment: definition and scope. In: Bartuska T. and Young, G. (Eds), *The Built Environment: A Creative Inquiry into Design and Planning*. Menlo Park, CA: Crisp Publications.

Douglas, M. and Wildavsky, A.B. (1982) *Risk and Culture: An Essay on the Selection of Technical and Environmental Danger*. Berkeley, CA: University of California Press.

Dynes, R. (2003) Finding order in disorder: continuities in the 9-11 response. *International Journal of Mass Emergencies and Disasters*, **21**, 9–23. Research Committee on Disasters, International Sociological Association.

Griffiths, R. (2004) Knowledge production and the research-teaching nexus: the case of the built environment disciplines. *Studies in Higher Education*, **29**, 709–726.

Haigh, R., Amaratunga, D. and Keraminiyage, K. (2006) An Exploration of the Construction Industry's Role in Disaster Preparedness, Response and Recovery. Proceedings of the Annual International Research Conference of the Royal Institution of Chartered Surveyors (RICS COBRA 2006), The RICS and The Bartlett School, University College London.

Haigh, R. and Amaratunga, D. (2010) An integrative review of the built environment discipline's role in the development of society's resilience to disasters. *International Journal of Disaster Resilience in the Built Environment*, **1**, 11–24.

Holling, C. (1973) Resilience and stability of ecological systems. *Annual Review of Ecology and Systematics*, **4**, 1–23.

Horne III, J. and Orr, J. (1998) Assessing behaviors that create resilient organizations. *Employment Relations Today*, **24**, 29–39.

Kendra, J. and Wachtendorf, T. (2003) Creativity in Emergency Response to the World Trade Center Disaster . . . Beyond September 11th: An Account of Post-Disaster Research. Special Publication No. 39. Boulder, CO: Natural Hazards Research and Applications Information Center, University of Colorado.

Kelman, I. (2009) Natural Disasters Do Not Exist (Natural Hazards Do Not Exist Either). Version 2, 10 September 2009 (Version 1 was 26 July 2007). http://www.ilankelman.org/miscellany/NaturalDisasters.rtf

Lengnick-Hall, C. and Beck, T. (2003) *Beyond Bouncing Back: The Concept of Organizational Resilience*. Paper presented at the Academy of Management, Seattle, WA, 1–6 August.

Lengnick-Hall, C. and Beck, T. (2005) Adaptive fit versus robust transformation: how organizations respond to environmental change. *Journal of Management*, **31**, 738–757.

Lettieri, E., Masella, C. and Radaelli, G. (2009) Disaster management: findings from a systematic review. *Disaster Prevention and Management*, **18**, 117–136.

Lloyd-Jones, T. (2006) *Mind the Gap! Post-disaster Reconstruction and the Transition from Humanitarian Relief*. London: RICS.

Longstaff, P. (2005) *Security, Resilience, and Communication in Unpredictable Environments Such as Terrorism, Natural Disasters and Complex Technology*. Report for the Center for Information Policy Research, Harvard University, Cambridge, MA.

McEntire, D., Fuller, C., Johnston, C. and Weber, R. (2002) A comparison of disaster paradigms: the search for a holistic policy guide. *Public Administration Review*, **62**, 267–281.

Mileti, D. (1999) *Disasters by Design: A Reassessment of Natural Hazards in the United States*. Washington DC: Joseph Henry Press.

O'Keefe, P., Westgate, K. and Wisner, B. (1976) Taking the naturalness out of natural disasters. *Nature*, **260**, 566–567.

Paton, D., Johnston, D., Smith, L. and Millar, M. (2001) Community response to hazard effects: promoting resilience and adjustment adoption. *Australian Journal of Emergency Management*, **16**, 47–52.

Sutcliffe, K. and Vogus, T. (2003) Organizing for resilience. In: Cameron, K. (Ed.), *Positive Organizational Scholarship*. San Francisco, CA: Berrett-Koehler Publishers Inc. pp. 94–110.

Swiss Reinsurance Company (2000) *Natural Catastrophes and Man-Made Disasters in 1999: Storms and Earthquakes Lead to the Second–Highest Losses in Insurance History*. Zurich: Sigma.

Swiss Reinsurance Company (2010) *Natural and Man-Made Catastrophes in 2009*. Zurich: Sigma.

Wildavsky, A. (1988) *Searching for Safety*. New Brunswick, CT: Transaction Books.

2 Capacity Development for Post-Disaster Reconstruction of the Built Environment

Kanchana Ginige and Dilanthi Amaratunga

2.1 Introduction

'The built environment is the aggregate of human-constructed "physical plant", with its myriad of elements and systems. It includes the buildings where we live, work, learn, and play; the lifelines that connect and service them; and the community and region that they are a part of. It is the roads, utility lines and the communication systems we use to travel, receive water and electricity or send information from one place to another. The pipes and transmission lines that carry vital supplies and wastes for use or treatment are other essential elements. Very simply, the built environment comprises the substantive physical framework for human society to function in its many aspects – social, economic, political, and institutional' (Geis, 2000, page 8).

Geis' definition ably demonstrates the critical role that the built environment plays in the functioning of society. As a consequence, destruction of a community's built environment during a disaster, regardless of that disaster's origin will, in all likelihood, significantly hinder the regular functioning of that society. In this context, rapidly reconstructing the built environment after a disaster becomes critical. Post-disaster reconstruction of the built environment primarily focuses on the repair and reconstruction of physical infrastructure and buildings. Although reconstruction is usually urgent in order to minimise the social and economic disruption to a society, it is important that this understandable desire for haste, does not compromise goals for sustainability and greater resilience that would reduce a society's vulnerability to future disasters. It is now widely acknowledged that the characteristics of the built environment – issues such as location, form and type – will significantly impact

Post-Disaster Reconstruction of the Built Environment: Rebuilding for Resilience, First Edition.
Edited by Dilanthi Amaratunga and Richard Haigh.
© 2011 Blackwell Publishing Ltd. Published 2011 by Blackwell Publishing Ltd.

disaster vulnerabilities (Duque, 2005). It is therefore vital that any reconstruction programme adequately addresses mitigation and long-term sustainability, rather than merely reinstate what has been destroyed.

The challenge of sustainable reconstruction is wide-ranging in nature. This chapter focuses on one aspect: how to analyse, create, utilise and retain capacities for construction of the built environment so that communities are more resilient and less vulnerable to the threat posed by local hazards. This chapter presents a framework that has been developed to develop capacity for post-disaster reconstruction in the built environment. In particular, the framework aims to strengthen the knowledge, abilities, skills and behaviour of individuals responsible for the built environment, and improve institutional structures and processes to ensure that post disaster reconstruction meets its mission and goals in a sustainable way.

2.2   Capacity needs for post-disaster reconstruction

Capacities exist in different forms; they may be knowledge, skills, technology and resources. According to the United Nations International Strategy for Disaster Reduction (UN/ISDR) (2009), capacity is the combination of all the strengths, attributes and resources available within a community, society or an organisation, and that can be used to achieve agreed goals. They can exist in the forms of infrastructure and physical means, institutions, societal coping abilities, human knowledge, skills, and collective attributes such as social relationships, leadership and management.

Effective post-disaster reconstruction may require the affected society to possess or have access to a range of capacities: organisations with clear responsibilities and knowledge for managing disasters; well-developed disaster plans and preparedness; coping mechanisms; adaptive strategies; memory of past disasters; good governance; ethical standards; local leadership; physical capital; human capital; and, resilient buildings and infrastructure that cope with and resist extreme hazard forces (Benson *et al.*, 2007). It is worth taking a moment to consider in more detail the important role that some of these capacities play in the reconstruction process.

Governance is the exercise of economic, political and administrative authority to manage a country's affairs at all levels through bringing together the actions of state, non-state and private sector actors (United Nations Development Programme [UNDP], 2004). The characteristics of good governance – participation, rule of law, transparency, responsiveness, consensus orientation, equity, effectiveness, efficiency, accountability and strategic vision – are therefore essential for sustainable development and disaster risk reduction in a country. Related capacities include: the authorities, institutions and plans to manage post-disaster reconstruction activities and coordinate its stakeholders; policy and legislation to regulate disaster reconstruction; and, human resources to manage the authorities and implement the policies and legislation. In this context, governments and local governments have an important role to perform by ensuring all the necessary capacities are in place. They are responsible for: the process of decision making to formulate policies; the subsequent implementation of those policies; and, initiation of organisations for disaster management at the central and local levels.

In the case of post-disaster reconstruction, the national and local government is, for example, responsible for the enforcement of building codes, land-use planning, environmental risk and human vulnerability monitoring and safety standards (UNDP, 2004). Building codes, for example, are the critical frontline defence for achieving more strongly engineered structures, including large private buildings, public sector buildings, infrastructure, transportation networks and industrial facilities. Similarly, the adverse effects of disasters can be greatly reduced if it is possible to avoid the hazardous areas being used for settlements or as sites for important structures. In this context, most urban masterplans involving land-use zoning attempt to separate hazardous industrial activities from major population centers (Nateghi-A, 2000). Thus, land-use policies perform an important role in reducing disaster vulnerabilities of the built environment. However, as Nateghi-A notes, building codes, land-use policies or design standards are unlikely to result in a more resilient built environment unless the professionals responsible for their implementation accept their importance and endorse its use, and understand the code and the design criteria required of them. Similarly, it is unlikely to be effective unless the code is fully enforced by authorities checking and penalising designs that do not comply.

'A code has to fit into an environment and there has to be the establishment of an effective administration to check code compliance in practice: the recruitment of ten new municipal engineers to enforce an existing code may have more effect in increasing construction quality in a city than proposing higher standard building codes' (Nateghi-A, 2000, page 207).

Such examples highlight the importance of considering capacity for the built environment in the broadest sense, as so many stakeholders are involved in its development and maintenance. In this context, education and training are vital in developing the necessary human capital for reconstruction. Training must address the needs of all those involved in the process, from government policy makers and enforcers, to private sector and non-government organisations that are directly engaged in the reconstruction activity.

Supporting this view, Bosher *et al.* (2007a) assert that risk and hazard training should be systematically integrated into the professional training and development of architects, planners, engineers and developers. They also recognise the importance of cross-disciplinary training for construction professionals and emergency managers. Similarly, the Royal Institution of Chartered Surveyors (RICS) President's Commission on Major Disaster Management (MDMC), which was convened after the 2004 Indian Ocean tsunami in response to members' concerns on what they and the RICS could do to help, has explored the strategic and practical ways it could bring the skills of RICS members and others involved in the built environment to provide help in the return to normality for those affected by disasters each year (Lloyd-Jones, 2006). In doing so, they highlight the wide range of property and construction skills that are required for effective mitigation and reconstruction in the built environment. Some examples include:

- Aiding local government land administration, cadastral mapping
- Knowledge of land and property legislation, providing support on land rights and claims
- Knowledge of local regulatory frameworks and ways they could be improved
- Training and knowledge transfer
- Disaster risk assessment
- Links with other built environment professions; inter-disciplinary and team working
- Contacts with local business and industry; networking
- Knowledge of appropriate forms of disaster-resistant construction and engineering

It is also important that the capacity of those outside the usual professions is not ignored. Many studies have recognised the need to include local community participation in the reconstruction process (see for example: Jayaraj, 2006; Pardasani, 2006; Owen and Dumashie, 2007). The involvement of the beneficiaries and the wider community in reconstruction can lead to more sustainable outcomes. The more the affected people are engaged in the process and the greater the level of stakeholder engagement, the more they are able to influence and take ownership of the outcomes (Lawther, 2009). They may also bring vital local knowledge and experience that is not available through external capacity: locations that are less vulnerable to hazards based on past disasters; an understanding of locally available material that can be used for construction; and, special community needs that should be addressed. In particular, capacity of a community represents the internal strengths of the community and its external opportunities. McGinty (2002) identifies five major elements of capacity in the community through a report on capacity building for regional Australia by Professor S. Garlick (1999): knowledge building as the capacity to enhance skills, utilise research and development and foster learning; leadership as the capacity to develop shared directions and influence what happens in regions; network building as the capacity to form partnerships and alliances; valuing community by recognising the community; and, the capacity of the community to work together to achieve their own objectives.

While local capacity is often deemed desirable, and perhaps preferable for use in reconstruction, it is often insufficient to meet the huge challenges that disasters leave behind. In many countries, the lack of local knowledge, resources and expertise following a disaster can be overcome by adequate global cooperation. The supportive role of international agencies can assist countries in mitigation and reconstruction by applying their existing knowledge and resources (El-Masri and Tipple, 2002). International aid through finance, technology, expertise and human resource development become significant capacities in this context. The following example from www.worldwildlife.org is indicative of the type of capacity that can be offered by major international agencies. Significantly, although external capacity is provided in this instance by the international agency, a vital role in this engagement is the development of long-term local capacity (World Wildlife Fund).

'In 2008, Cyclone Jokwe, a Category 3 cyclone with peak winds of 120 mph, formed over the South-West Indian Ocean on March 2 and passed over northern Madagascar before slamming hard into the coast of Mozambique in the World Wildlife Fund's (WWF) Coastal East Africa priority place. The cyclone affected 200 000 people, destroyed 9000 homes, damaged 3000 others and caused widespread damage to crops, including the destruction of over 150 000 cashew trees, one of Mozambique's most significant agricultural resources. Many of the areas affected by Cyclone Jokwe are part of a proposed three-year natural resource management and livelihoods joint program of the humanitarian aid agency CARE and the WWF. At the request of CARE, WWF's Humanitarian Partnerships team traveled to Mozambique to conduct a rapid environmental assessment (REA) to examine ways to ensure the sustainability of CARE's reconstruction activities and materials. In addition to performing the assessment, WWF conducted training and capacity-building activities with CARE and WWF staff on performing post-disaster rapid environmental assessments for future response activities.'

In addition, international agencies such as the United Nations fulfil important capacity needs by composing policies, frameworks and guidelines that formulate disaster mitigation and reconstruction internationally. According to Schipper and Pelling (2006), there are three key domains in the global arena which should be integrated in order to achieve sustainability in disaster risk management. These are climate change, natural hazards and development. Climate change policy is based on a specialised UN convention that requires global cooperation in order to function: the United Nations Framework Convention on Climate Change (UNFCCC). Disaster risk reduction is mainly guided by an international framework called Hyogo Framework for Action 2005–2015 (HFA) that is governed by United Nations secretariat of the International Strategy for Disaster Reduction (ISDR). Finally, development aims are linked to the internationally agreed United Nations Millennium Development Goals, which were introduced in the United Nations Millennium Declaration. All three of these domains, which are coordinated internationally, inform and guide regional and national practices.

It is evident from the examples provided, that the capacities required for effective reconstruction following a disaster can be found at every level: from the local to the international; from qualified professionals to volunteers; and from the public to the private sectors. Any holistic attempt to consider capacity for reconstruction must address these wide-ranging perspectives. The following section examines some of the major capacity gaps in respect of reconstruction.

2.3 Capacity gaps in post-disaster reconstruction

2.3.1 Policy environment

Disaster reconstruction requires certain organisational and procedural measures to regulate activities. These measures that guide reconstruction have to be sustained over a number of years and have to survive any changes in political administration that are likely to happen within that time, since the changes in physical planning in the built environment, upgrading to structures, and changes in the characteristics of the building stock, are processes which need a long time scale (Nateghi-A, 2000). Likewise, different levels of authorities – local, provincial, national and international – must cooperate in order to ensure sustainable and equitable development (El-Masri and Tipple, 2002). In order to promote effective cooperation, the different organisations must redefine and readjust their roles in order to establish adequate communication networks and warning systems; to disseminate existing and new knowledge; to help in effective technology transfer; and to mobilise adequate resources. Ambiguously defined roles between local and national levels can have serious implications for effective disaster reconstruction. One of the most widely reported problems, particularly in developing countries, is the centralised systems that make it difficult for decision makers to relate to communities because of spatial and socio-economic distance. In light of this, El-Masri and Tipple (2002) suggest a comprehensive decentralisation of decision making to sub-national and local levels. They suggest this will enhance local initiatives, maximise the use of resources, respond to the real needs of the people, and build appropriate systems for defining responsibilities and accountability in the administrative system. Although this decentralisation is not a simple task it could be achieved by building consensus and capacity at different levels.

In addition, lack of coherence among different policies towards a common goal is also a crucial problem in post-disaster reconstruction. As Mileti (1999) elaborates, to facilitate sustainable disaster mitigation, all policies and programs related to hazards and sustainability should be integrated and consistent. In this context, Schipper and Pelling (2006) observe that on a policy level, influence of the three spheres of development goals, responding to climate change and reducing risk of natural disasters have become three realms of action. 'Unfortunately, action often remains segregated both institutionally and from a disciplinary standpoint, thus not taking advantage of the interrelated nature of the realms' (Schipper and Pelling, 2006, page 19).

Elsewhere, evaluations following disasters have revealed shortcomings in construction techniques, code enforcement, and the behavior of structures under stress (Mileti, 1999). Thus, codes, standards, and practices for all hazards must be re-evaluated in light of the goal for sustainable mitigation while communities must improve adherence to them. Such problems are not restricted to developing countries: the post-windstorm damage surveys carried out by Henderson and Ginger (2008) in Australia revealed houses that did not conform to the relevant standards. Problems related to the use of unconservative design parameters, poor or faulty construction practices, and use of products that had not been designed, tested or installed in a manner appropriate for a region subjected to such winds. Henderson and Ginger suggest that education is required to raise awareness of the consequences in making inappropriate design assumptions or faulty construction. They emphasise that this education must address every step of the building process, from regulation, through design, construction, certification and maintenance, and must reach all parties, including the designer, builder, certifier, and owner.

Further, there are other gaps in the policy environment such as inadequately sensitive policies to hazards and their consequences. According to Le Masurier *et al.* (2006) the consequences of large disasters have not been taken into account adequately when developing policies. As Le Masurier *et al.* highlight, such policies demands a greater degree of coordination in reconstruction programmes. As an example, land-use planning policies need to be integrated with awareness of natural hazards and disaster risk mitigation in order to minimise the disaster vulnerabilities of new developments. However, as El-Masri and Tipple (2002) suggest, integration of disaster risk reduction awareness itself only is not sufficient for the land-use policies to be effective in post-disaster reconstruction, but they need to be complemented by appropriate building design, construction methods and use of building materials as well.

2.3.2   *Human resources, institutions and community*

As noted in the study by Lloyd-Jones (2006), chartered surveyors and other construction professions have key roles to play during all pre- and post-disaster phases. However, coordination between professions and other stakeholders is vital. Evidence from the field suggests that this does not always happen. For example, in the South Asian subcontinent the built environment has become highly specialised with a range of disciplines, professionals and decision makers, each with distinct approaches for mitigation and rehabilitation within their own disciplinary field (Laverack, 2005).

It is also critical that training and development is not restricted to the professional level. Developing the skills and knowledge of the major construction trades is essential to ensure that satisfactory levels of quality can be achieved. Similarly, whilst high levels of community involvement are accepted as preferable in post-disaster reconstruction, there are difficulties in implementing this approach (Lawther, 2009). Community members may not have sufficient technical and construction project management skills or an understanding of appropriate standards. The community may also have problems in interacting with the other stakeholders. Such problems are only likely to be addressed by an appropriate organisational management structure that can address these shortcomings and facilitate implementation for longer term benefits. Further, economic constraints and poverty in disaster-affected communities may also hinder effective community involvement in reconstruction. According to Martin (2002) poverty brings day to day living to a higher priority than disaster planning.

However, the RICS (Lloyd-Jones, 2006) suggests that in the longer term, improved governance, policies, planning, management and capacity-building can provide the framework for better access by households and local communities to the professional expertise and knowledge within business, local government and civil society. In doing so, it will help communities to reduce their risk to natural disasters, and build their properties, villages and neighborhoods to withstand the threat posed by hazards, when they cannot be avoided. In this context, the chapter introduces a framework for capacity development for post-disaster reconstruction in its next section.

2.4   Capacity development framework

2.4.1   *The concept of capacity development*

Capacity development is the process by which people, organisations or society systematically stimulate and develop their capacities over time through improvement of knowledge, skills, systems and institutions (UN/ISDR, 2009). According to Bolger (2000) the objectives of capacity development are to:

- Enhance, or more effectively utilise, skills, abilities and resources
- Strengthen understandings and relationships
- Address issues of values, attitudes, motivations and conditions in order to support sustainable development

Capacity development is recognised as a relatively new concept, which emerged in the 1990s and is complimentary to previously introduced ideas in development thinking such as institution building, institutional development, human resources development, development management and institutional strengthening. Lusthaus *et al.* (1999) see capacity development as providing a link to take previously isolated approaches into a coherent strategy with a long-term perspective and vision of social change. In a similar vein, UN/ISDR (2009) note that the concept of capacity development extends the term capacity building to encompass all aspects of creating and sustaining capacity growth over time.

Current views on capacity development show a process which runs across several stages. The United Nations Development Programme introduces five stages in its capacity development process (UNDP, 2008).

- Step 1: Engage stakeholders on capacity development
- Step 2: Assess capacity assets and needs
- Step 3: Formulate a capacity development response
- Step 4: Implement a capacity development response
- Step 5: Evaluate capacity development

Accordingly, capacity development for post-disaster reconstruction can be described as the ability of relevant stakeholders to identify constraints and to plan and manage reconstruction in the built environment effectively, efficiently and sustainably in this chapter. This definition involves both the development of capacities of human resources, institutions and communities, and also a supportive policy environment. It encompasses the process by which individuals, communities and institutions develop, utilise and retain their skills, abilities and knowledge individually and collectively, to identify their problems and constraints, set mitigation and reconstruction objectives, formulate policies and programmes, perform functions required to solve those problems, and achieve a set of mitigation and reconstruction objectives. In this context, a framework has been developed to elaborate the process of capacity development for post-disaster reconstruction and it is introduced in detail in the subsequent section.

2.4.2 Why a framework?

Developing required capacities for post-disaster reconstruction is a long and slow process that requires a significant commitment from all stakeholders involved in reconstruction. However, there is a dearth of literature on this process and it has not been clearly identified and described, especially from a stakeholder perspective. Table 2.1 attempts to describe this complex process and includes both the stakeholder groups whose capacity must be considered, and the various stages of capacity development.

Table 2.1 Capacity development framework

		Stages of capacity development		
		Analysis	Creation	Utilisation	Retention
Stakeholders	National and local governments International community Community Civic society Private and corporate sector Academia and professional associations				

2.4.3   *Four phases of capacity development*

The horizontal axis of the framework illustrates the stages of capacity development for post-disaster reconstruction. These stages are analysis, creation, utilisation and retention.

Analysis

It is increasingly recognised that the term capacity building is misleading, as building has the undertone of starting something from the beginning, whereas in practice, improving capacity must take account of the current context. One common and problematic theme of capacity development is the individual nature of capacity development interventions. Capacity development is highly influenced by local context and thus is unique in each of its applications. The first stage of capacity development focuses on the analysis of existing capacity, and identification and prioritisation of capacity gaps.

Creation

The second stage focuses on the need to create capacity in order to address the identified gaps. Creating capacity requires enormous efforts and time in understanding the local context and finding appropriate means to build capacity. Effective human and institutional capacity rests on a strong foundation that facilitates the creation of new capacities through learning opportunities as well as by putting in place processes that enhance the adaptability required for dealing with a dynamic environment. Such a foundation is created through formal training, informally through on-the-job training, as well as through accumulation of norms, routines and processes, which promote capacity creation on a continuous basis.

Utilisation

The third stage considers how developed capacities are mobilised and deployed under realistic conditions. A failure to make effective use of existing capacities can undermine mitigation and reconstruction activities. Efficient and effective use of existing capacities is an important aspect of capacity building as it recognises the need to make use of the affected community's own assets, thereby reducing its sense of helplessness. Making the best use of existing capacities will involve mobilisation of all the creative and innovative capacities that can be found in existing human and institutional capacities.

Retention

The final stage addresses the need to retain and sustain capacity over time. Capacity development must be designed in such a way that it is sustainable beyond any initial external intervention. Sustaining capacity is more likely to occur in the context of stable political, institutional and economic conditions that provide an atmosphere of support for the capacity building efforts in society. Sources of funding are an important element of sustainability and capacity retention. In the long-run, the key to sustaining capacity will be the availability of local sources of funding. Sustainable capacity building will need to address the capacity to mobilise domestic resources.

2.4.4   *Stakeholders*

Stakeholders of an organisation are usually identified as the groups who have a direct relevance to the organisation's core economic interests. A broader view of stakeholders also considers those who may be affected, positively or negatively, by the organisation (Mitchell *et al.*, 1997). With this broader view, individuals, groups, neighbourhoods, organisations, institutions, societies and even the natural environment can be considered as actual or potential stakeholders (Mitchell *et al.*, 1997). However, when a process is taken into account, stakeholders can generally be defined as the individuals or organisations that gain or lose from the success or failure of the particular process (Nuseibeh and Easterbrook, 2000). Thus, in the process of capacity development for post-disaster reconstruction, they are the entities which gain or lose from the success or failure of the process.

According to Newcombe (2003) stakeholders of a construction project are groups or individuals who have a stake in, or expectation of, the project's performance and include clients, project managers, designers, subcontractors, suppliers, funding bodies, users and the community. Similarly, stakeholders of post-disaster reconstruction are the groups or individuals who can affect or are affected by the achievement of a reconstruction project's objectives (Siriwardena *et al.*, 2009). Incidentally, capacities necessary for reconstruction emerge from different parties in the society, such as government and local governments, international agencies, academia and professional bodies, private organisations and non-government organisations, and the local community. All these parties might be considered as stakeholders to the post-disaster reconstruction process. In this context, the capacity of all these stakeholders must be considered if the goals of reconstruction are to be achieved efficiently, effectively and sustainably.

National and local government

Government performs a critical role in development since it has a unique capacity as a mediator between private and public interests and as an actor with local, national and international connections (UNDP, 2004). National and local governments of a country hold the main responsibility for the coordination of different stakeholders at different levels. Coordination among different stakeholders and different levels of authorities such as local, provincial, national and international should be ensured in order to achieve success in disaster management of a country. Government's capability as a mediator and coordinator can facilitate a country by transferring technical know-how and good practice that are useful for post-disaster reconstruction of the built environment from other countries. Similarly, it can bring finances and other resources to a country through its international relations.

Further, national and local governments of a country have the authority to develop and enforce rules, laws and regulations. In relation to disaster risk reduction in the built environment, governments have administrative and legislative power to enforce regulations and policies on construction operations. Thus, regulating the process of construction to ensure that necessary disaster mitigation and prevention measures are integrated into the practice to reduce disaster risk through enforcement of policy and legislation becomes a duty of national and local government. According to Bosher *et al.* (2007b), building codes, proper engineering design and construction practices, and land-use plans and regulations are critical for disaster mitigation in the built environment. Thus this group of stakeholders is defined as public and semi-public entities that have the authority to make and enforce rules, laws and regulations pertaining to the built environment in this categorisation.

International community

Non-profit making organisations which possess membership of more than one country and set up as intergovernmental organisations or international non-governmental organisations are considered as the international community under this categorisation. International organisations can be seen in two main types, namely, intergovernmental organisations and international non-governmental (Iriye, 2002). The first type is set up by intergovernmental agreements such as United Nations whilst the second, which is voluntary and open to anyone who wishes to join, is formed by private individuals and groups (Iriye, 2002).

International community also performs a significant role in formulation of policies, guidelines and regulations for disaster risk reduction. International entities such as the UN/ISDR and the UNDP are some leading performers in relevant policy making. The policy guidelines such as Hyogo Framework for action and United Nations Millennium Development Goals (MDGs) are globally agreed international agendas for integrating disaster risk reduction for development activities around the world. Thus, international community attempts to bring different nations together to achieve disaster risk reduction under common agendas.

In addition, international community provides necessary assistance for post-disaster reconstruction for nations in need. Incidentally, capacities and experiences of international community in the field of disaster management are vital mainly because, in many developing countries, the lack of knowledge, resources and expertise can be overcome by adequate global cooperation in tackling natural disasters (El-Masri and Tipple, 2002). Thus, in the context of post-disaster reconstruction of the built environment, international community is able to provide necessary skills and knowledge, technology and financial aid.

Community

Community is identified as the individuals and groups sharing a natural and built environment that is vulnerable to hazards. In other words, community is the general public; the users and occupants of the built environment and the beneficiaries of post-disaster reconstruction. Thus, special community needs and concerns require to be necessarily integrated into reconstruction activities in the built environment.

According to Lawther (2009), involvement of the beneficiaries and the wider community in the reconstruction can lead to more sustainable outcomes of projects. As UN/ISDR states, disaster risk reduction strategies need to build on people's local knowledge and cultural practices, and apply tools and approaches that people can easily understand and integrate into their lives. In particular, members of local communities represent the greatest potential source of local knowledge of hazardous conditions, and are the repositories of traditional coping mechanisms suited to their individual environment (UN/ISDR).Thus, experience and participation of local community are extremely important for post-disaster reconstruction of the built environment. Their leadership and involvement in the relevant initiatives, and their knowledge and experiences related to the disaster vulnerabilities of the area, safe locations for construction and local resources that can be used for construction are significant in this context.

Civic society

This group includes the non-governmental organisations (NGOs) that participate in disaster risk reduction activities, including not-for-profit and voluntary groups that are organised on a local, national or international level. These are the voluntary and social organisations that are non-state owned who appear for the purposes of post-disaster reconstruction. The UN/ISDR Secretariat believes that building the resilience of nations and communities to disasters cannot be done without the active participation of NGOs. In this context, UN/ISDR is determined to build a Global Network of NGOs for Community Resilience to Disasters, with the aim of addressing disaster risk reduction issues at sub-national and community levels. UN/ISDR highlights the following activities as the most important of NGOs' role in disaster risk reduction.

- NGOs can operate at grassroots level with communities and local organisations as partners, and take a participatory approach to development planning. This allows them to respond better to local people's priorities and build on local capacities.

- NGOs enjoy higher operational flexibility as they are relatively free from bureaucratic structures and systems, and better able to respond and adapt quickly and easily.
- NGOs often work with and on behalf of the most needy groups: the poorest and the most vulnerable.

Thus, the assistance extended by the civic society towards post-disaster reconstruction of the built environment can be through policy development and advocacy, education and awareness raising, technical assistance and human resources for risk and vulnerability assessments and enhancing community participation into local construction activities.

Private and corporate sector

This category consists of privately owned profit-orientated business and industrial groups. In most societies, the private sector has been the driving force behind socio-economic development (UN/ISDR, 2008). According to UN/ISDR, the private sector adversely suffers from the consequences of disasters and therefore it has a role to play in reducing disaster risk. In general, the private sector has a role to play in moving towards sustainable development that incorporates an awareness of disaster risk (UNDP, 2004). In this context, the UN Global Compact, launched in 2000 to bring businesses together with UN agencies, labour, civil society and governments, requests businesses to integrate disaster prevention into their decision-making.

Incidentally, there are various private and corporate sector institutions which have a direct interest in post-disaster reconstruction of the built environment. In particular, the majority of the developers, consultants, contractors and subcontractors, banks and finance institutions that design, construct, maintain and finance the built environment can be categorised under this stakeholder group. They are also the entities who are responsible for implementation of policies, regulations or guidelines including building codes and construction standards for disaster risk reduction in actual practice to minimise disaster vulnerabilities in the built environment.

Academia and professional associations

This group of stakeholders are universities, research organisations, and professional associations engaged in research, and training and development of individuals and organisations involved in post-disaster reconstruction. Academia and various professional associations are the prominent stakeholders in the built environment for post-disaster reconstruction-related education, training, and research and development, including invention of methods and techniques for reconstruction and development of reconstruction standards and guidelines.

2.4.5   *Advantages of the capacity development framework*

The capacity development framework provides an over-arching structure by which capacity for reconstruction can be created, utilised and sustained. It also identifies the major stakeholder groups that are involved in or affected by the reconstruction activity. In doing so it provides a means for identifying capacity gaps, and a way to prioritise development of those capacities.

2.5   Summary

Post-disaster reconstruction in the built environment refers to repairing and reconstructing physical infrastructure and buildings in the aftermath of a disaster. Post-disaster reconstruction depends on various capacities and these capacities arise from a number of different stakeholders since the built environment is a combination of many stakeholders. In particular, availability of authorities and institutions, policies and legislation, frameworks and guidelines, human and other physical resources, projects and programmes, knowledge, skills and experience of people and their proper implementation are significant for post-disaster reconstruction. In this context, national and local governments, international community, community, civic society, the private and corporate sector, and academia and professional associations have been identified as the prominent stakeholder groups in post-disaster reconstruction.

Although it has been noted that there are many capacities which exist in relation to post-disaster reconstruction in the built environment, there are number of capacity gaps that need to be addressed in order to make reconstruction more effective. Lack of appropriate policies, deficiencies in policy implementation, deficiencies in state of the art technology for rapid and sustainable reconstruction, deficiencies in trained and skilled human resources for post-disaster reconstruction, and lack of disaster reconstruction related knowledge, experience and coordination among stakeholders can be highlighted as key capacity gaps in relation to post-disaster reconstruction in the built environment. Thus, bridging these gaps is extremely important in achieving successful post-disaster reconstruction. Development of necessary capacities in a sustainable approach is essential to overcome the capacity gaps. Therefore, a framework has been introduced for capacity development for post-disaster reconstruction in the built environment with a stakeholder perspective in this chapter. Capacity development for post-disaster reconstruction is referred to as the ability of relevant stakeholders to identify constraints and to plan and manage reconstruction in the built environment effectively, efficiently and sustainably.

The framework identifies four stages of capacity development; namely, analysis, creation, utilisation and retention. These stages are identified against the aforementioned stakeholder groups that are involved in post-disaster reconstruction in the built environment. The capacity development framework brings several advantages by integrating a stakeholder perspective to capacity development and by fragmenting the process of capacity development for post-disaster reconstruction in the built environment into smaller elements whilst emphasising the relationships among those elements.

References

Benson, C. and Twigg, J. with Rossetto, T. (2007) *Tools for Mainstreaming Disaster Risk Reduction: Guidance Notes for Development Organisations*. Switzerland: ProVention Consortium.

Bolger, J. (2000) Capacity development: why, what and how. *Capacity Development – Occasional Series*, **1**, 1–8.

Bosher, L., Dainty, A., Carrillo, P. and Glass J. (2007a) Built-in resilience to disasters: a pre-emptive approach. *Engineering, Construction and Architectural Management*, **14**, 434–446.

Bosher, L., Dainty, A., Carrillo, P., Glass, J. and Price, A. (2007b) Integrating disaster risk into construction: a UK perspective. *Building Research and Information*, **35**, 163–177.

Duque, P.P. (2005) *Disaster Management and Critical Issues on Disaster Risk Reduction in the Philippines*. International Workshop on Emergency Response and Rescue, 31 October–1 November 2005, Taipei, Taiwan.

El-Masri, S. and Tipple, G. (2002) Natural disaster, mitigation and sustainability: the case of developing countries. *International Planning Studies*, **7**, 157–175.

Garlick, S. (1999) *Capacity Building in Regional Western Australia: A Regional Development Policy for Western Australia: A Technical Paper*. Perth: The Department of Trade and Commerce, Government of Western Australia.

Geis, D.E. (2000) By design: the disaster resistant and quality-of-life community. *Natural Hazards Review*, **1**, 151–160.

Henderson, D. and Ginger, J. (2008) Role of building codes and construction standards in windstorm disaster mitigation. *The Australian Journal of Emergency Management*, **23**, 40–46.

Iriye, A. (2002) *Global Community: The Role of International Organisations in the Making of the Contemporary World*. Berkeley, CA: University of California Press.

Jayaraj, A. (2006) Post disaster reconstruction experiences in Andhra Pradesh, in India. In: IF Research Group (Ed.), *International Conference on Post-Disaster Reconstruction – Meeting Stakeholder Interests*. 17–19 May 2006, Florence, Italy. The IF Research Group, Université de Montréal.

Laverack, G. (2005) Evaluating community capacity: visual representation and interpretation. *Community Development Journal*, **41**, 266–276.

Lawther, P.M. (2009) Community involvement in post disaster reconstruction – case study of the British Red Cross Maldives recovery programme. *International Journal of Strategic Property Management*, **13**, 153–169.

Le Masurier, J., Rotimi, J.O.B. and Wilkinson, S. (2006) *A Comparison between Routine Construction and Post-disaster Reconstruction with Case Studies from New Zealand*. 22nd ARCOM Conference on Current Advances in Construction Management Research, organised by Association of Researchers in Construction Management (ARCOM), 4–6 September, Birmingham, UK.

Lloyd-Jones, T. (2006) *Mind the Gap! Post Disaster Reconstruction and the Transition from Humanitarian Relief*. London: Royal Institution of Chartered Surveyors.

Lusthaus, C., Adrien, M.-H. and Perstinger, M. (1999) Capacity Development: Definitions, Issues and Implications for Planning, Monitoring and Evaluation. Universalia Occasional Paper, No 35, 1–21.

Martin, B. (2002) Are disaster management concepts relevant in the developing countries: the case of the 1999–2000 Mozambican floods. *The Australian Journal of Emergency Management*, **16**, 25–33.

McGinty, S. (2002) Community capacity building. In P. L. Jefferies (Ed.), *Australian Association of Research in Education 2002 International Education Research Conference Proceedings*, Brisbane.

Mileti, D.S. (1999) Disasters by design. In: Britton, N.R. (Ed.), *The Changing Risk Landscape: Implications for Insurance Risk Management*. Sydney: Southwood Press, pp. 1–16.

Mitchell, R.K., Agle, B.R. and Wood, D.J. (1997) Toward a theory of stakeholder identification and salience: defining the principle of who and what really counts. *Academy of Management Review*, **22**, 853–886.

Nateghi-A., F. (2000) Disaster mitigation strategies in Tehran, Iran. *Disaster Prevention and Management*, **9**, 205–211.

Newcombe, R. (2003) From client to project stakeholders: a stakeholder mapping approach. *Construction Management and Economics*, **21**, 841–848.

Nuseibeh, B. and Easterbrook, S. (2000) *Requirements Engineering: A Roadmap*. Proceedings of the Conference on The Future of Software Engineering, organised by ICSE, June 4–11, 2000, Limerick, Ireland.

Owen, D. and Dumashie, D. (2007) Built Environment Professional's Contribution to Major Disaster Management. FIG Working Week – Strategic Integration of Surveying Services, organised by International Federation of Surveyors, 13–17 May 2007, Hong Kong.

Pardasani, M. (2006) Tsunami reconstruction and redevelopment in the Maldives. *Disaster Prevention and Management*, **15**, 79–91.

Schipper, L. and Pelling, M. (2006) Disaster risk, climate change, and international development: scope for and challenge to integration. *Disasters*, **30**, 19–38.

Siriwardena, N., Haigh, R. and Ingirige, M.J.B (2009) *Identifying and Classifying Stakeholders of Post Disaster Reconstruction Projects in Sri Lanka*. Salford Postgraduate Research Conference 2009, 6 May 2009, The University of Salford, Salford, UK.

UN/ISDR (United Nations International Strategy for Disaster Reduction) (2008) *Private Sector Activities in Disaster Risk Reduction – Good Practices and Lessons Learned*. Geneva: UN/ISDR.

UN/ISDR (United Nations International Strategy for Disaster Reduction) (2009) *2009 UN/ISDR Terminology on Disaster Risk Reduction*. Geneva: UN/ISDR.

UNDP (2004) *Reducing Disaster Risk: A Challenge for Development*. Geneva: UNDP.

UNDP (2008) *Capacity Assessment Practice Note*. New York: UNDP.

World Wildlife Fund. *Humanitarian Partnerships – Post Disaster Reconstruction Projects*. www.worldwildlife.org/what/partners/humanitarian/postdisaster.html [accessed 02/11/2009].

3 Capacity of the Construction Industry for Post-Disaster Reconstruction: Post-Tsunami Sri Lanka

Krisanthi Seneviratne

3.1 Introduction

Against the increasing frequency of disasters, the construction industry has a very important role to play. As disasters damage and destroy the built environment, the construction industry is engaged in providing temporary shelter, restoring public infrastructure and services such as hospitals, schools, water supply, power, communications and securing income earning opportunities for vulnerable people. Beyond rebuilding, reconstruction must integrate risk reduction to ensure that the work takes place in safer locations, according to robust building codes and safety standards. Furthermore, it has been identified that buildings need to be made more resilient and this should be systematically integrated into the planning, design, construction and operational processes. Therefore it appears to be highly relevant to review the capacity of the construction industry in the context of disaster management.

The tsunami that struck the coastal area of northern Sumatra in Indonesia and affected Indonesia, Thailand, Sri Lanka, India, Maldives, Bangladesh, Malaysia, Myanmar and Somalia has been identified as one of the deadliest and costliest disasters in history. Though historically, Sri Lanka had experienced few, large scale natural disasters other than the occasional flooding of rivers, the country has recently been recognised as a tsunami-prone area. As a tsunami may pose a significant hazard in a number of coastal areas throughout the world, coastal countries must examine and evaluate the impact of tsunamis and develop plans to mitigate and reduce any apparent hazards. Worldwide communities face an increase in the frequency and variety of disasters that can cause direct and indirect effects. However, disasters particularly affect

Post-Disaster Reconstruction of the Built Environment: Rebuilding for Resilience, First Edition.
Edited by Dilanthi Amaratunga and Richard Haigh.
© 2011 Blackwell Publishing Ltd. Published 2011 by Blackwell Publishing Ltd.

developing countries rather than the developed world, mainly due to their lack of capacity to cope. Thus, it is important to develop these countries' capability at all levels. Within these contexts, this chapter will focus on the capacity of the construction industry in post-disaster reconstruction by paying particular attention to post-tsunami Sri Lanka.

The chapter commences with an introduction to the impact of tsunamis in terms of physical, socio-economic, human and environmental aspects. It then provides an introduction to the concept of disaster risk management followed by a discussion on the role of the construction sector in post-disaster recovery. The next section focuses particularly on the role of the construction industry in post-tsunami Sri Lanka. A comprehensive analysis is then provided on the capacity of the construction industry in post-tsunami reconstruction, in which the emphasis is on addressing the capacity gaps and strategies to minimise them.

3.2 Impact of tsunami

Tsunamis are caused mainly by earthquakes, volcanic eruptions or by landslips over the ocean. From time to time tsunamis have caused destruction mainly in littoral countries of the Pacific Ocean, Indonesian region and eastern USSR (Bradford, 2001 cited in Arya *et al.*, 2006). Although a few cases of tsunamis have also been reported in the Indian Ocean, these coastal impacts were not as severe as the December 2004 tsunami (Arya *et al.*, 2006). In December 2004, a massive earthquake of magnitude 9.0 struck the coastal area of northern Sumatra in Indonesia and this triggered a tsunami that affected Indonesia, Thailand, Sri Lanka, India, Maldives, Bangladesh, Malaysia, Myanmar and Somalia (Pheng *et al.*, 2006; Sonak *et al.*, 2008; Srinivas and Nakagawa, 2008). It is recognised as one of the deadliest and costliest disasters in history (Hansen, 2005; Morin *et al.*, 2008).

Tsunami is a Japanese term meaning wave (nami) in a harbour (tsu) and a series of travelling waves of extremely long length and period (Camilleri, 2006; UNESCO-IOC, 2006). Because of their destructiveness tsunamis have a huge impact on the human, social and economic parts of societies. The effects of tsunami destruction are described by two types of measurements: run-up elevation and inundation distance (UNESCO-IOC, 2006).

- *Run-up elevation:* This is the maximum vertical height above mean sea level at the farthest point on the shore that the sea surface attains during a tsunami. UN Special Envoy (2005 cited in UNESCO-IOC, 2006) indicates that, while any tsunami run-up over 1 metre can be dangerous, the December 2004 tsunami generated a run-up of nearly 34 metres in local Sumatra Island killing 170 000 people and destroying US$4.5 billion worth of property.

● *Inundation distance:* This is the maximum horizontal distance that a tsunami run-up can penetrate inland (UNESCO-IOC, 2006). The extent of inundation is an indication of the volume of water carried and the extent of damage along the coastline. Usually tsunamis may penetrate inland around 300 metres or more, flooding many areas with water and debris (UNESCO-IOC, 2006). The inundation distance is not the same throughout the shoreline and depends on the geographical context of the land. The steep-sided islands with barrier reefs are better protected from inundations and the lower grounds can be protected by developing closely spaced structures or planting dense stands of trees (UNESCO-IOC, 2006).

Sri Lanka is a country with an area of 65 610 square kilometres and 1340 kilometres of coastline. The tsunami hit the eastern, southern and south-western coasts of Sri Lanka which is more than two-thirds of its coastline (Oxfam, 2005; Yamada *et al.*, 2006). The impact of the tsunami can be explained in terms of physical, socio-economic, human and environmental effects as follows.

3.2.1   *Physical effects*

The 2004 Indian tsunami caused an estimated US$9.9 billion worth of damage (Koria, 2009). Destruction of transport and infrastructure such as roads, bridges, power and telecommunications caused logistical challenges and hampered the efforts of the relief agencies (Perry, 2007). During the immediate aftermath of the tsunami, the roads in the tsunami affected areas were not accessible and clearing them became an immediate priority (Yamada *et al.*, 2006). In Sri Lanka, the cost of power lines and transformers destroyed was about US$10 million. In the field of sanitation and water supply damage the costs amount to US$42 million. Total damages in the transport sector including railways, roads, bridges amount to US$85 million (Klem and Frerks, 2005). Apart from these, around 183 schools, four universities and 18 vocational centres were damaged causing losses of about US$26 million in Sri Lanka (Klem and Frerks, 2005; Reconstruction and Development Agency [RADA], 2006). The damage to the health sector is about US$60 million (Klem and Frerks, 2005).

People who worked in the fishing industry lived along the shoreline and as these houses were constructed poorly, many of them were damaged completely or partially. In Sri Lanka, housing constructed of brick was largely eroded at the base of the structures which could then collapse at any moment (Rodriguez *et al.*, 2006). The tsunami resulted in the total or partial destruction of nearly 100 000 homes in Sri Lanka (Koria, 2009). In Indonesia it was assumed that 101 000 housing units needed to be replaced and some 95 000 units rehabilitated. Overall reconstruction and rehabilitation investment was estimated to be above US$7 billion (Steinberg, 2007).

Though many non-governmental organisations (NGOs) provided large numbers of tents as temporary shelters, most of them were not customised to the environmental conditions in Sri Lanka and due to the high humidity, the tents were unbearable to live in (Yamada *et al.*, 2006). In India tents were not used by villagers due to the extreme heat generated inside these structures (Rodriguez *et al.*, 2006).

🖊 3.2.2 *Socio-economic effects*

Although the macroeconomic impact of tsunami-affected countries was limited, the socio-economic impact of affected areas is found to be huge (Lyons, 2009). The Asian Development Bank (ADB) (2005 cited in Lyons, 2009) estimated that the bulk of Sri Lanka's loss was in assets (US$970–1000 million or 4.4%–4.5% of GDP) rather than income (1.5% of GDP for 2005 and 2006) and the greatest share of this was in housing (US$341 million). However, according to Papathoma *et al.* (2003), loss of life, destruction of property and engineered structures and coastal infrastructure had led countries to experience major losses in economic and business interruption. Also, disasters have particularly affected lower-income countries (Srinivas and Nakagawa, 2008). For instance, one-third of the population in the areas affected by the tsunami in Sri Lanka live below the poverty line (Oxfam, 2005; Perry, 2007).

According to McMahon *et al.* (2006) an estimated 275 000 jobs were lost in Sri Lanka. Central bank sources predicted the rebuilding cost of housing and townships would amount to US$2 billion (IPS, 2005 cited in Pathiraja and Tombesi, 2009). Fishing communities experienced a substantial loss in terms of boats, motors and nets that would directly affect their livelihoods. In Sri Lanka it was estimated that 19 000 fishing boats were damaged and most of the deaths were among the fishing communities (Oxfam, 2005).

In addition to fishing, the agriculture, tourism and salt industries were impacted as a consequence of the tsunami (Rodriguez *et al.*, 2006). The damage to the assets in the tourism sector is estimated at US$250 million (Klem and Frerks, 2005). The agricultural sector was damaged due to debris from the tsunami waves, sedimentation, erosion, decrease in land fertility and salt pollution (Sonak *et al.*, 2008).

The tsunami killed people and not only destroyed homes and infrastructure but also community structures and community cohesion (Steinberg, 2007). Most people lost their loved ones, neighbours and friends and the sense of loss was widespread, affecting most communities. Natural disasters are not gender neutral and it is reasonable to assume that women are more vulnerable than men in disasters and their ability to recover is less favourable (Tobin, 1999). The female death rate in the tsunami was higher and sexual abuse in refugee settings was an issue, impacting on their role as an economic provider (Oxfam, 2005; Rodriguez *et al.*, 2006; Sonak *et al.*, 2008). As a result of the high death rate of women, men are facing the challenge of raising and educating their children (Rodriguez *et al.*, 2006). According to RADA (2006), the 2004 tsunami orphaned 1500 children.

3.2.3 *Human effects*

The victims of the tsunami suffered two successive waves and many people with minor injuries from the first wave were unable to move or swim and did not survive the second wave (Yamada *et al.*, 2006). Usually there are three to five times more injured than dead but, in contrast, in the tsunami disaster, it was reported that there were fewer injured than dead. For example, there were 200 injured people and 6000 deaths in Nagapattinam, India (Yamada *et al.*, 2006). The death toll of the 2004 tsunami is estimated to be between 200 000 and 300 000 (Poisson *et al.*, 2009). Srinivas and Nakagawa (2008) indicate that the majority were women and children. Indonesia was hit most severely with around 145 000 deaths (Steinberg, 2007). The death toll of the Indian disaster is estimated at 13 500 (Achuthan, 2009); 5792 people were reported missing with 6913 injured (Sonak *et al.*, 2008). In Sri Lanka, over 35 000 lives were lost, and over half a million people were displaced (Koria, 2009). Many bodies were buried in mass graves made necessary by rapid decomposition from prolonged water immersion. Identification of dead bodies had to be abandoned (Yamada *et al.*, 2006).

Discarded waste from densely populated camps combined with wet weather conditions spread malaria and dengue fever. This was controlled through fogging around camps, which is a temporary method to control mosquitoes, flies and other pests using pesticides and diesel to create fog in the affected area. Many people, including children, suffered great psychological trauma. Sonak *et al.* (2008) claim that there is considerable potential that residents of the tsunami-affected Islands face risks from kidney damage and various diseases like cholera, typhus, diphtheria and enteric fever if they consume ground water with high levels of saline and faecal coliforms (Sonak *et al.*, 2008).

Foreign medical personnel encountered linguistic and cultural barriers that needed considerable local resources to bridge the gap. As the tsunami damaged some pharmaceutical storehouses, it resulted in a shortage of medication. International support was required, which caused further problems due to different brand names, expired medication and unavailability of appropriate storage facilities.

3.2.4 *Environmental effects*

While the initial focus is on humanitarian aspects of the disaster, longer term rehabilitation and reconstruction should concentrate on the recovery of environmental assets such as water, land, forests, agricultural areas, fisheries, eco systems such as mangroves and coral reefs, etc. that may be damaged during disasters (Rodriguez *et al.*, 2006; Srinivas and Nakagawa, 2008).

The generation of debris, erosion along the coast, sedimentation along lagoons and water ways, salinisation of agricultural land and fresh water resources, impacts on fishery resources and plantations, are just some of the environmental impacts caused by a tsunami (Sonak *et al.*, 2008).

Among the environmental impacts caused by a tsunami, solid waste and disaster debris remains the most critical problem face. The debris may be mixed with hazardous materials and toxic substances such as asbestos, oil, fuel and other industrial chemicals (Srinivas and Nakagawa, 2008). Also, inappropriate dumping and air burning of debris can cause secondary impacts on the environment.

Another key impact is the contamination of soil and water. Seawater flooding, spreading of pollutants from waste, and untreated sewage and chemicals, resulted in the deterioration of the quality of ground and surface water resources. An estimated 62 000 ground water wells were contaminated by seawater, waste water and sewage, affecting access to drinking water and agriculture (Srinivas and Nakagawa, 2008).

3.3   Disaster risk management

As shown in Figure 3.1, the disaster risk management and response spiral identified two main phases: pre-disaster risk reduction phase and post-disaster recovery phase. The pre-disaster risk reduction phase includes risk and vulnerability assessment, risk reduction and disaster preparedness. The post-disaster recovery phase comprises relief, transition and reconstruction. Similarly, in 2004, the United Nations summarised the concept of disaster risk management

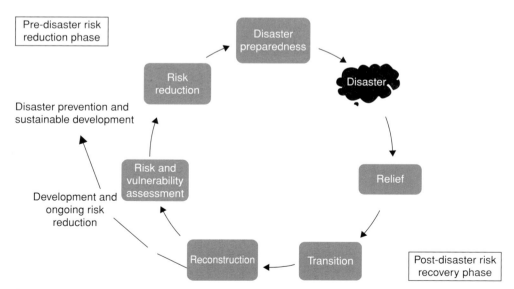

Figure 3.1   The risk management and response spiral

into four phases: hazard identification, mitigative adaptations, preparedness planning and recovery. The United Nations advocates the avoidance of threats and hazards within the disaster risk management framework, emphasising the mainstreaming of disaster risk management into national development strategies (Bosher *et al.*, 2007a).

Risk and vulnerability assessment involves identifying the nature and magnitude of current and future risks from hazards to people, infrastructure and buildings (RICS *et al.*, 2009). According to Papathoma *et al.* (2003), tsunami vulnerability analysis is fundamental to effective disaster plans as it is not possible to develop or implement sensible mitigative measures without a meaningful analysis. They indicate that both the population and the infrastructure within any given tsunami flood zone are not uniformly at risk and depend on a variety of parameters which includes the built environment, sociological data, economic data and environmental or physical data. Table 3.1 demonstrates the parameters and their characteristics.

Through vulnerability analysis it is possible to identify which public and private buildings should be reinforced or relocated and which buildings are likely to contain large numbers of trapped survivors (Papathoma *et al.*, 2003). It would be unrealistic to prevent or limit building and occupation of the coastal environment and reinforce every building within the tsunami flood hazard zone due to the economic costs. Also, it would not be possible to construct large and hard engineered coastal barriers such as breakwaters, walls and revetments. Therefore, detailed information on which buildings, structures and groups of people are vulnerable to the impact of tsunamis helps to develop cost-effective mitigation measures (Papathoma *et al.*, 2003).

Risk reduction refers to measures that can be taken to minimise the destructive effects of hazards and therefore minimise the scale of a possible disaster. As it is realised many disasters are not unexpected, the impacts of such disasters can be reduced through hazard mitigation (Bosher *et al.*, 2007a). Two types of mitigating hazards have been identified as follows (Tobin, 1999; Atmanand, 2003; Bosher *et al.*, 2007a; RICS *et al.*, 2009):

- *Structural mitigation:* This means the strengthening of buildings and infrastructure exposed to hazards (via building codes, engineering design and construction practices, etc.).
- *Non-structural mitigation:* This includes directing new developments away from known hazard locations through land use plans and regulations, relocating existing developments to safer areas and maintaining protective features of the natural environment (such as sand dunes, forests and vegetated areas that can absorb and reduce hazard impacts).

Preparedness involves dealing with the activities and measures taken in advance to ensure effective response to the impact of hazards, including the issuance of timely and effective early warnings and the temporary evacuation of people and property from threatened locations (Atmanand, 2003).

Table 3.1 Parameters and their characteristics. Reproduced by permission of Maria Papathoma Köhle (source: Papathoma *et al.,* 2003)

Parameter	Classification
Built environment	Number of storeys in each building: • Only one floor (vertical evacuation is impossible – high vulnerability) • More than one floor (vertical evacuation is possible – low vulnerability) Description of ground floor: • Open plan with movable objects like tables, chairs, etc. (high vulnerability to injury/damage) • Open plan or with big glass windows without movable objects (moderate vulnerability) • None of the above (low vulnerability) Building surroundings: • No barrier (very high vulnerability) • Low/narrow earth embankment (high vulnerability) • Low/narrow concrete wall (moderate vulnerability) • High concrete wall (low vulnerability) Building material, age, design: • Buildings of fieldstone, unreinforced, crumbling and or deserted (high vulnerability) • Ordinary brick buildings, cement mortar, no reinforcement (moderate vulnerability) • Precast concrete skeleton, reinforced concrete (low vulnerability) Movable objects: • Moveable objects (high vulnerability) • Non movable objects (low vulnerability)
Sociological data	Population density: • Population density during the night • Population density during the day • Population density during the summer • Population density during the winter Number of people per building: • High (high vulnerability) • Low (low vulnerability)
Economic data	Land use: • Business (shops, storage rooms, taverns, hotels) • Residential • Services (schools, hospitals, power stations)
Environmental/ physical data	Physical or man-made barriers: • Natural sandy or marsh (high vulnerability) • Soil embankment (moderate vulnerability) • Concrete stone wall (low vulnerability) Natural environment: • Wide intertidal zone (low vulnerability) • Intermediate intertidal zone (moderate vulnerability) • Narrow intertidal zone (high vulnerability) Land cover/vegetation: • No vegetation (high vulnerability) • Scrub and low vegetation (moderate vulnerability) • Trees and dense scrub (low vulnerability)

3.4   Role of the construction sector in post-disaster recovery

It is claimed that the construction sector should play an important role in the structural elements of mitigation, while developers and planners should be able to influence non-structural elements (Wamsler, 2006 cited in Bosher *et al.*, 2007a).

Activities that are planned and conducted before the disaster impact are called a proactive approach. In contrast, activities that are conducted after the disaster impact are called a reactive approach (Moe and Pathranarakul, 2006). The government of India recognised the need for a shift from the reactive approach to the proactive approach as it minimises the damage and loss to the people (Atmanand, 2003). Pheng *et al.* (2006) too emphasise the fact that the construction industry must set their sights towards improving local mitigation and preparedness instead of recovery and reconstruction in countries where tsunamis are infrequent but are a risk.

The emergency operations phase takes place immediately after the disaster to save and protect lives. Soon after, the relief work starts and this plays a crucial role during the first weeks and months following a disaster and includes the delivery of minimum safety, stability and confidence building beyond the materials supply (Regnier *et al.*, 2008). Relief activities include medical attention, body identification, clearing away rubble, debris, providing transport access, providing survival requirements (water purification kits, cooking utensils, foods, safe areas, relocation, shelter), and general living and psychological support (Perry, 2007).

The transition phase involves activities such as community surveys, needs assessment, land survey and acquisition and the provision of transitional shelter (RICS *et al.*, 2009). Care and maintenance of transitional shelters is required until permanent housing is constructed. However, the delays caused in reconstruction prolonged the transitional shelter usage (RADA, 2006).

Commencement of the recovery phase begins with the restoration of essential buildings and infrastructure services destroyed in the disaster and rehabilitation to assist the victims in returning to their pre-disaster livelihood (Pheng *et al.*, 2006). Koria (2009) indicates that a continuum exists between emergency relief operations, recovery initiatives and development activities. Accordingly, relief organisations are set up for rapid response and development organisations are built for long-term response, while recovery initiatives are situated in the middle ground to support local stakeholders to recover long-term capability.

Recovery is usually known to be slow, expensive and complex in terms of coordination and management (Koria, 2009). Recovery was set up across seven programme areas in Sri Lanka, three of which dealt with construction including shelter, health and water, and sanitation and used over 80% of the total available funds (Koria, 2009).

Despite the tsunami's enormous negative impact, the reconstruction process presented an opportunity for re-building better than original through addressing the failures of the past (RADA, 2006). RICS *et al.* (2009) also emphasised the feeding back of recovery experience to improve the resilience of vulnerable communities and inform the disaster management process thereby reducing future risks.

3.5 Post-tsunami Sri Lanka: the role of the Sri Lankan construction industry

It is certainly not possible to eliminate all disasters and many communities will always remain hazard prone (Sonak *et al.*, 2008). Therefore recovery remains important and it is pertinent to focus on recovery and factors that are conducive to facilitating recovery (Sonak *et al.*, 2008). Bosher *et al.* (2007a) highlighted the fragility and vulnerability of the built environment to both natural and man-made disasters. Special characteristics of construction that make built items more vulnerable are shown in Table 3.2.

Aside from this, the construction process itself can lead to disasters through environmental impacts that it may cause, e.g. through material extraction, processing, transportation, storage, use of land, faults of design and construction, methods of construction, extent of use of energy and waste (Ofori, 2002).

Disasters always hit human settlements, people and property. In December 2004 the tsunami destroyed hundreds of thousands of buildings around the Indian Ocean causing severe devastation to infrastructure such as roads, railways and bridges (Pathiraja and Tombesi, 2009). According to UN-HABITAT (2007), human settlement and constructed items are vitally important for the economic activities of a country and provide social and welfare benefits. For instance, they provide the space needed for the production of all goods and services, and fulfil the basic needs of people by providing shelter and offering them the opportunity to improve their living standards (Ofori, 2002). One

Table 3.2 Special characteristics of construction that make built items more vulnerable. Reproduced by permission of Gonzalo Lizarralde (source: Ofori *et al.*, 2002)

Characteristics	Resultant vulnerability
Immobile	Built items cannot be moved as a precaution and are exposed to disasters which occur where they are located
Highly expensive to test	Not possible to test the completed built item, by exposing it to the full force of a possible disaster and tests applied may not fully reflect the real situation
Long development process	Planning, design and construction involve various operations with dispersed control
Durable	Though durability is an essential requirement and a feature, items may weaken when exposed to disasters
Usage	Built items are occupied and utilised for various purposes and disasters may lead to loss of life as a result

of the most visible consequences of many disasters is the widespread destruction of houses (Barenstein and Pittet, 2007). According to Kennedy (2008), settlement and shelter construction is considered as an essential component of post-disaster reconstruction because housing reconstruction and rehabilitation is seen as central to the reconstruction of communities which needs to be integrated with recovery of economic and social sectors (Steinberg, 2007). Therefore the construction industry has a vital role to play in reconstruction after disasters.

Professionals, practitioners, volunteers from international institutions, NGOs that specialise in building, civil engineering, architecture, urban planning and environmental studies are all required to respond in various stages of disaster management (Atmanand, 2003; Pheng *et al.*, 2006). A guide for humanitarian agencies (RICS *et al.*, 2009) highlights the importance of the contribution of built environment professionals in achieving the long-term goal of sustainable recovery and development. It identifies the role of the different disciplines in disaster risk management and response.

At the same time it is claimed that disasters have a greater impact on the built environment of developing countries than on industrialised ones (Ofori, 2002). While disasters cause huge damage to people and property, it takes a long time to reconstruct after a disaster. Loss of wealth that took a long time to accumulate damages the economical and physiological conditions of people in developing countries. They are not in a position to replace or recover the damages through insurance or other means. On the other hand, the governments of these countries are not capable of compensating their people for damages due to budgetary constraints (Ofori, 2002). These issues hamper the development of countries. Waves destroyed the fragile homes of the less well-off, made of wood, clay thatch, wattle and daub, leaving more expensive brick or concrete built buildings standing. For example, in Vaharai, the poorest area of Batticaloa district, Sri Lanka, almost 70% of houses were made of clay and thatch (Oxfam, 2005). In Sri Lanka, many of those people who lived in flimsy fishing shanties were affected by the waves (Perry, 2007).

Koria (2009) claimed that long-term recovery has not progressed as expected and a discussion follows on the influence of political factors, coordination issues, beneficiary identification and participation issues, policy issues and lack of expertise and knowledge on implementing projects. Due to these issues, three years after the tsunami, homeless people still had to reside in refugee camps or temporary structures built close to their destroyed settlements. In addition most of the new buildings have failed to meet standards in quality and character.

- *Political factors*: Impractical political pledges on completion of the recovery led to a hostile media environment and some internal and external political agendas contributed to additional delays to the projects (Koria, 2009). Recovery work in the north of the country also had to stop due to lack of access.

- *Coordination issues*: Availability of extensive resources created a significant number of players within the context of recovery. This made coordination more difficult and hampered the optimum use of resources. As cultural differences exist between the international agencies and the local community, it is recommended they use local expertise and strengthen the local partnership (Perry, 2007).
- *Beneficiary identification and participation*: The beneficiaries have been difficult to identify and because of this their participation has been minimal. The affected people have different needs and therefore it is important to understand and cater to their social, cultural and environmental characteristics.
- *Public policies*: Public policies on relocation were identified as inconsistent with regards to the creation of a buffer zone. These changes caused delays in the projects. For instance a 200 metre coastal buffer zone was created which was later revised to a significantly narrower zone. Buffer zone policy is exemplified as a policy which was generated under urgent pressure and failed to address the root causes of vulnerability which may amplify the social, economic and environmental weaknesses that turn natural hazards into large-scale disasters (Ingram *et al.*, 2006). Though the 200 metre buffer zone was short lived it has amplified pre-tsunami socio-economic disparities, threatened coastal livelihoods, resulted in unfavourable social conditions and presented additional environmental threats (Ingram *et al.*, 2006). While relocation has emerged as central to disaster recovery in Sri Lanka and India, it is argued that if relocation efforts are to be successful, the communities should be actively involved in the decision-making process. People have complained about relocation as transferring communities from one risk area to another, e.g. to areas that would be affected by floods, to areas with a high prevalence of wild elephants, or to high-rise buildings that to some people seem inconceivable as dwellings.
- *Lack of technical and managerial expertise*: The lack of appropriate technical and managerial expertise and knowledge within the practicing organisations was rated and identified as a major constraint within the recovery context. This has led to significant failure in terms of delivering the projects within time, cost and scope limits. An inability to foresee market conditions caused the knock-on effects of escalated materials, labour and services costs and consequent delays.
- *Information and knowledge dissemination*: Most organisations were unable to capture, retain and re-use the knowledge from previous similar projects and various parties to the projects had to rediscover this knowledge (Koria, 2009). Management roles within projects were not clearly defined and insufficient or unsuitable resources were allocated to the projects.
- *Lack of guidance*: Bosher *et al.* (2007b) found that there is a lack of guidance on how to deal with unexpected disaster events and how to use this information to improve the design of buildings and infrastructure to cope with such risks and dangers. Further, it is revealed there is a lack of 'joined up'

thinking regarding how the expertise is being integrated into the process of design, construction and operation (Bosher *et al.*, 2007a).

- *Financial gaps*: Donors preferred to fund short-term relief operations rather than long-term reconstruction, and as a result funds available for reconstruction were sparser. Also, problems associated with aid utilisation and accountability were evident. Scarcity of building materials, shortage of skilled carpenters, electricians and plumbers all contributed to delays in rebuilding (Yamada *et al.*, 2006).

3.6 Capacity of the construction industry in post-tsunami reconstruction

It is important for reconstruction to be more sensitive to adaptations required to mitigate the impacts of disasters. For this to be achieved the local construction industry needs to appreciate the importance of a building delivery process and its life cycle of planning, design, construction, operation and maintenance. According to Ofori (2002), the ability of the built facility to withstand damage or to prevent or reduce loss of life can be enhanced through appropriate design which involves the structure, installations, layout, dimensions, selected materials and method of construction. For these to be realised, the construction industry needs to be provided with the requisite capacity and capability to design and construct, which can only be achieved through deliberate, planned, strategic and systematic efforts.

These should include not only the countries where disasters are frequent but also the countries where disasters are infrequent but vulnerable to future events. As an example, though natural- and human-induced hazards have caused minor disruption to the economy, infrastructure and people of the UK, it has been argued that the increasing magnitude and frequency of these events is due to climate changes and hence the vulnerability of the built environment is increased (Cabinet Office, 2006 cited in Bosher *et al*, 2009). According to Bosher *et al.* (2009), floods during June and July 2007 in the UK were wake-up calls to this vulnerability. Existing and future threats to building and infrastructure in the UK have been identified and these emphasise the input of all stakeholders in construction (Bosher *et al.*, 2007a). Accordingly they emphasise the importance of embedding flood hazard mitigation into pre-construction decision making. According to Camilleri (2006), though tsunamis in the Pacific Ocean are more frequent and devastating than in the Mediterranean, due to the vast development that has occurred around the Mediterranean shoreline over the past century, it is now understood that economic measures to reduce the risks from a tsunami are a necessity

Designing and constructing a resilient built environment demands an in-depth understanding of the expertise and knowledge required to avoid and mitigate the effects of threats and hazards (Hamelin and Hauke, 2005; Lorch, 2005; Bosher *et al.*, 2006, 2007 cited in Bosher *et al.*, 2007a). The pre-construction phase emerges as the most critical phase for integrating disaster risk management into the design, construction and operations process. Apart from this, the stakeholders involved in the preliminary phase should consider what materials they propose to use, where they plan to build the development, what they plan to develop and how the development will be built. When designing service networks (infrastructure), their location should be considered carefully.

It is important to ensure that up-to-date and secure building schematics are made available to the emergency services as the project progresses during the construction phase. Existing buildings and structures at risk should be considered and the impact of disasters must be identified, e.g. providing necessary defences for seaports, extra protection for nuclear power stations and transport infrastructure situated near to the coast (Bosher *et al.*, 2007a). Designers are perceived to be the most important construction stakeholders who should provide essential input into disaster risk management activities. Further, civil engineers, clients, developers and emergency risk managers are also key stakeholders who should provide essential input to disaster risk management within the context of design, construction and operation processes (Bosher *et al.*, 2007a).

Bosher *et al.* (2009) suggests a range of strategies that will be required to address the issue of built-in resilience for hazards, including innovation and knowledge, operations, planning, legislative and regulatory. A discussion of the major omissions that should be addressed to enhance the capacity of the construction industry to respond to disaster management including post-disaster reconstruction follows:

3.6.1 Competence development

Professionals involved in reconstruction work must be trained/advised and understand how recovery efforts can help reduce vulnerabilities and make communities more resilient to disasters (UN-HABITAT, 2007). Human resources or construction professionals should be provided with the knowledge and skills required to undertake appropriate designs and construction. Educational and professional institutions should be a powerful force for change in the industry. Educational institutions need to redevelop their curricula in order to include disaster management modules, while professional institutes can include various disaster management courses and training programmes as part of their continuous professional development schemes (Ofori, 2002). It is suggested that they develop necessary training programmes and accreditation schemes for recovery practices. In Sri Lanka it was found that higher educational institutions showed clear gaps in planning, implementing and managing these required programmes and associated research (Amaratunga and Haigh, 2008). Networking with international partners too can provide support to achieve the necessary competence. It is recommended that professionals should go through certified training as an indication of expertise. Kennedy *et al.* (2008) suggests all parties should undergo formal training, read published materials, and discuss and learn from the experiences of others to improve and perform better next time.

As many reconstruction housing projects fail due to inadequate infrastructure provision and livelihood support (Steinberg, 2007), there is a need to prioritise the planning exercises to limit the gaps in planning skills. As an example, a case study carried out in Sri Lanka illustrates that the selection of inappropriate locations for housing reconstruction as one of the main factors that affected the success of post-tsunami housing reconstruction. Often the donors were provided with inappropriate land to build upon and as a result of this, 5000 homes were built in the wrong place. This resulted in a misspend of US$40 million (Amaratunga and Haigh, 2008).

As the formation of construction labour and the consequent transfer of knowledge occurs through informal relationships, evaluating the true capacity of the construction industry in Sri Lanka has proven to be a difficult task (Pathiraja and Tombesi, 2009). According to the 2003 figures of the Institute of Construction Training and Development (ICTAD) the construction industry in Sri Lanka provides direct employment for around 300 000 people, which in turn becomes more than 1 000 000 when considering informal linkages. These numbers include the staff of around 2000 ICTAD registered contractors, 100 consultancy organisations and approximately 200 private sector property development entities. In addition there are over a dozen major institutions that provide assistance to the construction industry (Pathiraja and Tombesi, 2009). However, when compared with the scale of construction work required after the disaster, the industry's involvement in public sector development has been minimal in recent times (Pathiraja and Tombesi, 2009).

3.6.2 Innovation

There is a need to develop new, more resilient technologies and materials (Bosher *et al.*, 2009), e.g. the need for proper sewerage systems and cost-effective sewerage treatment plants. Where technological innovation is required to increase the resilience of reconstructed items, it is recommended they adapt existing methods and extend existing skills through training programmes where possible (RICS *et al.*, 2009).

For example, engineers and researchers could design a 40 square metre house for the coastal areas of Sri Lanka that they believe could withstand a tsunami and which costs between US$1000 to US$1500 (Hansen, 2005). It is simply designed with gaps between the walls that will enable water to flow through the structure without destroying it. Designers suggest that these houses would be approximately five times stronger than a conventional house of the same size (Hansen, 2005).

Development of decision support tools could enable stakeholders to make informed decisions regarding the integration of hazard mitigation measures during the process of planning, design, construction and operation (Bosher *et al.*, 2009).

3.6.3   *Information and knowledge dissemination*

Information is vital for planning, early warning and rehabilitation and reconstruction (Moe and Pathranarakul, 2006). Sri Lanka has demonstrated the need for proper information and knowledge dissemination as this has been identified as one of the major reasons behind the unsuccessful post-tsunami recovery (Haigh *et al.*, 2006). Many claim logistics management and information sharing and management is highly demanding in successful disaster management operations (McMahon *et al.*, 2006; Moe and Pathranarakul, 2006; Perry, 2007). Loss of information and loss of collective memory can be reduced by supporting initiatives such as libraries (Kennedy *et al.*, 2008).

3.6.4   *Planning regulations and building codes*

Statutes that relate to construction are formulated to ensure the health and safety of the occupants and their neighbours, e.g. land use planning and regulations determine the zoning, density, massing, and setbacks. Building design codes ensure the safety of users and their neighbours. In the context of disasters there is a need to develop these statutes based on professionally established hazard and vulnerability assessment (Ofori, 2002; Pheng *et al.*, 2006; Bosher *et al.*, 2009). These standards will help to establish a proper framework of documentation, procedures and practices to guide and facilitate continuous development of companies in the various aspects of tsunami management.

Kennedy *et al.* (2008) indicates spatial and urban planning can reduce the disaster risk and integrate development. This includes evacuation routes in the settlement design, tsunami barriers, buildings with vertical evacuation facilities and warning systems, using shelter and infrastructure shape and orientation to reduce exposure to wind and water forces and spatial planning to direct rain runoffs or to provide fire breaks through post-disaster programmes. The presence of housing and other buildings very close to the coast and the lack of appropriate planning and building codes/standards increased the damage and destruction of infrastructure and property (Srinivas and Nakagawa, 2008). Therefore proper land use planning is required for locating people and infrastructure in areas less prone to damage (Sonak *et al.*, 2008). Accordingly there is a need to identify and map zero construction zones, vulnerable zones and evacuation sites. Further it is accepted as prudent to move all non-essential structures further inland and to protect the shoreline with suitable vegetation (Camilleri, 2006). It is evident that natural barriers like sand dunes, coral reefs and mangroves have provided protection from the tsunami, e.g. in Sri Lanka's Yala and Bundala national parks. Arya *et al.* (2006) highlighted the plantation of mangroves and shelterbelt along coastal areas to minimise the effects from waves, wind, etc. Therefore it is essential to balance the coastal zone developments of harbours, buildings and other infrastructure with coastal zone management and restoring coastal ecosystems to enhance resilience (Srinivas and Nakagawa, 2008).

In addition there should be an effective and efficient enforcement framework to give practical effect to the regulations (Pheng *et al.*, 2006). For example, in Sri Lanka most houses constructed did not meet the requirements of the official building code issued by the Government and were considered unsafe. Since Sri Lanka has recently been identified as a tsunami prone area, it is important that the country pays more attention to developing research in the field of building materials in order to cater for future needs (Amaratunga and Haigh, 2008).

In the UK, despite an ever increasing range of guidance, information and legislation, there is a lack of suitable guidance specifically focused on proactive flood mitigation measures (Bosher *et al.*, 2009). The availability of such guidance, but also the awareness of it, is essential for key stakeholders. For instance, in Indonesia, despite the publication of improved building codes for anti-seismic housing construction in May 2005, many reconstruction agencies were not aware of the codes and did not follow them. As a result, the quality of construction and earthquake resistance of dwellings is poor in many projects and communities will not live in them (Steinberg, 2007).

3.6.5   Institution building

A disaster management authority should be appointed for managing disasters at a national level. Developing a national disaster management plan would clarify roles, responsibilities and streamline coordination across administrative levels and various stakeholders (Amaratunga and Haigh, 2008).

3.6.6   Waste management

Management of waste created by natural hazards is crucial and the need for clear guidelines is highlighted. It is important to explore ways of recycling and reusing debris such as concrete, blocks, bricks, etc. For example, an average weight of debris (bricks, concrete and roofing material) per house destroyed was in the range of 3000 kilograms. For approximately 100 000 houses destroyed and damaged there would have been about 300 million kilograms of debris alone in Sri Lanka (Amaratunga and Haigh, 2008). Implementation of prevailing rules, scarcity of land, unawareness of new waste management strategies, poor local expertise and poor attitudes are identified as among the reasons for failure in waste management.

3.6.7   Research and development

As the built environment is highly vulnerable to tsunami disasters, research and development programmes should be conducted to investigate the relationships between tsunamis and the built environment in terms of developing appropriate designs and finding suitable materials that are tsunami resistant (Pheng *et al.*, 2006). Since Sri Lanka has recently been identified as a tsunami prone area, more attention should be focused on developing research on building materials to cater for future needs in this country (Amaratunga and Haigh, 2008).

Depending on local construction skills, techniques and resources is essential to avoid the risk of introducing imported materials and components that are locally and culturally unfit. Furthermore, the importance of using local skills and resources in long running economic developments will reinforce the local building sector (UN-HABITAT, 2007). Therefore it is necessary to identify, through research, disaster-resistant materials which are suited to the local context, and are of good quality, durable and affordable.

The 'bottom up' and multidisciplinary approach to construction will enhance built-in resilience (Bosher *et al.*, 2009). To avoid the risk of introducing unfamiliar forms of construction based on imported materials and components, it is necessary to account for local skills, techniques and resources (RICS *et al.*, 2009). For example, it is found that the delivery of housing based on the community approach is faster and is of a higher quality, satisfaction and accountability than the traditional contractor-based methods in Sri Lanka (Amaratunga and Haigh, 2008).

3.6.8   Raising awareness in poor and vulnerable people

Loss of official documents and those who held land rights caused problems of ineligibility for official grants. For example, in Sri Lanka, with the introduction of buffer zones, 45 000 houses needed to be built and government policy was to provide those people affected with a new, permanent house in a new location. However, to claim a house, they were required to have a title or deed for their old house and those who rented, squatted or shared missed out on the opportunity (Oxfam, 2005). Therefore, it is important to educate the poor and less educated people on their rights and how to claim them. Lack of knowledge considerably increases the vulnerability of the population and raising awareness through road shows, films, exhibitions, leaflets and posters is identified as a key tool to reduce such vulnerability (Morin *et al.*, 2008).

3.7   Summary

The tsunami of December 2004 caused significant and extensive destruction of lives, houses, public infrastructure and services across the whole region. Though Sri Lanka had experienced a few large scale natural disasters, it was hugely affected by this tragic tsunami. Sri Lanka, and many other countries, are facing an increasing frequency and variety of disasters that have direct and indirect effects. Poor countries are found to be affected in particular. The built environment is highly vulnerable to both natural and man-made disasters. In these contexts the construction industry has a vital role to play in post-disaster reconstruction and disaster management. Reconstruction efforts must integrate risk reduction to reduce the future vulnerabilities and impacts of disasters by considering the process of reconstruction as an opportunity for 'building back better'. Unfortunately, reconstruction has not progressed as expected and is hampered by poor coordination, political restraints, poor beneficiary identification and participation, public policies, poor expertise and knowledge, poor information and knowledge dissemination, lack of guidance and financial control.

The ability of the built facility to withstand damage or to prevent or reduce loss of life can be enhanced through appropriate design which involves the structure, installations, layout, dimensions, selected materials and method of construction. For these to be realised the construction industry must be provided with the requisite capacity and capability for design and construction. In order to enhance the capacity of the construction industry to respond to disaster management it is important to address the issues of competence development, innovation of technologies and materials, information and knowledge dissemination, planning regulations and building codes, institutional building, waste management, research and development and to raise the awareness of poor and vulnerable people.

References

Achuthan, N.S. (2009) Four years beyond tsunami: contours of a roadmap for a coordinated 'multi-hazard (including tsunami) risk management action plan for tsunami-affected villages in Tamail Nadu': overview of ongoing/projected initiatives. *Disaster Prevention and Management*, **18**, 249–269.

Amaratunga, D. and Haigh, R. (2008) *Statement of Research Gaps in Post-tsunami Sri Lanka. Research to Enhance Post-tsunami Reconstruction Efforts*. EURASIA, University of Salford, UK.

Arya, A.S., Mandal, G.S. and Muley, E.V. (2006) Some aspects of tsunami impact and recovery in India. *Disaster Prevention and Management*, **15**, 51–66.

Atmanand (2003) Insurance and disaster management: the Indian context. *Disaster Prevention and Management*, **12**, 286–304.

Barenstein, J.D. and Pittet, D. (2007) *Post-Disaster Housing Reconstruction. Current Trends and Sustainable Alternatives for Tsunami-affected Communities in Coastal Tamil Nadu*. Institute for Applied Sustainability to the Built Environment, University of Applied Sciences of Southern Switzerland.

Bosher, L., Dainty, A., Carrillo, P. and Glass, J. (2007a) Built-in resilience to disasters: a pre-emptive approach. *Engineering, Construction and Architectural Management*, **14**, 434–446.

Bosher, L., Dainty, A., Carrillo, P., Glass, J. and Price, A. (2007b) Integrating disaster risk management into construction: a UK perspective. *Building Research and Information*, **35**, 163–177.

Bosher, L., Dainty, A., Carrillo, P., Glass, J. and Price, A. (2009) Attaining improved resilience to floods: a proactive multi-stakeholder approach. *Disaster Prevention and Management*, **18**, 9–22.

Camilleri, D.H. (2006) Tsunami construction risks in the Mediterranean – outlining Malta's scenario. *Disaster Prevention and Management*, **15**, 146–162.

Haigh, R., Amaratunga, D. and Kerimanginaye, K. (2006) An exploration of the construction industry's role in disaster preparedness, response and recovery. In: Sivyer, E. (Ed.), *COBRA 2006*. London: University College London, RICS.

Hansen, B. (2005) Simple, economical house design to resist future tsunamis. *Civil Engineers* (August), 13–14.

Ingram, J.C., Franco, G., Rio, C.R. and Khazai, B. (2006) Post-disaster recovery dilemmas: challenges in balancing short-term and long term needs for vulnerability reduction. *Environmental Science and Policy*, **9**, 607–613.

Kennedy, J., Ashmore, J., Babister, E. and Kelman, I. (2008) The meaning of 'build back better': evidence from post-tsunami Aceh and Sri Lanka. *Journal of Contingencies and Crisis Management*, **16**, 24–36.

Klem, B. and Frerks, G. (2005) Tsunami response in Sri Lanka: report on a field visit from 6–20 February 2005, Clingendael Institute.

Koria, M. (2009) Managing for innovation in large and complex recovery programmes: tsunami lessons from Sri Lanka. *International Journal of Project Management*, **27**, 123–130.

Lyons, M. (2009) Building back better: the large-scale impact of small-scale approaches to reconstruction. *World Development*, **37**, 385–398.

McMahon, P., Nyheim, T. and Achwarz, A. (2006) After the tsunami: lessons from reconstruction. *McKinsey Quarterly*, (1), 94–105.

Moe, T. L. and Pathranarakul, P. (2006) An integrated approach to natural disaster management: public project management and its critical success factors. *Disaster Prevention and Management*, **15**, 396–413.

Morin, J., Coster, B.D., Paris, R., Flohic, F., Lavigne, D.L. and Lavigne, F. (2008) Tsunami-resilient communities' development in Indonesia through educative actions lessons from 26 December 2004 tsunami. *Disaster Prevention and Management*, **17**, 430–446.

Ofori, G. (2002) Construction industry development for disaster prevention and response, keynote paper. *Proceedings of the Conference on Improving Post Disaster Reconstruction in Developing Countries*, 23–25 May 2002, organised by the Information and Research for Reconstruction and IF Research Group, University of Montreal, Canada. Available at: http://www.GRIF.UMontreal.ca/pages/papersmenu.html

Oxfam (2005) *Targeting Poor People. Rebuilding Lives after the Tsunami*. Oxfam International.

Papathoma, M., Dominey-Howes, D., Zong, Y. and Smith, D. (2003) Assessing tsunami vulnerability, an example from Herakleio, Crete. *Natural Hazards and Earth System Sciences*, **3**, 377–389.

Pathiraja, M. and Tombesi, P. (2009) Towards a more 'robust' technology? Capacity building in post-tsunami Sri Lanka. *Disaster Prevention and Management*, **18**, 55–65.

Perry, M. (2007) Natural disaster management planning: a study of logistics managers responding to the tsunami. *International Journal of Physical Distribution and Logistics Management*, **37**, 409–433.

Pheng, L.S., Raphael, B. and Kit, W.K. (2006) Tsunamis: some pre-emptive disaster planning and management issues for consideration by the construction industry. *Structural Survey*, **24**, 378–396.

Poisson, B., Garcin, M. and Pedreros, R. (2009) The 2004 December 26 Indian Ocean tsunami impact on Sri Lanka: cascade modelling from ocean to city scales. *Geophysics Journal International*, **177**, 1080–1090.

RADA (2006) *Post-Tsunami Recovery and Reconstruction: Progress, Challenges, Way Forward*. Colombo: Reconstruction and Development Agency.

Regnier, P., Neri, B., Scuteri, S. and Miniati, S. (2008) From emergency relief to livelihood recovery lessons learned from post-tsunami experiences in Indonesia and India. *Disaster Prevention and Management*, **17**, 410–429.

RICS, ICE, RIBA and RTPI (2009) *The Built Environment Professions in Disaster Risk Reduction and Response. A Guide for Humanitarian Agencies*. London.

Rodriguez, H., Wachtendorf, T., Kendra, J. and Trainer, J. (2006) A snapshot of the 2004 Indian Ocean tsunami: societal impacts and consequences. *Disaster Prevention and Management*, **15**, 163–177.

Sonak, S., Pangam, P. and Giriyan, A. (2008) Green reconstruction of the tsunami-affected areas in India using the integrated coastal zone management concept. *Journal of Environmental Management*, **89**, 14–23.

Srinivas, H. and Nakagawa, Y. (2008) Environmental implications for disaster preparedness: lessons learnt from the Indian ocean tsunami. *Journal of Environmental Management*, **89**, 4–13.

Steinberg, F. (2007) Housing reconstruction and rehabilitation in Aceh and Nias, Indonesia-rebuilding lives. *Habitat International*, **31**, 150–166.

Tobin, G.A. (1999) Sustainability and community resilience: the holy grail of hazards planning? *Environmental Hazards*, **1**, 13–25.

UN-HABITAT (2007) *Sustainable Relief and Reconstruction – Synopsis from World Urban Forum II and III. From Conceptual Framework to Operational Reality*. Nairobi, Kenya, UN Human Settlements Programme.

UNESCO-IOC (2006) *Tsunami Glossary*. IOC Information Document No. 1221. Paris: UNESCO.

Yamada, S., Gunatilake, R.P., Roytman, T.M., Gunatilake, S., Fernando, T. and Fernando, L. (2006) The Sri Lanka tsunami experience. *Disaster Management and Response*, **4**, 38–48.

4 Resourcing for Post-Disaster Reconstruction: A Longitudinal Case Study Following the 2008 Earthquake in China

Yan Chang, Suzanne Wilkinson, Regan Potangaroa and Erica Seville

4.1 Introduction

This chapter examines the large-scale and intensive reconstruction resource management process used following the Wenchuan earthquake in China in May 2008. In particular, the chapter focuses on the various post-disaster reconstruction resourcing problems encountered by the Chinese Government and affected communities and how these problems were solved. An explicit lack of resources required for housing reconstruction combined with implicit resourcing capacity bottlenecks significantly limits the degree to which a successful recovery could occur. Despite the inherent link between post-disaster resourcing and reconstruction performance, little research has been conducted which shows the interconnectedness of reconstruction and resource management. A longitudinal case study in China's Sichuan Province is discussed in this chapter, which follows the earthquake reconstruction and resource management activities from six weeks after the earthquake until to date. This chapter is based on the one and half years' case study looking back at post-earthquake resourcing policies, decisions and programmes and to examine their effectiveness in housing reconstruction. The chapter identifies changing practices and identifies the challenges which confront the construction professionals and industry in such an environment and assesses the transferability of the techniques applied in China to other disaster affected regions.

Post-Disaster Reconstruction of the Built Environment: Rebuilding for Resilience, First Edition.
Edited by Dilanthi Amaratunga and Richard Haigh.
© 2011 Blackwell Publishing Ltd. Published 2011 by Blackwell Publishing Ltd.

4.2   The impact of the 2008 Wenchuan earthquake

On 12 May 2008 the magnitude (M) 8.0[1] earthquake struck China's province of Sichuan and its neighbours, causing massive casualties and widespread destruction to buildings and infrastructure. The epicentre was in Wenchuan County, a rural and mountainous region in Sichuan Province, about 80 kilometres from capital city Chengdu. More than 15 million housing units collapsed during the earthquake and resulted in direct losses to buildings and infrastructure of over US$150 billion (Paterson *et al.*, 2008). Approximately 34 125 kilometres of highways, 1263 reservoirs, 7444 schools, 11 028 medical institutions and numerous urban, rural residences and factories were devastated by the earthquake with direct economic losses reaching US$123.66 billion (State Planning Group, 2008). Housing was the single greatest component of all losses in terms of economic value and in terms of buildings damaged.

The damage experienced by structures in the Wenchuan earthquake was largely contingent on the construction type.[2] In the mountainous terrain of Sichuan Province, most housing buildings are one or two storey masonry structures composed of bricks or concrete blocks that do not possess seismic-resistant elements such as reinforced concrete. These structural defects made housing highly vulnerable to earthquakes, and consequently, many collapsed. Tables 4.1 and 4.2 provide a general sense of the immense scale of housing reconstruction required in the rural and urban earthquake impacted regions.

Table 4.1   Rural housing retrofit and reconstruction (source: State Planning Group of Post-Wenchuan Earthquake Restoration and Reconstruction, 2008)

	Unit	Total	Sichuan	Gansu	Shaanxi
Strengthening	Household (10 000)	214.11	190.42	12.55	11.14
New construction	Household (10 000)	328.97	298.49	26.04	4.44
	Room number (10 000)	1315.87	1193.96	104.14	17.77

Table 4.2   Urban housing retrofit and reconstruction (source: State Planning Group of Post-Wenchuan Earthquake Restoration and Reconstruction, 2008)

	Unit	Total	Sichuan	Gansu	Shaanxi
Strengthening	Area (10 000 m^2)	5807.09	5517.7	220.47	68.92
New construction	Suite (10 000)	85.98	84.29	1.28	0.41
	Area (10 000 m^2)	6620.10	6490.03	98.50	31.57

4.3   Wenchuan earthquake reconstruction process

4.3.1   *Legislative and policy guidelines for Wenchuan earthquake reconstruction*

After the earthquake, the State Council of China took swift legislative action to establish a multi-governmental management framework for the recovery.

On June 4, 2008, the State Council issued *The Regulations on Post-Wenchuan Earthquake Restoration and Reconstruction*[3] which was the first law with respect to post-disaster recovery and reconstruction in China's history. The 'Regulations' provides a legal basis for various departments and government entities both in and out of quake-hit regions to assist with recovery and reconstruction, and establishes guiding principles for damage assessment, temporary housing, reconstruction planning, financing, implementation, and management.

As a supplementary policy to aid the full implementation of the 'Regulations', *One-on-one Assistance Program for Post-Wenchuan Earthquake Restoration and Reconstruction* became a key constituent of China's post-disaster management framework. In this programme, the earthquake-stricken areas of Sichuan, Gansu and Shaan-xi provinces were divided into 24 districts and twinned with 24 relatively developed localities across China. Sister localities have been tasked over the next three years with funding 1% of their annual GDP, provision of human resources and temporary housing units, and in-kind support from planning institutions and other departments in association with disaster reconstruction.

Four months after the earthquake *The State Overall Planning for Post-Wenchuan Earthquake Restoration and Reconstruction* was released for reconstruction implementation in the earthquake-impacted region. As the recovery steps entered into an overall rebuilding stage, the role and responsibilities of the Chinese government shifted to technical support and supervision of reconstruction implementation with less administrative intervention.

The State Overall Planning for Post-Wenchuan Earthquake Restoration and Reconstruction was aimed at a more integrated and balanced reconstruction and redevelopment of earthquake affected areas. Reconfirming people's safety as a top priority, the document is committed to a stringent 'quality control' consistently incorporated into housing reconstruction. In addition, reconstruction planning was oriented towards accelerating and streamlining the reconstruction process by improving layout and land use without detriment to the sustainability of quake-affected areas.

In all, due to the centralised government system, China was capable of mobilising and deploying the necessary materials, resources and funds by assembling the forces of governments, organisations and communities. It is important to note that a series of recovery policies and regulations were introduced after the earthquake to guide people in disaster relief and those participating in the long-term reconstruction and rehabilitation process.

4.3.2   *Housing reconstruction approach*

One of the most complex tasks facing recovery managers is to determine and implement the most appropriate ways to reconstruct buildings and infrastructure (International Recovery Platform, 2007). There is a growing body of literature regarding the pros and cons of different approaches to post-disaster housing reconstruction (e.g. Barakat, 2003; Barenstein, 2006; Twigg, 2006; Barenstein, 2008). Several elements have to be considered when choosing the appropriate reconstruction method, including the wider political and social contexts (Oliver-Smith, 1990), operational requirements (Harvey, 2005), as well as the expectations and preferences of the people affected (Bolin and Stanford, 1998). In addition, rebuilding approaches may also be determined by factors such as the agency's available resources, overall mandate, experience and capacities (Barenstein, 2006). Even within the same disaster area, different agencies or individuals may adopt varied reconstruction methods. In spite of the various categories in the literature, four main approaches are widely recognised and applied in past disaster reconstruction practice:

- *Government-driven approach*: Government takes on the entire programme of post-disaster reconstruction.
- *Contractor-driven approach*: External expertise or professional building contractors are contracted with large-scale design and construction of housing for communities affected.
- *Donor-driven approach*: Non-governmental organisations (NGOs) or other designated organisations are tasked with housing rebuilding programmes for beneficiaries.
- *Owner-driven approach*: House owners are responsible for rebuilding their own houses through self-maintenance with limited external financial, technical and material assistance.

The paternalistic pure government-driven reconstruction approach could only be expected in cases of massive devastation, such as the rebuilding of China's Tangshan city after the catastrophic earthquake in 1976. Regarding post-disaster reconstruction in developing countries, various international aid agencies had been dedicated to providing technical and financial assistance for disaster-stricken areas. This donor-driven model was modified in the 1980s to mix the provision of rebuilding housing by outside contractors with locally managed self-help building programmes. In 1982, the United Nations Disaster Relief Organisation (1982) advocated the community participatory approach as a key to success in post-disaster reconstruction for developing societies, as the involvement of external interventions from both aid agencies and contractors brought about other problems during reconstruction. In contrast, in developed nations, the housing rebuilding is to a great extent dependent on personal savings, insurance and bank loans to finance recovery of residential properties. In this system, the reconstruction model developed and applied in developed countries can be categorised as an owner/community-driven approach. As Davidson *et al.* (2007) pointed out, construction in general and reconstruction in particular are rooted in their socio-politico-economic contexts and there is no single 'best' approach for community participation.

With the widespread housing reconstruction commencing in China's earthquake impacted areas in August 2008, all the discussed reconstruction approaches and other forms[4] can be seen with a few variations in different locations and programmes. The self-construction with a combination of governmental subsidies, partnership assistance, and social help and support was advocated in rural areas. Many examples of local reconstruction planning, land readjustment, joint housing and actual rebuilding participation, show how people affected leverage their value into collective reconstruction solutions. However, there are challenges associated with a community participatory role in terms of ensuring that new construction meets building standards and is seismically resistant. Some counties in the affected rural areas in Sichuan Province adopted innovative approaches to address this, encompassing accessible education and training for farmers not only on the technical specifics of building codes, but on the basic reasoning that underlie them, and linking the provision of subsidies to inspection and monitoring at each stage of the house reconstruction process (Paterson *et al.*, 2008). In contrast, in urban cities impacted, housing reconstruction projects were mainly packed with professional building contractors or other social organisations. Hence, although house owners came to the fore in reconstruction after the Wenchuan earthquake – assuming more design and planning at the upfront stage, and control and monitoring responsibilities at the late stage, the real construction work of their house was still procured and tasked with construction expertise, builders or contractors.

4.3.3 Housing reconstruction progress

The earthquake-impacted areas, in conjunction with officials from the Earthquake Relief Headquarters of the State Council (ERHSC) and the Ministry of Housing and Urban-Rural Development (MOHURD), and with the participation of the Sponsoring Locality, launched a series of policies and initiatives to expedite housing retrofit assistance for residents of damaged buildings and to enhance permanent housing reconstruction programmes. In spite of different forms of these initiatives, the key element in common was to stimulate and promote the use of resources both nationwide and locally.

One year on from the Wenchuan earthquake, Sichuan Provincial People's Government Information Office held a press conference on 7 May 2009 informing the public on the progress of '5.12' Wenchuan earthquake restoration and reconstruction (see Table 4.3). As of 4 May 2009, the province had started rebuilding 1.248 million permanent rural housing units, accounting for 98.8% of the reconstruction task, 241 000 of them under construction and 1 008 000 delivered. The repair and retrofit for rural housing was completed by the end of 2008. Maintenance and reinforcement of 719 000 (50.7%) of the province's urban housing was completed. Among 138 000 of initiated urban housing units, 105 000 were under construction, whereas 33 000 were built and occupied. In February 2010, by the time the researchers had conducted the fourth field visit to the earthquake zone in China, rural housing reconstruction was basically completed as well as the rebuilding of the vast majority of urban housing (Table 4.3).

Table 4.3 Housing reconstruction progress as of 4 May 2009 in Sichuan Province (source: Sichuan Provincial People's Government Information Office, 2009)

Housing type	Damaged/ destroyed units	Retrofit		Rebuilding	
		Units to be strengthened	Completion rate	Units to be rebuilt	Completion rate
Rural housing	3.476 million	2.213 million	100%	1.263 million	79.8%
Urban housing	1.732 million	1.418 million	50.7%	0.314 million	10.5%

4.4   Resourcing for Wenchuan earthquake reconstruction

4.4.1   Post-disaster reconstruction resourcing context

In comparison with pre-event project construction, the post-disaster reconstruction environment is chaotic, dynamic and complex (Berke et al., 1993; Alexander, 2004; Birkland, 2006; Davidson et al., 2007). Many factors contribute to this difficulty such as the variety and severity of disaster impacts, including physical, economic, social and psychological (International Federation of Red Cross and Red Crescent Societies [IFRC], 2006; Waugh and Smith, 2006), the diversity and multiplicity of tasks that each working unit is to execute (Berke et al., 1993), and volatility and instability of the market in disaster-stricken areas (Jayasuriya and McCawley, 2008).

In the wake of a disaster, the majority of manufacturing-supply facilities and operational systems in up-stream industries in disaster-stricken areas are likely to be damaged and the construction transaction market tends to be in disorder and adversarial. This, if compounded with disruption of transportation and energy supply, and historical problems of local industry, could significantly exacerbate the difficulty in project sourcing within the construction industry (Cho et al., 2001; Singh, 2007; Jayasuriya and McCawley, 2008). Consequently, the urgent housing reconstruction projects are susceptible to numerous resourcing bottlenecks such as lack of resources and alternatives (Russell, 2005; Zuo and Wilkinson, 2008), lack of access to available resources (Green et al., 2007) and limited sources of resources (Brunsdon et al., 1996; Oxfam Australia et al., 2007), which significantly impede the reconstruction process in disaster-affected countries. These, together with an ineffective resourcing approach (United Nations Development Programme, 2005; IFRC, 2006) and poor resource management (Steinberg, 2007), lead to the recurrent concerns following a disaster such as project failure and rework (Steinberg, 2007), 'Dutch disease' (Corden, 1984; Adam and Bevan, 2004), and 'cost surge' (Rodriguez et al., 2007; Jayasuriya and McCawley, 2008) undermining the effectiveness of post-disaster reconstruction performance.

The pressure to acquire resources for post-disaster reconstruction is even higher for poorer countries as they have to rely on external assistance, such as NGOs, international NGOs (INGOs), World Bank, etc.; or reallocate resources from existing projects to rehabilitation or reconstruction to meet their recovery needs (Freeman, 2004; Jayasuriya and McCawley, 2008). As a result, dependence on external aid is likely to suppress the local self-production capacity and reduce the likelihood of the reconstruction programme succeeding (Cuny, 1983). These resource reallocation tactics disrupt markets and economic order (Makhanu, 2006), adversely affecting sustainable productivity layout and economic and social development goals in the long run (Work Bank Operations Evaluation Department, 2005).

4.4.2 Resource availability in Wenchuan earthquake housing reconstruction

Resource management for post-disaster reconstruction projects requires a focus on building materials, plant, human power, skills, resilient infrastructure and good disaster preparedness plans. The availability of resources is not only concerned with the quantity required for long-term disaster reconstruction, but also their accessibility in procurement to available resources and usability during the construction process.

The Wenchuan earthquake serves as an exemplar for the disproportionate volume between resource demand and supply during the post-quake reconstruction. The supply shortfalls of building materials in Table 4.4 indicate that the reconstruction demand was sufficient to pose difficulty to reconstruction practitioners in procurement of these materials. Labour scarcity was another problem leading to wage increases undermining the functionality of local construction markets.

Based on our longitudinal investigation, in the peak of housing reconstruction in December 2008, the most needed construction materials in quake-affected areas of Sichuan Province were brick, aggregate, cement with a 127%, 125%, and 30% price increase respectively and steel with a 30% price fall[5]. In June 2009, one year after the Wenchuan earthquake, the local material production could basically meet the needs of local reconstruction and prices of brick and cement in affected areas of Sichuan decreased whereas that of steel and aggregate increased (see Figures 4.1–4.4[6] and Table 4.5).

Much of the supply shortage of these materials at the early stage of reconstruction came from large-scale and intensive building demand. In addition, the inadequacy of local production capacity, increased transportation fees, and the scarcity of local raw minerals also contributed to the subsequent supply

Table 4.4 Supply shortfalls of cement, brick and steel in earthquake-stricken areas (source: http://www.sc.gov.cn)

Construction materials	2008–2009	2009–2010	2010–2011
Cement (million tons)	53	39	31
Brick (billion pieces)	35.5	17.8	3
Steel (million tons)	3–3.6	3–3.6	3–3.6

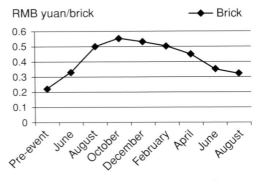

Figure 4.1   Brick price trend (source: authors' own market investigation)

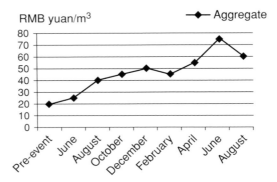

Figure 4.2   Aggregate price trend (source: authors' own market investigation)

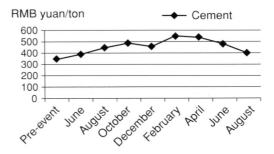

Figure 4.3   Cement price trend (source: authors' own market investigation)

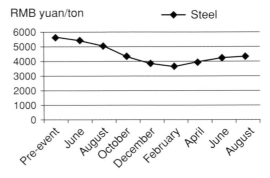

Figure 4.4   Steel price trend (source: authors' own market investigation)

Table 4.5   Most needed building materials for housing reconstruction

Order	December 2008	June 2009
1	Brick	Aggregate
2	Cement	Steel
3	Steel	Cement
4	Aggregate	Brick

disruption and inflation in disaster-affected areas. Brick supply pressure was caused by oil and raw coals prices rising in China, a monetary incentive from the government and a traditional value on the New Year. For example, the local government promised that for house owners who embarked on rebuilding before 12 May 2009, the government would subsidise 5000 Chinese dollars per household. This financial encouragement along with people's intention to welcome traditional Chinese Spring New Year in their new house caused competition among house owners for limited resources, leading to the brick price soaring.

The global economic crisis in 2008 was less insidious, yet equally destructive, to China's construction market with the greatest impact on the steel industry. According to China Iron and Steel Association (2009), in the second half of 2008, steel producers in China were cutting their production due to the reducing demand for and dropping prices of steel. Nevertheless, the heavy overcapacity of the steel industry made steel makers in China stock the raw materials they purchased at a high price and unwilling to output steel products. This profit-driven risk aversion of steel makers to some extent exacerbated the steel supply shortage in the disaster-impacted areas.

According to Figures 4.1–4.4 and Table 4.5, the price decline of brick and cement started from April 2009. The reasons for this are that nearly 70% of rural housing was completed and the production of 70% of brick factories and plants were restored and functional, supplemented by newly-built production facilities. By comparison, the price of aggregate and steel increased from April 2009. The possible explanation for price elevation of aggregate may be that the governmental authorities of China implemented a strict control on the exploitation of sands and gravels in early 2009. The ever increasing standards for environmental protection in association with exploitation of natural resources had further heightened pressure to push up the production cost of aggregate. In response to the global financial crisis, in November 2008, the Chinese government launched a 4 trillion Chinese dollars (approximate US$600 billion) investment programme aimed at boosting the economy. Many investment funds will be used for the construction of railways and airport infrastructure which require large amount of steel and cement. In 2009, the reviving signs of the steel industry appeared in the Chinese market accompanied by the nationwide steel product price escalation.

4.5   Key resourcing problems and solutions adopted by the Chinese reconstruction teams

Resourcing broadly encompasses a wide range of activities that have a bearing on resource provision for construction projects, embracing resource planning and preparedness, resource procurement, resource delivery, and development of resource alternatives. A desirable resourcing approach calls for an integrated connection with all stakeholders and a cohesive driving force (Voordijk, 2000; Cox and Ireland, 2002; Yeo and Ning, 2002). In a disaster context, this could be accomplished by embracing those actors into an adaptive and cooperative process for attaining a common aim of resource management. With this in mind, China's resourcing practice in the housing sector after the Wenchuan earthquake is reviewed and discussed below.

4.5.1   Resourcing problems arising during post-earthquake reconstruction

Inflation of construction materials after the earthquake

The construction market is a source of resources for reconstruction. Post-disaster reconstruction activities tend to see inflation and profiteering after a catastrophic event (Jayasuriya and McCawley, 2008). In response, governmental authorities generally tend to resort to 'hard intervention' solutions through price restrictions and transaction regulations which, however, might cause resources to be allocated inefficiently and disaster relief delayed (McGee, 2008). This standpoint can also be substantiated by reconstruction examples in the aftermath of the Wenchuan earthquake. Apart from legal support for resource allocation nationwide, Chinese authorities took a series of measures including:

- A range of temporary price restrictive interventions
- Directing resource supply to the most severely earthquake-stricken areas
- Assigning inspectors on the ground to monitor the selling price of suppliers

In line with the top priority of reconstruction, there was a specific provision of construction materials for reconstruction of rural residences in earthquake-hit areas to first meet rural communities' needs. Furthermore, the supplementary 'price monitoring system' was set up in a number of material production factories to ensure the effective implementation of temporary price restriction. However, these 'hard' interventions rightly focused on the supply side to directly contain the profiteering behaviour in the construction market. Although these actions to some degree eased the temporary tension of material supply in earthquake-affected areas, they also provided a major disincentive to other suppliers from actively engaging in post-disaster resource provision efforts.

Intensive reconstruction demands and supply disruption

When coping with the enormous gap between large-scale reconstruction demands and resource supply, the Provincial Government of Sichuan Province (PGSP) enhanced supplementary subsidies for transportation costs and exempted highway tolls for vehicles delivering materials to the earthquake-affected zone. 0.1 billion Chinese dollars (approximate US$14.65 million) of financial aid was assigned to corporations tasked with specific material production and delivery to designated areas in 2008. In additions, the PGSP sanctioned 75 cement production lines, 760 brick factories and two large-scale high-performance steel factories to meet the reconstruction demand for the following three years. Meanwhile, the PGSP established a transparent supply–demand information platform embracing databases varying from demand quantities on construction materials, directories of 95 cement and steel production companies in neighbouring provinces, to detailed daily prices of steel, cement and brick in 51 affected counties (PGSP, 2008).

However, this action brought about another concern that with supply gradually outpacing demand at the late stages of reconstruction, excess productivity was likely to have a negative effect on the local economy. This concern was verified in the longitudinal study by the fact that with the majority of housing (80%) in rural areas rebuilt one year after the earthquake, the supply of bricks exceeded the real demand, and the local brick production industry suffered a 'crisis of excessive production'.

The on-going activities of housing reconstruction in China underscore that although governments launched a variety of supporting programmes for resource provision in post-earthquake reconstruction, in reality, the availability, accessibility and affordability of resources still posed a significant obstacle to the housing reconstruction sector after the Wenchuan earthquake especially in the most severely quake-hit areas of Mianzhu, Qingchuan and Beichuan. Resource shortages and cost increases, along with supply disruption, affected the process and performance of housing reconstruction in these areas in terms of project time and budget.

Accessibility of resources: transportation and logistics

The Wenchuan earthquake also highlighted the vulnerability of the transportation system in China. Road and railway systems in earthquake-affected areas were damaged and cut off by a large number of secondary hazards such as landslides, landslips, mud-rock flow and 'quake lakes'. Reopening access was a slow process due to the particularly mountainous terrain, constant aftershocks and persistent rain. Lack of access and consistency in delivery, together with the volatile price fluctuation of fuel, inhibited the recovery process.

All modes of transport were utilised in such a disaster situation, including human-power and animal carrying methods to deliver building materials to construction sites. In order to improve the transportation of construction materials from outside earthquake-impacted areas, the Transport Ministry of China approved four river–land joint routes for large cargo delivery. The Transport Network Planning Programme was devised and launched incorporating the concept of the Lifeline Highway Network. It required at least two highly earthquake-resistant highways to reach other jurisdictions in each quake-affected town and the total length of the network was 4900 kilometres. Three railway construction programmes, such as the Dujiangyan–Chengdu express line, were also put on the agenda (State Planning Group, 2008).

Undoubtedly, these transport capacity-extending strategies for reconstruction in China to a great extent relieved the tension of resource transport. Nevertheless, it also led to a conflict between normal delivery activities and reconstruction needs, adding difficulty in overall coordination and deployment of resources. In comparison with the pre-event situation, the increasing cost of new transport alternatives had an adverse effect on efficient and timely resource delivery to rebuilding projects.

4.5.2   Stakeholders in China's resourcing experience

China's efforts for post-earthquake reconstruction resource availability were mainly driven by governmental agencies and other authorities. Apart from assistance from the National Government and sponsoring localities, Provincial and local governments played a critical role through the enforcement of building codes, implementation of policies from the Central Government, as well as through regulation of the construction market in the aftermath of the Wenchuan earthquake to contain the cost surge and inflation of building materials. Other stakeholders, especially the construction industry, appeared to be less involved initially. This fact suggests the key aspects of a successful resourcing approach including input requirements, market linkage, infrastructure settings and key stakeholder interactions are not yet being addressed. There is an absence of understanding that the government is one of a complex set of institutions that the nation utilises for dealing with natural disasters. As the main operators in physical reconstruction and rehabilitation after a disaster, the construction sector needs to take steps to improve resourcing capability rather than rely on policies and administrative interventions from governmental authorities. Large-scale industry players should take a lead responsibility as key stakeholders to enhance the overall resourcing capability in the industry by providing information and knowledge sharing programmes for small and medium businesses.

The construction industry is not only a critical component of the nation's economy, but also a fundamental contributor in disaster management and mitigation (Spence and Kelman, 2004; Pheng *et al.*, 2006; Zuo *et al.*, 2006). However, recent natural disasters have highlighted the weaknesses of the construction industry in the disaster-stricken areas and their inadequate engagement in the mitigation of such events (Pribadi *et al.*, 2003; Gharaati, 2006; Pheng *et al.*, 2006). For example, in Banda Aceh, Indonesia, the quality of post-tsunami reconstruction projects was largely impaired by the low quality of local construction practitioners. The use of building materials of poor quality and low-skilled labour further added risks to future disasters, led to costly financial waste and a dysfunctional recovery system in the tsunami-impacted region.

Likewise, the construction industry in China has not been sufficiently involved in disaster planning and management. In spite of resourcing facilitation efforts made by the Chinese government, the local contractors and reconstruction organisations in earthquake-affected areas appeared to be less reactive in reconstruction resource procurement. Except for a few large construction companies which had resourcing contingency plans integrated into the overall project plan, there were no set schemes and strategies regarding post-disaster reconstruction, and the awareness of engagement in disaster management was poor.

The supply chain mechanisms traditionally used for resourcing construction projects in China's construction history turned out to be inadequate for post-disaster reconstruction needs. The present system of such resourcing for earthquake housing rebuilding relied much on mandated policies and legislation by which the material production and transaction was directed towards the priority of reconstruction. Changing the culture within the industry is a long-term goal, which necessarily depends on the establishment of on-going partnerships and trust along the supply chain. In dealing with the post-disaster reconstruction situation, the supply chain maintenance and management should be achieved by a combination of government-facilitation with market forces-driven leadership. In challenging conventional thinking and recognising the need to balance relationships and commercial issues in the construction market, there appears to be a much greater chance of fulfilling the aims and objectives set out by the leading advocates in the construction industry. There is also certainly a need to expand the training of architects, engineers and builders in safer ways of building.

4.5.3 Resourcing challenges to post-quake reconstruction practitioners

Heightening participatory awareness of construction practitioners for proactive resourcing

There is a need within the construction industry to design programmes to heighten the awareness of industry players to engage in disaster management, planning and mitigation. Crucially, the critical awareness and practices to enable the proactive and active participation of the construction industry can only be developed and sustained through developing capabilities before a disaster occurs. This requires the Chinese construction industry to establish a collective mindset to introduce changes in both structural and non-structural risk reduction practices and to formulate an adaptive procurement strategy when dealing with possible resourcing problems after an event. A well-conceived resourcing plan for a rebuilding scenario in a disastrous environment is needed which shows that contractor organisations are prepared for managing the resourcing needs during recovery. Involving construction-related stakeholders in local emergency management departments and other disaster mitigation programmes could improve construction industry awareness.

Supervision and management of owner-driven reconstruction

The policy-driven commitment to ensure that local communities take a leading role in rebuilding their own houses is an important strategy China's authority adopted for Wenchuan earthquake recovery and reconstruction. This decision recognised that even participatory approaches would be slower than authoritative, officially government-driven or typical 'top-down' models of implementation, and that they would be more effective over the long-term and more firmly rooted in community ownership. In some earthquake-hit areas where the engineering technical assistance was less accessible, under the self-rebuilding housing reconstruction model, there was a lack of trust between the house owners and construction teams. For instance, house owners wanted to procure building materials on their own due to the lack of trust in builders, but they had little knowledge about materials and purchased materials inadequate for the design.

Due to the shortage of local building professionals, construction teams from neighbouring cities and towns flooded into the earthquake-impacted communities. Some of them won bids for the rebuilding work with a lower price; however, many cheated house owners during reconstruction regarding material selection and purchase as well as construction quality. In some other areas, the shortage of construction materials made the house owners turn to inferior building materials due to the urgency caused by the government monetary incentive and the 'New Year – New Start' expectations. However, a number of the defects of poorly supervised or rushed reconstruction buildings could be seen after the housing was completed in these affected places.

The stakeholder problems faced posed a challenge to the local construction authorities on how to manage, supervise and regulate the behaviour of industry players as well as house owners during reconstruction and how to educate the public to raise their legal, contractual and quality requirements without affecting their participation.

Authoritative interventions vs market-driven resource production and supply

Material price restrictions from the government to some degree eased the tension of inflation in China's post-earthquake construction market. However, the reliance on mandated resolutions and increasing production methods was unable to solve the root problems of resource availability in a post-disaster environment.

The trade-off between levels of macro control and market self-regulation is needed for Chinese policy makers to settle different and conflicting interests of stakeholders without detriment to the disaster-affected areas. To deal with disaster situations, it is important to assess the adequacy of material supply and the availability of labour and skills before embarking on large building programmes to ensure a reasonable phasing of reconstruction processes.

Resourcing-related environmental concern

Another concern with post-disaster resourcing activities lies in their impact on the local natural environment. Other specific problems are becoming increasingly prominent environmentally during reconstruction after the Wenchuan earthquake. Raw material exploitation for making building components and products poses a great threat to the natural environment system. For example, as a conventional masonry material, shale-made brick was the most needed building material in mountainous affected areas in Sichuan Province. However, the excessive use of shale mineral was likely to degrade the river banks, damaging the eco-system and livelihood nearby. Inappropriate sourcing approaches are likely to induce secondary hazards. Deforestation for satisfying reconstruction logging needs, for example, exacerbated the negative earthquake impact, contributing to the incidence of flooding and landslides. In addition, the high volume of waste and debris from earthquake-destroyed buildings and from secondary hazards like rock-mud flow, also served as a problem because of the need to dispose and remove debris.

4.6 Summary

Throughout the housing reconstruction and repair process following China's Wenchuan earthquake Chinese authorities sought to adopt a range of policies and measures that could make the resourcing task smooth. Apart from the legal act issued after the earthquake, the one-on-one programme was a policy put forward which ties post-disaster recovery and reconstruction to the availability of public assistance nationwide. The administrative intervention into transaction of building products along with capacity expansion of local supplies and services in the construction market to some extent eased the tension between resource demand and supply in the earthquake-impacted region. The examples and evidence provided in the chapter illustrate largely successful efforts or partial accomplishments of resourcing for Wenchuan earthquake reconstruction. But each suggests that even positive experiences are not without their challenges and limitations.

This chapter provides a view of the elements that made resourcing for housing reconstruction such a complex undertaking in the aftermath of a disastrous event. Different resourcing efforts for housing reconstruction are needed with an emphasis on enabling technical and institutional support from not only the government, but also other stakeholders involved, the construction industry in particular. The observations incorporated into this chapter demonstrate that while there are few absolute solutions to resourcing needs, the difficulties can be transferred and applied in other national contexts and disaster circumstances. The impacts of the Wenchuan earthquake, and the related lessons and experiences, provide a variety of approaches and considerations pertinent to where resourcing bottlenecks are likely to exist in the aftermath of an event and where efforts are needed to align resources more closely to reconstruction requirements. In order to improve resource management for the next disaster, the policies and programmes developed for the Wenchuan earthquake will need to be assessed and redesigned to accommodate lessons learnt and best practice approaches.

Notes

1. M 8.0 represents surface wave magnitude (Ms), the national standard used by the Chinese government for earthquake magnitude. M 7.9 for the Wenchuan earthquake from the United States Geological Survey (USGS) reports represents moment magnitude (Mw).
2. A selection of the survey publications on the Wenchuan earthquake is available from the website of the organisations: Earthquake Engineering Field Investigation Team, Earthquake Engineering Research Institute, Multidisciplinary Center for Earthquake Engineering Research (MCEER), United States Geological Survey (USGS), and National Center for Research on Earthquake Engineering.
3. Also known as *Wenchuan Earthquake Disaster Recovery and Reconstruction Act*, came into effect on June 4, 2008.
4. For a thorough analysis on the housing reconstruction approaches, see the paper by Barenstein (2006) referenced in this chapter.
5. Price fluctuation rate is based on the contrast between April 2008 and December 2008. Data was obtained through the researchers' personal communication with officers in the Price Bureau of People's Government of Mianzhu City, Sichuan Province and field-based market investigation.
6. In Figures 4.1–4.4, in order to present a clearer effect, the price was measured in the Chinese currency unit RMB, rather than converted into US$.

References

Adam, C. S. and Bevan D.L. (2004) Aid, public expenditure and Dutch disease. *Growth, Poverty Reduction and Human Development in Africa.* St Catherine's College, Oxford: University of Oxford.

Alexander, D. (2004) Planning for post-disaster reconstruction. *I-Rec 2004 International Conference 'Improving Post-Disaster Reconstruction in Developing Countries',*

Coventry, UK, I-Rec information and research for reconstruction, Coventry University, online publication prepared in 2008 by the IF Research Group. http://www.grif.umontreal.ca/pages/papersmenu2004.htm

Barakat, S. (2003) *Housing Reconstruction after Conflict and Disaster*. Network Paper Number 43. London: Humanitarian Practice Network at Overseas Development Institute.

Barenstein, J.D. (2006) *Housing Reconstruction in Post-Earthquake Gujarat: A Comparative Analysis*. London: Humanitarian Practice Network at Overseas Development Institute.

Barenstein, J.D. (2008) From Gujarat to Tamil Nadu: owner-driven vs contractor-driven housing reconstruction in India. *I-Rec 2008 International Conference 'Building Resilience: Achieving Effective Post-Disaster Reconstruction'*. Christchurch, New Zealand. http://www.resorgs.org.nz/irec2008/i-rec2008_papers.shtml

Berke, P.R., Kartez, J. and Wenger, D. (1993) Recovery after disaster: achieving sustainable development, mitigation and equity. *Disasters*, **12**, 94–109.

Birkland, T.A. (2006) *Lessons of Disaster: Policy Change After Catastrophic Events*. Washington, DC: Georgetown University Press.

Bolin, R. and Stanford, L. (1998) The Northridge earthquake: community-based approaches to umet recovery needs. *Disasters*, **22**, 21–38.

Brunsdon, D.R., Charleson, A.W., King, A.B., Middleton, D.A., Sharpe, R.D., Shephard, R.B. *et al.* (1996) Post-earthquake co-ordination of technical resources: the need for a unified approach. *Bulletin of the NZ National Society for Earthquake Engineering*, **29**, 280–283.

China Iron and Steel Association (2009) 2008 Report of China's Steel Industry. China Iron and Steel Association (CISA).

Cho, S., Gordon, P., Moore, J.E., Richardson, H.W., Shinozuka, M. and Chang, S. (2001) Integrating transportation network and regional economic models to estimate the costs of a large urban earthquake. *Journal of Regional Science*, **41**, 39–65.

Corden, M.W. (1984) Booming sector and Dutch disease economics: survey and consolidation. *Oxford Economic Papers*, **36**, 359–380.

Cox, A. and Ireland, P. (2002) Managing construction supply chains: the common sense approach. *Engineering, Construction and Architectural Management*, **9**, 409–418.

Cuny, F.C. (1983) *Disasters and Development*. New York: Oxford University Press.

Davidson, C.H., Johnson, C., Lizarralde, G., Dikmen, N. and Sliwinski, A. (2007) Truths and myths about community participation in post-disaster housing projects. *Habitat International*, **31**, 100–115.

Freeman, P.K. (2004) Allocation of post-disaster reconstruction financing to housing. *Building Research and Information*, **32**, 427–437.

Gharaati, M. (2006) An overview of the reconstruction program after the earthquake of Bam, Iran, *I-Rec 2006 International Conference on Post-Disaster Reconstruction: Meeting Stakeholder Interests*, Florence, Italy. I-Rec information and research for reconstruction, Coventry University, online publication prepared in 2007 by the IF Research Group. http://www.grif.umontreal.ca/pages/papersmenu2006.htm

Green, R., Bates, L.K. and Smyth, A. (2007) Impediments to recovery in New Orleans' Upper and Lower Ninth Ward: one year after Hurricane Katrina. *Disasters*, **31**, 311–335.

Harvey, P. (2005) Cash and Vouchers in Emergencies. HPG Discussion Paper. London: Humanitarian Policy Group at Overseas Development Institute.

International Federation of Red Cross and Red Crescent Societies (IFRC) (2006) Annual Report 2006. International Federation of Red Cross and Red Crescent Societies.

International Recovery Platform (2007) *Learning from Disaster Recovery: Guidance for Decision Makers*, I. Davis (Ed.): Geneva, Kobe: United Nations International Strategy for Disaster Reduction (UNISDR).

Jayasuriya, S. and McCawley, P. (2008) Reconstruction after a Major Disaster: Lessons From the Post-Tsunami Experience in Indonesia, Sri Lanka, and Thailand. ADBI Working Paper 125. Tokyo: ADB Institute.

Makhanu, S.K. (2006) Resource mobilization for reconstruction and development projects in developing countries: case of Kenya, *I-Rec 2006 International Conference on Post-Disaster Reconstruction: 'Meeting Stakeholder Interests'*, Florence, Italy. I-Rec information and research for reconstruction, Coventry University, online publication prepared in 2007 by the IF Research Group. http://www.grif.umontreal.ca/pages/papersmenu2006.htm

McGee, R.W. (2008) An economic and ethical analysis of the Katrina disaster. *International Journal of Social Economics*, **35**, 546–557.

Oliver-Smith, A. (1990) Post-disaster housing reconstruction and social inequality: a challenge to policy and practice. *Disasters*, **14**, 7–19.

Oxfam Australia, Australian Red Cross, and World Vision (2007) Emergency Response Supply Chain Assessment. Australian Agency for International Development (AusAID).

Paterson, E., Re, D.D. and Wang, Z. (2008) *The 2008 Wenchuan Earthquake: Risk Management Lessons and Implications*. RMS Publications. S. Ericksen, Risk Management Solutions (RMS). http://www.rms.com/Publications/2008_Wenchuan_Earthquake.pdf

Pheng, L.S., Raphael, B. and Kit, W.K. (2006) Tsunamis: some pre-emptive disaster planning and management issues for consideration by the construction industry. *Structural Survey*, **24**, 378–396.

Pribadi, K., Hoedajanto, D. and Boen, T. (2003) Earthquake disaster mitigation activities in Indonesia Jan 1999 – Nov 2003. In: Meguro, K., Dutta, D. and Katayama, T. (Eds), *Seismic Risk Management for Countries of the Asia Pacific Region*, Proceedings of the 3rd Bangkok Workshop, December 2003, ICUS 2004-01.

Rodriguez, H., Quarantelli, E.L. and Dynes, R.R. (2007) *Handbook of Disaster Research*. New York: Springer.

Russell, T.E. (2005) *The Humanitarian Relief Supply Chain: Analysis of the 2004 South East Asia Earthquake and Tsunami*. Massachusetts Institute of Technology, Engineering Systems Division. Master Thesis.

Sichuan Provincial People's Government (2008) *Press Conference for Housing Building Materials in Earthquake-stricken Areas*. Retrieved 15 December 2008.

Sichuan Provincial People's Government Information Office (2009) The progress of 5.12 Wenchuan earthquake restoration and reconstruction. Sichuan Provincial People's Government Press Conference. www.chinanews.com.cn

Singh, B. (2007) *Availability of Resources for State Highway Reconstruction : A Wellington Earthquake Scenario*. Civil and Environmental Engineering Department, The University of Auckland. Master Thesis.

Spence, R., and Kelman, I. (2004) Editorial: Managing the risks from natural hazard. *Building Research and Information* **32**, 364–367.

State Planning Group of Post-Wenchuan Earthquake Restoration and Reconstruction (2008) The State Overall Planning for Post-Wenchuan Earthquake Restoration and Reconstruction. National Development and Reform Committee.

Steinberg, F. (2007) Housing reconstruction and rehabilitation in Aceh and Nias, Indonesia – rebuilding lives. *Habitat International*, **31**, 150–166.

Twigg, J. (2006) Technology, Post-Disaster Housing Reconstruction and Livelihood Security. Disaster Studies Working Paper 15. London: Benfield Hazard Research Centre, University College London.

United Nations Development Programme (2005) *Survivors of The Tsunami: One Year Later*, Shepard, D. (Ed.). Regional Bureau for Asia and the Pacific with the assistance of the Communications Office of the Administrator, United Nations Development Programme (UNDP).

United Nations Disaster Relief Organisation (1982) *Shelter after Disaster: Guidelines for Assistance*. New York: UNDRO.

Voordijk, H. (2000) The changing logistical system of the building materials supply chain. *International Journal of Operations and Production Management*, **20**, 823–841.

Waugh, W.L. Jr and Smith, R.B. (2006) Economic development and reconstruction on the Gulf after Katrina. *Economic Development Quarterly*, **20**, 211–218.

Work Bank Operations Evaluation Department (2005) Lessons from Natural Disasters and Emergency Reconstruction, The World Bank Group. http://www.worldbank .org/ieg/disasters/lessons_from_disasters.pdf

Yeo, K.T. and Ning, J.H. (2002) Integrating supply chain and critical chain concepts in engineering-procurement-construct (EPC) projects. *International Journal of Project Management*, **20**, 253–262.

Zuo, K. and Wilkinson, S. (2008) Supply chain and material procurement for post disaster construction: the Boxing Day tsunami reconstruction experience in Aceh, Indonesia. In: Haigh, R. and Amaratunga D. (Eds), *CIB W89 International Conference on Building Education and Research BEAR 2008*. Sri Lanka: Heritance Kandalama.

Zuo, K., Wilkinson, S., Le Masurier, J. and Van der Zon, J. (2006) Reconstruction procurement systems: The 2005 Matata flood reconstruction experience. *I-Rec 2006 International Conference on Post-Disaster Reconstruction: Meeting Stakeholder Interests*. Florence, Italy. I-Rec information and research for reconstruction, Coventry University, online publication prepared in 2007 by the IF Research Group. http://www.grif.umontreal.ca/pages/papersmenu2006.htm

5 Empowerment in Disaster Response and Reconstruction: Role of Women

Nirooja Thurairajah

5.1 Introduction

The seeming randomness of the occurrence of disaster, its impacts and the uniqueness of these events demand dynamic, real-time, effective and efficient solutions from the field of disaster management, thus making this field of study necessary. In less than half a decade, the world has witnessed numerous catastrophes which took away many hundred thousands of lives and caused huge damage to economies and unimaginable human suffering. In addition to man-made disasters, experts predict that the rising global temperature from climate change is yet to cause severe natural disasters around the world.

The impact of disasters is influenced by the prevailing socio-economic conditions of the particular community in addition to the scale of the cause itself. The extent of impact of a disaster on the affected communities mainly depends on hazards and vulnerabilities that exist in those affected areas. Hence, there is a need to include local knowledge on existing hazards and vulnerabilities and local needs to enhance the resilience of communities against future disasters. Although many governments take preventive action to face disasters the lack of incorporation of communities' engagement into these initiatives lead to further vulnerabilities in those areas. The lack of involvement of both men and women in disaster management has exposed them to more potential dangers. Recent studies have reflected the need for gender consideration in disaster management and emphasised its importance in building disaster resilient communities (International Labour Organisation, 2003; Ariyabandu and Wickramasinghe, 2003). However, the inclusion of women's contributions to post-disaster reconstruction in building disaster-resilient communities and its effect on women's development are of major concern which need to be further addressed by the field of the built environment.

Post-Disaster Reconstruction of the Built Environment: Rebuilding for Resilience, First Edition.
Edited by Dilanthi Amaratunga and Richard Haigh.
© 2011 Blackwell Publishing Ltd. Published 2011 by Blackwell Publishing Ltd.

5.2   The concept of empowerment

5.2.1   The emergence of empowerment

The origin of empowerment as a form of theory was traced back to the Brazilian humanitarian and educator, Paulo Freire (1973 cited in Hur, 2006) when he proposed a plan for liberating subjugated people through education. Although Paulo did not use the term empowerment, his emphasis on education as a means of inspiring individuals and group challenges to social inequality provided an important background for social activists who were concerned about empowering marginalised people (Parpart *et al.*, 2003). The concept is conceived as the idea of power since it is closely related to changing power by gaining, expending, diminishing and losing (Page and Czuba, 1999).

While explaining about the origin of the concept, Shackleton (1995) says that there is no single cause or origin of the empowerment movement, rather, it emerges from the increasing specialisation of some work, the changing shape of organisations and a shift towards placing greater value on human beings at work. Kinlaw (1995) views empowerment as a concept which is applied to improve the management of employees within organisations. Kinlaw relates empowerment to McGregor's theory Y and participative management. Among the studies on empowerment within organisations, the concept was referred to as employee involvement, participative management and quality circles. During earlier days, the concept was part of improvement programmes in organisations (Kinlaw, 1995). Employees' performance was boosted through empowerment. Although the term was initially not part of these practices, the interest of organisations and scholars on this construct made it form as a concept. Nesan and Holt (1999) state that empowerment is more a philosophy than a set of tools or management principles to be readily applied to business organisations. Though the term empowerment has been used frequently in management literature, it has been defined in several ways by organisations and scholars. Accordingly, empowerment is a diverse concept which is open to a number of different interpretations. During the last decade the term has become a widely used word in the social sciences across many disciplines such as community psychology, management, political theory, social work, education, women studies and sociology (Lincoln *et al.*, 2002).

5.2.2   Meanings of empowerment

Empowerment has been defined in several ways by many authors for different contexts. Even though the meaning of the terms delegation and empowerment may look similar they are different to each other. Shackleton (1995) states that in delegation a leader or manager decides to pass on a task or a specific part of his or her job to another individual for a specific reason. However, empowerment is a philosophy of management which widens the responsibility associated with the current task or role without necessarily changing the task or role itself. Handy (1993) simply explains empowerment as encouraging people to make decisions and initiate actions with less control and direction from their manager.

In a study by Loretta and Polsky (1991), empowerment is about giving up some control and sharing of additional knowledge of company goals and achievements from the organisation's management stand. On the other hand, it is about the acceptance of risk by taking more responsibility from the employees' stand. Avrick *et al.* (1992) state empowerment as giving authority commensurate with their responsibilities to initiate positive change in their organisation. This demands total commitment, involvement, support and trust from the management. While explaining about empowerment, Rubinstein (1993) states that every individual is responsible for acceptance or rejection of the quality of prior work; self inspection and control of current work; and acceptance or rejection of finished work. Within the above studies the authors have considered the concept with the view of enhancing the employee's position while delivering the required output for the management of organisations.

Ripley and Ripley (1992) explain empowerment from four dimensions: as a concept; as a philosophy; as a set of organisational behavioural practices and as an organisational programme. While elucidating about empowerment under each dimension, they state that:

- Empowerment as a concept is the vesting of decision making or approval authority to employees where, traditionally, such authority was a prerogative.
- Empowerment as a philosophy and as a set of behavioural practices means allowing self-managing teams and individuals to be in charge of their own career destinies, while meeting and exceeding company and personal goals through shared company vision.
- Empowerment as an organisational programme involves providing the framework and permission to the total workforce in order to unleash, develop and utilise their skills and knowledge to their fullest potential, for the good of the organisation, as well as for themselves.

In the above definition of empowerment, Ripley and Ripley (1992) identify the possible means of including empowerment into the organisation. In other words it provides a guide to practitioners and scholars to investigate the roles and implications of the concept within the management.

According to Wang and Burris (1994) empowerment is about increasing power-to, especially for marginalised people and groups who are farthest down the ladder in the power-over hierarchy and who have least access to knowledge, decisions, networks and resources. According to a study in the United States, empowerment is viewed as expanding assets and capabilities of poor people to participate in, negotiate with, influence, control and hold accountable institutions that affect their lives. Within this study, empowerment of the local community was achieved through community participation where it emphasised the need to consider the community's perceptions about indicators and indicators' context specificity as it may delay the impacts (Moser and Moser, 2003).

In a study on empowering Russian adults at individual and household level, empowerment is viewed as taking actions that selectively empower those with little power to redress power inequality (Lokshin and Ravallion, 2003). This project was aimed at addressing inequality of personal power and inequality of economic welfare. Under this study, the Cantil ladder (nine steps for power and welfare) was used to rank respondents' power by themselves. This study found that there is strong agreement in how perceptions of power and welfare react to differences in individual and household characteristics.

In a study within the construction industry, Nesan and Holt (1999) collectively define empowerment as the process of giving employees the authority to take decisions, relating to their work processes and functions, and within the limits provided by management, but requiring them to assume full responsibility and risk for their actions. Further, it was stated that empowerment is not an act or incident that can visibly or physically happen, but it is employees' perception or realisation that they believe in, and control, what happens to their work processes; and that they are capable of controlling those processes efficiently. In a gender-related study, even though Eylon and Bamberger (2000) view empowerment from two different perspectives, i.e. a cognition (psychological approach) or social act (sociological approach), they accept that empowerment cannot be neatly conceptualised as either a cognition or social act.

5.2.3   *The path towards empowerment*

Empowerment is multidimensional and occurs within sociological, psychological, economic, political and other dimensions. Earlier studies on empowerment state that empowerment can occur at individual level or collective level (Boehm and Staples, 2004; Hur, 2006). Empowerment can be illustrated as a social process since it occurs in relation to others and as an outcome that can be enhanced and evaluated against expected accomplishments (Parpart *et al.*, 2003). According to Hur (2006), the process of empowerment can be synthesised into five progressive stages as illustrated in Figure 5.1: an existing social disturbance, conscientising, mobilising, maximising and creating a new order. As shown in Figure 5.1, Hur (2006) considers empowerment as a process that has two interrelated features: personal empowerment and collective empowerment. Each feature has its own components. A set of four components including meaning, competence, self-determination and impact were found in personal empowerment. In addition, a set of four components including collective belonging, involvement in the community, control over organisation in the community and community building are explored under collective

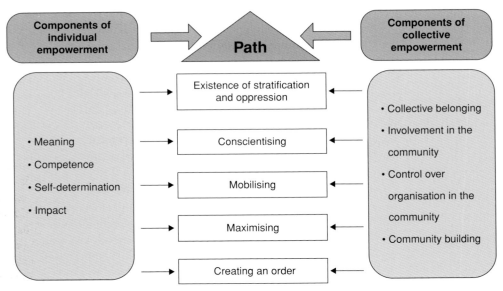

Figure 5.1   Path towards empowerment (adapted from Hur [2006]. Reproduced by permission of John Wiley & Sons, Inc.)

empowerment. Hur (2006) and Nilsen (2007) describe the components of individual empowerment as below:

- *Meaningfulness*: This is related to how the employees find the work relevant to their own values and ideas. It is also about internal commitment.
- *Competence*: It is a belief that one possesses the skills and abilities necessary to perform a job or task well (Gist, 1987 cited in Hur, 2006).
- *Self-determination*: Is about autonomy in personal working tasks. It concerns having the ability and authority to try one's own solutions.
- *Impact*: Impact is the perception of the degree to which an individual can influence strategic, administrative or operating outcomes at work.

Further, Fetterson, (2002) and Hur (2006) illustrate the components of collective empowerment as below:

- *Collective belonging*: Taking part in community activities or events that may lead to effecting change/affecting the power structure in communities.
- *Involvement in the community*: Taking part in community activities or events that may lead to effecting change in /affecting the power structure in communities.
- *Control over organisation in the community*: Component of gaining forces to influence representative groups, plus efficacy of those organisations like group support and advocacy.

- *Community building*: Creating a sense of community among residents that will increase its ability to work together, solve problems, and make group decisions for social change.

Further, the goal of individual empowerment is to achieve a state of liberation strong enough to impact one's power in life, community and society. On the other hand, the goal of collective empowerment is to establish community building, so that members of a given community can feel a sense of freedom, belonging and power that can lead to constructive social change. Although this research focuses on the process of empowerment it considers empowerment as both a process and as an outcome.

The concept of empowerment has been used widely within social studies and organisational development. Within social studies, the concept of empowerment is mainly used to address the problems of vulnerable and oppressed people. Worell and Remer (2003) viewed empowerment as a feminist therapy approach that resulted from the feminist movement of the 1960s. The increasing use of the concept of empowerment within social developmental studies leads the concept to be part of millennium development goals. This indicates the importance of the concept in the developmental process.

5.3   Women's empowerment

The concept of women's empowerment is a *fuzzy* concept as used by many organisations and researchers. Historical textual analysis and interviews with researchers and officials in development organisations disclose its flexibility and capacity to carry multiple meanings. In the efforts to utilise the concept for a broader social change agenda, policy makers and other officials juggle with these different meanings in order to keep that agenda alive. This indicates the need for a proper description of the concept in order to achieve its primary objectives without manipulation.

5.3.1   *Conceptualising women's empowerment*

The notion of women's empowerment is from the understanding of a route by which those who have been denied the ability to make strategic life choices can acquire such ability (Kabeer, 1999). Amartya Sen's (1999) 'development as freedom' approach has been a starting point for many recent definitions of empowerment. Sen (1999) argues that the goal of development is not to achieve a certain set of indicators, but to increase choices.

According to Magar (2003) women's empowerment is an outcome of a process whereby individual attitudes and capabilities, combined with collaborative actions, and reciprocally influenced by resources, results in a transformation to the desired achievements. On the other hand, Kabeer (1999) describes women's empowerment as a process by which women acquire the ability to make strategic life choices in terms of three interrelated dimensions that include resources (preconditions), agency (process) and achievements (outcomes).

Magar, in her study on empowerment approaches to gender-based violence, constructed a framework using the findings from earlier studies (Stein, 1997; Kabeer, 1999). This framework highlights individuals' attitudes and capabilities, which allow participation in various types of collaborative behaviour that leads to empowerment. The empowerment process comprises of two levels: the level of individual capacities observed in individual attitudes and capabilities and the level of group capacities (Magar, 2003) as shown in Figure 5.2. Individual attitudes (self-esteem and self-efficacy) along with specific types of skills, knowledge, and political awareness, are key ingredients in achieving empowerment at these two levels. Self-efficacy or agency is defined as the experience of oneself as a cause agent, not in terms of skills but rather in terms of one's judgement of what one can do with whatever skills one has (Bandura, 1995).

Although women's empowerment focuses on enhancing women's positions, the report from the United Nations division for the Advancement of Women from the Department of Economic and Social Affairs, the Beijing Declaration 75 and Platform for Action 76 encouraged men to participate fully in all actions towards gender equality and urged the establishment of the principle of shared power and responsibility between women and men at home, in the community, in the workplace and in the wider national and international

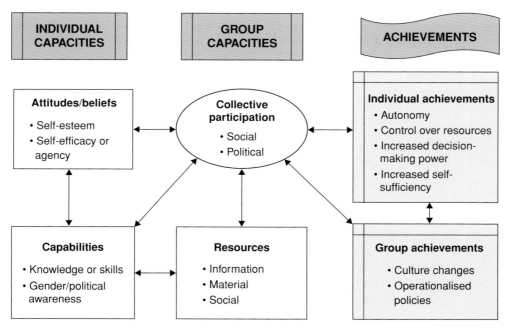

Figure 5.2   Women's empowerment conceptual framework (adapted from Magar [2003]. Reproduced by permission of John Wiley & Sons, Inc.)

communities (United Nations, 2006). This emphasises the need to bring about change in attitudes, relationships and access to resources and decision making, which are critical for the promotion of gender equality and the full enjoyment of all human rights by women. Even though in theory the concept seems well organised, in practice it faces many challenges in addition to women's own problems.

5.3.2 *Women's empowerment in different settings*

According to the report on poverty elimination and empowerment of women in the UK, empowerment was defined as individuals acquiring the power to think and act freely, exercise choice, and to fulfil their potential as full and equal members of society. This further emphasised the need to have equality of opportunity and equity of outcomes for women's empowerment. Further, the factors that influence empowerment of women were identified as: acquiring knowledge and understanding of gender relations and the ways in which these relations may be changed; developing a sense of self-worth, a belief in one's ability to secure desired changes and the right to control one's life; gaining the ability to generate choices and exercise bargaining power; developing the ability to organise and influence the direction of social change to create more justice in social and economic order. In this context, the construct of empowerment of women was aimed to address lack of commitment to improve opportunities for women, lack of income, limited access to services and opportunities for human development, lack of voice in political life and decision making and social subordination and exclusion (Department for International Development, 2000). It was highlighted that there is a need for support of gender equality and women's empowerment in terms of: economic and social policy; women's economic empowerment; education; women in public life; trade and globalisation; reproductive and sexual health; rights of the child; violence against women; basic services and infrastructure.

In another study within the UK, women's empowerment was viewed as a process by which women redefine and extend what is possible for them to be and do in situations where they have been restricted compared with men (Mosedale, 2005). It further identified the problems that can be addressed through empowerment such as women's gendered identities that disempowered them in public roles, and the need for change that expands options not only for themselves but also for women in general both now and in the future.

In a study within rural India, Roy and Tisdell (2002) refers to women's empowerment as a process by which women can gain power to diminish the forces of institutional deterrents to their development. Further, they state that the right to land is an important factor for women's empowerment as it is a more permanent source of income and it indicates that the person has a long-term interest in preserving the fertility of the land and therefore will be interested in investing in land. Furthermore, when income is higher this will increase the person's capacity to spend on consumption of food, housing, education, health and other necessities. It was found that when women work on someone else's land as paid labour or family land as unpaid labour they do not have an opportunity to invest and cannot have a significant voice in the family's expenditure plan.

In Nepal, government and non-governmental organisations funded projects are running to develop communities which include educational development, child welfare, empowerment of women and health services. A study on empowering women through a community development approach views empowerment as a multidimensional and interlinked process of change in power relations to expand individual choices and capacities for self-reliance (Mayoux, 2003 cited in Acharya *et al.*, 2005). In order to address female submission, silence, sacrifice, inferiority and obedience, problems in female illiteracy, and lesser mobility of women in employment in Nepal, a project considered the concept of empowerment of women through facilitating self-help group activities which are truly self-reliant, literacy programmes, group savings and credit programmes.

According to a study on rural women in Bangladesh, Parveen and Leonhäuser (2004) describe empowerment as an essential precondition for the elimination of poverty and upholding of human rights, in particular at the individual level as it helps to build a base for social change. This study found six indicators to measure a cumulative empowerment index: contribution to household income; access to resources; ownership of assets; participation in household decision making; perception of gender awareness; ability to cope with household shocks. The study addresses women's problems such as a limited role in household decision making, limited access and control over household resources, low level of individual resources, restricted mobility, inadequate skills and knowledge leading to vulnerability and heavy domestic workload.

A microfinance-based intervention on women's empowerment in South Africa (Kim *et al.*, 2007) reflected that violence against women is an explicit manifestation of gender inequality and increasingly being recognised as an important risk factor for a range of poor health and economic development outcomes. Violence perpetrated by a spouse or intimate partner is the most common form of gender-based violence. This causes direct injury or loss of life and it increases vulnerability to a range of health problems. Kim *et al.* found that reductions in violence occurred when women were enabled to: challenge the acceptability of such violence by expecting and receiving better treatment from partners; leave violent relationships; give material and moral support to those experiencing abuse; mobilise new and existing community groups; and raise public awareness about the need to address both gender-based violence and infections.

In Sri Lanka, the concept of empowerment is becoming an important concept to address women's difficulties especially to those who are more vulnerable in disaster situations. Many non-governmental organisations are working on women's empowerment in order to address women's safety and health-related issues and to enhance their status with regard to their assets, income and employment. A study about women who work in factories in the free trade zone, examined the violence against women through different elements of the concept of empowerment. Factory women who participated in a study (Hancock, 2006) rated violence against women as a major way in which to measure women's empowerment. This is a reflection of the problem itself at the societal level and provides new and constructive ways in which to conceptualise and measure women's empowerment.

In order to improve rural women's access to human, capital and information resources and their very limited role outside the confines of their homestead, improvement in women's participation was proposed as a solution in Kashmir within Pakistan (Weinberger, 2001). Weinberger (2001) states that participation means enabling poor people to take part in a process that strengthens their own abilities and possibilities, which is often regarded as a human right, allowing for equity and empowerment. In this context participation is viewed as a function of bargaining power. Further, incomplete information, cultural, ethnic and economic restrictions often influence the decision-making process of women and hence it has a bearing on women's empowerment. In addition, social networks play a major role in women's empowerment.

Weinberger (2001) states that two categories of participation approaches can be used in order to improve the participation of women: group-centred participation where costs and benefits of participation are connected to the uncertainty that surrounds the behaviour of fellow individuals and; decision to participate which is based on an individual analysis of cost and benefit. Prior to applying the concept of empowerment to enhance marginalised people's position in any situation, it is important to understand the prevailing conditions and needs of those people and others around them.

5.4   Women in a post-disaster setting

5.4.1   *The context of disaster response and reconstruction*

The report by the United Nations Environment Programme (cited Eilperin, 2009) stresses the need to protect the world from increasing global temperature which is currently projected to rise by up to 6.3 degrees Fahrenheit by the end of the century. The report further states that even if countries accomplish their most ambitious climate pledge they will not be able to reduce the temperature rise. This will increase the occurrence of tropical cyclones and heavy rainfall, and it is predicted that sea level may rise by nearly a metre. This indicates the need to build disaster-resilient communities in order to face the future. The current number of occurrences of natural disasters and the scale of their damage are drawing the attention of various sessions and meetings around the world. Within the disaster management setting, the need for building disaster-resilient communities has been increasingly highlighted since vulnerabilities and needs of communities can only be identified through a process of direct consultation and dialogue with the communities concerned as those communities can understand local realities and contexts better than outsiders (Haghebaert, 2007).

Although disasters and devastation are not new to the world, the extent of devastation to properties and lives in one single event shocks many not only within the country but around the world. This was evident during the Kashmir earthquake (2005), the Asian tsunami (2004), the Bam earthquake (2003), and the attack on the World Trade Center in the United States (2001). Post-disaster activities play a major part in rebuilding the community. Recovery activities, which include both the short- and long-term, continue until all systems return to normal or improved status. The rehabilitation phase includes medium-term interventions such as construction of transitional housing, provision of basic food to the affected population, provision of social services, road clearing, income generation and water system rehabilitation (Delaney and Shrader, 2001). As the emergency is brought under control, the affected community is capable of undertaking a growing number of activities aimed at restoring their lives and the infrastructure that supports them. There is no distinct point at which immediate relief changes to rehabilitation and then to long-term reconstruction development.

The reconstruction period includes the long term and often substantial investment in rebuilding the physical and social infrastructure of affected regions. Post-disaster reconstruction is a process that is the interaction of complex social, technological and economic factors and actions (Baradhan, 2006). There will be many opportunities during the reconstruction period to enhance prevention and increase preparedness, thus reducing vulnerability. However, though many organisations are involved in relief and rehabilitation most often they focus on the emergency phase and the reconstruction phase remains neglected (Jayaraj, 2006). Therefore, it is necessary for organisations that are concerned about disasters to use the opportunities and develop the community's capabilities. In the process of building disaster resilience, the built environment plays a major role during post-disaster reconstruction. Post-disaster reconstruction can provide many opportunities in rebuilding the social and economic status of the community (Thurairajah *et al.*, 2008).

5.4.2   *Role of women within a post-disaster context*

During post-disaster reconstruction many organisations fail to deliver the facilities according to the needs of affected communities. In many post-disaster situations although women belong to one of the most affected segments of communities, their needs and rights are not considered or are subject to gender inequality. A study has revealed the need for gender consideration within disaster management (Ariyabandu and Wickramasinghe, 2003). Most importantly, studies have recognised that disasters affect women and men differently due to the different roles and responsibilities undertaken by them and the differences in their capacities, needs and vulnerabilities (Ariyabandu and Wickramasinghe, 2003; Goonesekere, 2006). Women's local knowledge and expertise are essential assets for communities and households struggling to rebuild. In order to capture these capacities, disaster responders must work closely with women, remembering that the greatest need of survivors is for empowerment and self-determination. Women's under-representation in decision making in the management of construction organisations is one reason behind the unaccounted needs of those women in the planning of reconstruction activities. This affects the performance of the construction industry in post-disaster reconstruction and the industry fails to achieve its objective to satisfy its clients.

According to the International Labour Organisation report (International Labour Organisation, 2000), there are four general impacts that disasters have on the work of women. Firstly, women's economic insecurity increases. Since their productive assets are destroyed they often become sole earners, their household entitlements may decline, their small businesses are hard hit and they lose their work. In developing countries, after natural disasters, women lacking in land titles or farming small plots may be forced off their land (International Labour Organisation, 2000). Moreover, since land and employment arrangements are often negotiated through men, women may lose access to both without a man to represent them. Most importantly the gender stereotypes limit women's work opportunities especially in the post-disaster reconstruction stage. In addition, due to economic downturns after natural disasters women lose their jobs more quickly and in greater numbers than men (International Labour Organisation, 2000).

Women's workload increases significantly following a disaster and their working conditions in the household and paid workplace deteriorate (e.g. through lack of child care and increased work and family conflicts) (International Labour Organisation, 2000). The increased workload stems from damaged infrastructure, housing and workplaces; the need to compensate for declining family income and social services; and the responsibility of caring for orphaned children, the elderly and the disabled. This in turn limits women's mobility and time for income-generating work. Furthermore, this leads women to recover more slowly than men from major economic losses, as they are less mobile than male workers, are likely to return to paid work later and often fail to receive equitable financial recovery assistance from the government or external donors (International Labour Organisation, 2000). In certain communities, since women often take on more waged or other forms of income generating work and engage in a number of new forms of disaster work, they have also expanded their responsibilities.

Though decisions regarding resource allocation, enforcement of land and building regulations, and investment on economic and social development are made with an intention to satisfy both genders, many studies have highlighted the existing inequality in distribution chains and the implementation phase during post-disaster activities. Most importantly, women's contributions to post-disaster resilience have long been underestimated. Further, similar to using a generic term 'he' especially in written documents, linguistically females are subsumed under male. Pyles (2009) recognises that a core and often neglected element of disaster recovery has been the rebuilding and community development phase. Morrow and Peacock (1997) recognised that low income and marginalised communities are likely to suffer from a downward spiral of deterioration after a disaster. Further, Sundet and Mermelstein (1996) found that high poverty rates in communities were associated with the failure to survive. This can be seen in many examples within the research on disaster.

During the phase of planning and designing of shelters, women find that poor procedures in capturing women's demands and their ways of living lead to construction of inappropriate houses (Women's Coalition for Disaster Management, 2005). The guidelines used by the agencies/institutions were not clear about the definitions and the people to whom the support can be provided. For example, a government-initiated agency in Sri Lanka which worked on disaster reconstruction claimed that it would encourage 'household-driven housing reconstruction' while it does not clearly define the word household, especially where extended families live in the same house (Women's Coalition for Disaster Management, 2005).

A study found that time constraints on using the loans given for the reconstruction process added additional burden to people in affected families (Thurairajah *et al.*, 2010). The eligibility for special loans was based on the capacity to pay back the loan rather than on the vulnerability of people whose accommodation had been destroyed by the tsunami. The increased consumption of alcohol by men leads to misuse of funds allocated for reconstruction purposes (Women's coalition for Disaster Management, 2005). This shows the need to consider the equal distribution of funds to both men and women and to maintain a monitoring mission in order to provide effective distribution of funds for the purpose.

According to the study by the National Committee on Women (2006) on post-tsunami, it was found that female-headed households face discrimination in terms of their civil status, family and community support, property ownership, and access to resources. Patriarchal systems that exist within the community suppress women's legal rights such as property rights and land titles. Since land titles are allocated to the head of household, who is generally registered as male, there are concerns over the entitlements of women within the reconstruction phase. Although Sri Lankan law does not state that the male is the head of household, the patriarchal systems tend to locate women in a secondary position within the family-based household (National Committee on Women, 2006). However, government payments and interventions in the post-tsunami context target the family-based household as the unit that receives payments. Further, the head of household is eligible to receive these benefits, which puts women in a more marginalised position. A woman is usually recognised as a head of household in tsunami-affected families only when her spouse died or is unable to provide support to the family (National Committee on Women, 2006).

Women's participation in reconstruction of dwellings is not always antici-pated. Many women from certain parts of the affected communities mainly carry out their income-earning activities in their houses in tsunami-affected areas in Sri Lanka. Their lack of alternative housing, in addition to cultural fac-tors, force them to live in marginalised positions. Lack of experience/knowledge related to the construction of houses and their dependency on others to com-plete the projects made them more vulnerable as did the misuse of con-structed houses for women by others. In addition, their inability to use the new technology within their houses meant they could not benefit from them (Hidellage, 2008).

As the above problems illustrate, there is a need to address the role of women in post-disaster reconstruction. Bearing in mind the social conditions of women and the opportunities that post-disaster reconstruction can offer to the community (Thurairajah *et al.*, 2008), the concept of empowerment can be used to enhance women's positions and improve the performance of the construction sector.

5.5   Women's empowerment in post-disaster reconstruction

5.5.1   *Women's participation in disaster response and reconstruction*

Generally the role of women in the post-disaster stages can be categorised into three main areas: reproductive roles, community roles and productive roles (Ariyabandu and Wickramasinghe, 2003). Reproductive roles include roles within the household and the family: bearing, nurturing and rearing children; cooking; cleaning the house and yard; attending the market; caring for the sick and elderly. These roles may be expanded to include agricultural work with household stock and the long-term work of rebuilding family and community spirit, which is of no economic value. Women's community roles include: maintaining kinship relations; religious activities; social interactions and ceremonies; communal sharing and caring activities; communal survival activities (Ariyabandu and Wickramasinghe, 2003). These are generally done voluntarily and do not provide economic returns. Although these are usually related to reproductive function, there are instances where it includes work related to relief and reconstruction including the physical reconstruction of their homes (Enarson and Scanlon, 1999). Finally, productive roles give eco-nomic remuneration for manual labour, professional labour and subsistence activities. Although generally women's roles are classified under the above three categories, Fordham (1998) states that there is no simple distinction between the private and the public labour of women. This does not facilitate a frame to classify women's disaster responses into any discrete categories which could be used for any useful purpose.

The degree of vulnerability among women and their participation in post-disaster situations differs considerably. Disabled, elderly, pregnant and lactating women and widows often require assistance on a longer term or sustained basis, whereas other women can be supported up to the point where they achieve food and economic self-sufficiency. These distinctions are important in determining the types and levels of support to be provided to them. Therefore more local knowledge and wisdom needs to be incorporated into post-disaster recovery and development planning, particularly as they relate to women. According to the International Labour Organisation (2000), it is now widely accepted that women are not only responsible for attending to the basic needs of their children and families, but also account significantly for productive and income-generating activities. In addition, in a disaster situation women have demonstrated their capacity as income-earners, producers and managers of food production, providers of fuel and water, and participants in cultural, religious and political activities.

During the reconstruction phase and especially in temporary shelters, women take on a triple duty of reproductive work, community organisation and productive work in the informal economy, while men tend to return to their traditional role of waged work outside home. The tremendous impact of disasters on children and the elderly are largely shouldered by women. Generally, in post-disaster situations, the officials in charge of reconstruction activities find it difficult to obtain timely and accurate information. This is partly because decision making does not follow its usual procedure due to the urgency of the situation and the flow of information to lower ranks does not work as usual. Often, implementation is not effective. The contribution of women in these situations, therefore, would be very helpful.

Recent literature on post-disaster situations emphasises the dominance of men in development processes. Further, women remain marginal to the process and this indicates the need for diversified packages to women in post-disaster stages as not all women are from educated backgrounds. While women are severely affected by natural disasters this often provides women with a unique opportunity to challenge and change their gendered status in society. Women have proven themselves indispensable when it comes to responding to disasters (Bradshaw, 2002). Following Hurricane Mitch in 1998, women in Guatemala and Honduras were seen building houses, digging wells and ditches, hauling water and building shelters (Bradshaw, 2002). Although often against men's wishes, women have been willing and able to take an active role in tasks that were traditionally considered as male. This can have the effect of changing society's perceptions of women's capabilities. Women are most effective at mobilising the community to respond to disasters. They form groups and networks of social actors who work to meet the most pressing needs of the community.

The reconstruction phase is a significant period in disaster management as the results of the process are directly open to evaluation and criticism. The literature on disaster management (Fothergill, 1996) suggests that women's work around the household, on the job and in their neighbourhood contributes significantly to the social construction of daily life under extreme and routine conditions (Enarson and Morrow, 1998). However, the recognition of the importance of their work and their development still remains in doubt.

5.5.2 The state of women's empowerment in post-disaster response and reconstruction

In addition to poverty, environmental degradation and the different needs of men and women, the marginalised role of women within many organisations and their absence from the decision-making structures contributes to women's vulnerability in post-disaster situations. However, studies have recognised the concept of empowerment as a management philosophy which can help to overcome the problems that women face (Department of Economic and Social Affairs, 1999; UN-HABITAT, 2007). Further, they found that when women are empowered, they have the capacity and the inner will to improve their situation and gain control over their own lives. This can lead to an equal share in economic and political decision-making, and control of economic resources which will reduce their vulnerability in disaster situations. In reconstruction, the most vulnerable and marginalised sections of society, e.g. women, children, and the poor are the primary stakeholders and partners in the empowering process (Department of Economic and Social Affairs, 1999). This reflects the strong need to empower women from the affected community within post-disaster reconstruction to develop long-term disaster resilient communities.

According to a study on past disaster experiences, a pre-existing pervasive culture of acceptance or denial concerning violence against women, which includes no criminal legislation on domestic violence, presents compounded problems for organisations attempting to support women in the wake of the tsunami in South Asian countries. The denial or trivialising of violence against women by authorities only adds to the problem (Pheng et al., 2006). In certain developing countries male dominance and its negative implications is what underscores the importance of the longer-term vision for structural change to address gender inequality.

According to Navaratnaraja (2005), women were suddenly responsible for the whole family income after a war. This was due to identity problems faced by men and they were not able to provide support to the families during the civil war period in Sri Lanka. This led women to recognise their needs and capabilities in disaster settings. Further, women were trained in construction skills and were involved in reconstructing training centres under supervision. The Women's Coalition for Disaster Management (2005) highlighted the importance of providing compulsory criteria for including women in decision-making bodies in order to avoid dismal representation. Since certain organisations such as the Village Rehabilitation Committees and Divisional and District Grievance Committees play a very important role in reconstruction such as being responsible for making beneficiary lists, administration and disbursal of grants, and resolution of disputes, it is important to maintain representation from all in order to avoid any discrimination. Further, the Women's

Coalition for Disaster Management (2005) emphasised that tsunami recovery, rehabilitation and construction processes have to be based on the promotion and protection of rights rather than on a 'victim focus' which is limited to a welfare and dependency approach.

The impact of natural disasters and the consequent partial reconstruction efforts have presented challenges to women. Accordingly, post-disaster reconstruction strategies must anticipate these obstacles in order to minimise the future threat to women's safety and to facilitate the redevelopment of communities. The tsunami decimated Southeast Asia on 26 December 2004, killing more than 220 000 people in 12 countries and leaving 1.6 million people homeless. Although, the tsunami put an end to many opportunities and lives, the lives of survivors do not cease in disaster-affected areas. Although women's empowerment has been practised in livelihood activities during post-disaster reconstruction women's participation within the built environment must increase.

5.6   Summary

The increase in the occurrence of disasters has increased the urge of policy makers and researchers to focus on enhancement of society's capacity to withstand disasters in order to reduce damage to both human and material resources. Post-disaster recovery and reconstruction can provide windows of opportunity for physical, social, political and environmental development not only to reconstruct the impacted areas, but also to improve the socio-economic and physical conditions of the impacted population (Department of Economic and Social Affairs, 1999). However, in practice, too often disaster responses have not contributed to long-term development but actually subvert or undermine it, which results in lengthy post-disaster reconstruction activities and development opportunities being lost.

Despite many national and international agreements affirming human rights, women are still marginalised economically, politically and socially compared with men. In many developing countries women usually have less access than men to property ownership, investment options, training, employment and to information. Further, they are far less likely than men to be politically active and far more likely to be victims of domestic violence. Women are underrepresented in the political arena even in developed economies. Although international awareness of gender issues has improved, in reality no country has yet managed to eliminate the gender gap. Although some countries have made considerable progress in the recent past in removing obstacles to enable full participation of women in their respective societies, tremendous work in this area needs to be carried out.

Women's empowerment in developing countries has interconnected key issues, such as the role of culture, tradition, education, religion and economics. Despite the negative labelling on contextual factors, uneducated women are presented with many obstacles to empowerment in developing countries. Even with heightened awareness of women's issues, women who reside in rural and remote areas in developing countries are still voiceless compared with women in developed countries. Although many organisations are actively involved in women's empowerment, they tend to forget women's roles as policymakers rather than just 'policy takers'. Further, in order to validate and recognise the concept of empowerment the role of men in empowering women within the community should be recognised and considered in the process. The Feminist movement encourages all, including men, to consider a notion of empowerment with the potential to bring together all groups. However, this aspect needs to be considered in both developed and developing countries to make the concept effective instead of disempowering one group while empowering another.

During the post-disaster reconstruction phase some activities can be resuscitated. It is wise to search for the hidden resilience displayed by communities affected by disasters and then build upon it. This would entail a conscious strengthening of local knowledge and wisdom, applying appropriate solutions to crises. One of the main sustainable means for disaster victims to overcome their marginal condition is through an adjustment process of empowerment, allowing them to fulfil their basic human development needs. When designing protection and assistance programmes for women during and following emergencies it is essential for planners to broaden the concept of the status of women from the narrow conceptualisation as daughter or mother or wife. The capacity of women to mobilise people and manage change should not be underestimated. Instead of feeling that their voices cannot be safely heard, opportunities for women to engage in management and decision making related to all levels of disaster response and reconstruction should be offered. It can enable disasters to provide physical, social, political and environmental development opportunities that can be used during post-disaster reconstruction. This would eventually lead to not only the reconstruction of the affected areas, but also to improve the socio-economic and physical conditions of the affected communities in the long term.

References

Acharya, S., Yoshino, E., Jimba, M. and Wakai, S. (2005) Empowering rural women through a community development approach in Nepal. *Community Development Journal*, **42**, 34–46.

Ariyabandu, M.M. and Wickramasinghe, M. (2003) *Gender Dimensions in Disaster Management*. Colombo: ITDG South Asia Publication.

Avrick, A.J., Manchanda, Y.P. and Winter, E.F. (1992) Employee empowerment: world class test team. *Proceedings of American Production and Inventory Control Society*. APICS.

Bandura, A. (1995) *Self-Efficacy in Changing Societies*. New York: Cambridge University Press.

Baradhan, B. (2006) Analysis of the post-disaster reconstruction process following Turkish earthquakes, 1999. In: IF Research Group (Ed.), *International Conference on Post-Disaster Reconstruction: Meeting Stakeholder Interests*. 17–19 May 2006, Florence. The IF Research Group, University of Montreal.

Boehm, A. and Staples, L.H. (2004) Empowerment: the point of view of consumer. *Families in Society*, **85**, 270–280.

Bradshaw, S. (2002) Exploring the gender dimensions of reconstruction process post Hurricane Mitch. *Journal of International Development*, **14**, 871–879.

Delaney, P.L. and Shrader, E. (2001) *Gender and Post-Disaster Reconstruction: The Case of Hurricane Mitch in Honduras and Nicaragua*. Report prepared for the World Bank. www.anglia.ac.uk/geography/gdn [accessed 25/10/2007].

Department of Economic and Social Affairs (1999) *Women's Empowerment in the Context of Human Security*. Bangkok: Department of Economic and Social Affairs.

Department for International Development (2000) *Poverty Elimination and the Empowerment of Women*. London: DFID.

Eilperin, J. (2009) New analysis brings dire forecast of 6.3-Degree temperature increase. *The Washington Post*. http://www.washingtonpost.com/wp-dyn/content/article/2009/09/24/AR2009092402602.html [accessed 10/10/2009].

Enarson, E. and Morrow, B. (1998) Women will rebuild: a case study of feminist response to disaster. In: Enarson, E. and Morrow, B. (Eds) *The Gendered Terrain of Disaster: Through Women's Eyes*. Westport: Praeger Publishers. pp. 185–199.

Enarson, E. and Scanlon, J. (1999) Gender patterns in flood evacuation: a case studying Canada's Red River Valley. *Applied Behavioural Science Review*, **7**, 103–124.

Eylon, D. and Bamberger, P. (2000) Empowerment cognitions and empowerment acts. *Group and Organisation Management*, **25**, 354–372.

Fetterson, M.D. (2002) Empowerment evaluation: building communities of practice and a culture of learning. *American Journal of Psychology*, **30**, 89–102.

Fordham, M. (1998) Making women visible in disasters: problematising the private domain. *International Journal of Mass Emergencies Disasters*, **22**, 126–143.

Fothergill, A. (1996) Gender, risk and disaster. *International Journal of Mass Emergencies and Disasters*, **14**, 33–56.

Goonesekere, S.W.E. (2006) *A Gender Analysis of Tsunami Impact*. Colombo: Centre for Women's Research.

Haghebaert, B. (2007) Working with vulnerable communities to assess and reduce disaster risk. *Humanitarian Exchange*. London: Overseas Development Institute. pp. 15–18. http://www.odihpn.org/report.asp?id=2888 [accessed 25/10/2007]

Hancock, P. (2006) Violence, women, work and empowerment. *Gender, Technology and Development*, **10**, 211–228.

Handy, M. (1993) Freeing the vacuums. *Total Quality Management*, **11**, June.

Hidellage, V. (2008) Women's state within post tsunami reconstruction (Interview) (Personal communication, October 2008).

Hur, M.H. (2006) Empowerment in terms of theoretical perspectives: exploring a typology of the process and components across disciplines. *Journal of Community Psychology*, **34**, 523–540.

International Labour Organisation. (2000) *Crisis Response and Reconstruction*. http://www.ilo.org/public/english/employment/crisis/download/afghanwomen.pdf [accessed on 23/09/2007]

International Labour Organisation. (2003) *Gender in Crisis Response*. Report prepared for ILO InFocus Programme on Crisis Response and Reconstruction. http://www.ilo.org/public/english/employment/crisis/download/factsheet9.pdf [accessed 23/05/2008].

Jayaraj, A. (2006) Post disaster reconstruction experiences in Andhra Pradesh, in India. In: IF Research Group (Ed.), *International Conference on Post-Disaster Reconstruction: Meeting Stakeholder Interests*. 17–19 May, 2006, Florence. The IF Research Group, University of Montreal.

Kabeer, N. (1999) Resources, agency, achievements: reflections on the measurement of women's empowerment. *Development and Change*, **30**, 435–464.

Kim, J.C., Watts, C.H., Hargreaves, J.R., Ndhlovu, L.X., Phetla, G., Morison, L.A. *et al.* (2007) Understanding the impact of a micro-finance-based intervention on women's empowerment and the reduction of intimate partner violence in South Africa. *American Journal of Public Health*, **97**, 1794–1802.

Kinlaw, D.C. (1995) *The Practice of Empowerment*. Aldershot: Gower Publishing Limited.

Lincoln, N.D., Travers, C., Ackers, P. and Wilkinson, A. (2002) The meaning of empowerment: the interdisciplinary etymology of a new management concept. *International Journal of Management Reviews*, **4**, 271–290.

Lokshin, M. and Ravallion, M. (2003) Rich and powerful?: subjective power and welfare in Russia. *Journal of Economic Behavior and Organization*, **56**, 141–172.

Loretta, D. and Polsky, W. (1991) Share the power. *Personnel Journal*, September, 116.

Magar, V. (2003) Empowerment approaches to gender-based violence: women's courts in Delhi slums. *Women's Studies International Forum*, **26**, 509–523.

Morrow, B.H. and Peacock, W.G. (1997) Disasters and social change: Hurricane Andrew and the reshaping of Miami? In: Peacock, W.G., Morrow, B.H. and Gladwin, H. (Eds), *Hurricane Andrew: Ethnicity, Gender and the Sociology of Disasters*. New York: Routledge. pp. 226–242.

Mosedale, S. (2005) Assessing women's empowerment: towards a conceptual framework. *Journal of International Development*, **17**, 243–257.

Moser, C. and Moser A. (2003) *Background Paper on Gender-Based Violence*. Paper commissioned by the World Bank. Washington: Washington, D.C.

National Committee on Women. (2006) *Gender and the Tsunami*. Colombo: United Nations Development Fund for Women.

Navaratnaraja, A. (2005) *Poverty Reduction and Livestock Development in War Inflicted Northern Region of Sri Lanka*. Colombo: JICA.

Nesan, L.J. and Holt, G.D. (1999) *Empowerment in Construction*. Hertfordshire: Research Studies Press.

Nilsen, A.S. (2007) Tools for empowerment in local risk management. *Safety Science*, **46**, 858–868.

Page, N. and Czuba, C.E. (1999) Empowerment: what is it? *Journal of Extension*, **37**, 24–32.

Parpart, J.L., Rai, S.M. and Staudt, K. (2003) *Rethinking Empowerment: Gender and Development in a Global/Local World*. New York: Routledge.

Parveen, S. and Leonhäuser, I. (2004) Empowerment of rural women in Bangladesh – a household level analysis. *Conference on Rural Poverty Reduction through Research for Development and Transformation*, 5–7 October 2004. Berlin: Deutscher Tropentag.

Pheng, L.S., Raphael, B. and Kit, W.K. (2006) Tsunamis, some pre-emptive disaster planning and management issues for consideration by the construction industry. *Structural Survey*, **24**, 378–396.

Pyles, L. (2009) Community organising for post disaster social development: locating social work. *International Social Work*, **50**, 321–333.

Ripley, R.E. and Ripley, M.J. (1992) Empowerment, the cornerstone of quality: empowering management in innovative organisations in the 1990s. *Management Decision*, **30**, 20–43.

Roy, K.C. and Tisdell, C.A. (2002) Property rights in women's empowerment in rural India: a review. *International Journal of Social Economics*, **29**, 315–334.

Rubinstein, S.P. (1993) Democracy and quality as an integrated system. *Quality Progress*, September, 51–55.

Sen, A. (1999) *Development as Freedom*. Oxford: Oxford University Press.

Shackleton, V. (1995) *Business Leadership*. London: Routledge.

Stein, J. (1997) *Empowerment and Women's Health: Theory, Methods and Practice*. London: Zed Books.

Sundet, P. and Mermelstein, J. (1999) Predictions of rural community survival after a natural disaster: implications for social work practice. *Research in Social Work and Disasters*, **22**, 57–70.

Thurairajah, N., Amaratunga, D.G. and Haigh, R. (2008) Post Disaster Reconstruction As An Opportunity For Development: Women's Perspective, CIB W89. International Conference in Building Education and Research, 10–15 February 2008, Kandalama.

United Nations (2006) *Agreed Conclusions of the Commission on the Status of Women on the Critical Areas of Concern of the Beijing Platform for Action 1996–2005*. United Nations Publication (ST/ESA/304).

UN-HABITAT – Regional Office for Asia and the Pacific (2007) *Towards Women Friendly Cities*. Fukuoka: UN-HABITAT Regional Office for Asia and the Pacific.

Wang, C. and Burris, M. (1994) Empowerment through Photo Novella: portraits of participation. *Health Education and Behavior*, **21**, 171–186.

Weinberger, K. (2001) What role does bargaining power play in participation of women? A case study of rural Pakistan. *The Journal of Entrepreneurship*, **10**, 209–221.

Women's Coalition for Disaster Management – Batticaloa (2005) Concerns of displaced women's welfare and rights. *Options* 1. 25 September, pp. 38–41.

Worell, J. and Remer, P. (2003) *Feminist Perspectives in Therapy*. 2nd edn. New Jersey: John Wiley & Sons Inc.

6 Community-Based Post-Disaster Housing Reconstruction: Examples from Indonesia

Taufika Ophiyandri

6.1 Introduction

This chapter begins with a description of disaster figures in Indonesia and why Indonesia has become so vulnerable to disaster. It then presents the effort that the government of Indonesia has made to deal with this situation. Next, it introduces the meaning of community, discusses the importance of community participation in reconstruction and lists different types of community participation. It then analyses the housing reconstruction problems following the 2004 tsunami in Indonesia and compares the results of the community-based and contractor-based approach. Finally, it presents two case studies of good practices on community-based housing reconstruction.

6.2 Disaster vulnerability

6.2.1 A disaster prone country

The Republic of Indonesia is a country in Southeast Asia. It is the world's largest archipelagic country that consists of 17 508 islands, of which 6000 are inhabited. The total population of Indonesia given in the last population census in 2000 was 205 843 000, the fourth largest in the world after China, India, and the United States of America, and was projected to be 240 million in 2009.

Lying along the equator, Indonesia has a tropical climate, with two distinct monsoonal wet and dry seasons. Average annual rainfall in the lowlands varies from 1780–3175 millimetres, and up to 6100 millimetres in mountainous

Post-Disaster Reconstruction of the Built Environment: Rebuilding for Resilience, First Edition.
Edited by Dilanthi Amaratunga and Richard Haigh.
© 2011 Blackwell Publishing Ltd. Published 2011 by Blackwell Publishing Ltd.

regions. Humidity is generally high, averaging about 80%. Average temperatures are classified as follows: coastal plains, 28°C; inland and mountain areas, 26°C; higher mountain areas, 23°C (National Information Agency, 2004).

Indonesia is a beautiful country, but also prone to disaster, both natural or man-made. Some causes of disaster are related to geography, geology, climate or other factors related to social, cultural or political diversity. Types of disaster that occur in Indonesia are listed below (BAPPENAS and BAKORNAS PB, 2006):

- *Earthquakes and tsunamis*: Earthquakes occur relatively frequently in Indonesia because of tectonic plate movements and volcanic eruptions. Tectonic plate movements occur along Sumatra's west coast where the Asian Plate adjoins the Indian Ocean Plate; Java's south coast and the islands of Nusa Tenggara where the Australian Plate meets the Asian Plate; and Sulawesi and Maluku where the Asian Plate meets the Pacific Ocean Plate. They produce an earthquake belt in Indonesia with thousands of epicentres and hundreds of hazardous volcanoes. Tsunamis in Indonesia are mostly generated by tectonic earthquakes which occur along subduction zones and other seismic active areas. Between 1600 and 2000 there were 105 tsunamis of which 90% cent were generated by tectonic earthquakes, 9% by volcanic eruptions and 1% by landslides.
- *Volcanic eruptions*: There are 129 active volcanoes and 500 inactive volcanoes in Indonesia. They form a belt covering in one straight stretch Sumatra, Java, Bali and Nusa Tenggara before turning northwards to the Banda Sea and the northern part of Sulawesi. The belt's length is about 7000 kilometres and it contains volcanoes of mixed characteristics.
- *Floods:* Floods generally occur in the western part of Indonesia, which features a heavier rainfall than the eastern region. Indonesia's growing population and need for space to accommodate life supporting activities has indirectly contributed to flooding. Logging has increased to address the demand for space, which increases sedimentation in rivers which in turn produces uncontrolled runoff and high groundwater saturation.
- *Landslides*: Landslides frequently occur in areas with steep slopes and are normally triggered by heavy rainfall. Areas highly prone to landslides are Sumatra's Bukit Barisan mountains and mountain ranges in Java, Sulawesi and Nusa Tenggara. Fatal landslides also occur at drilling sites and in mining shafts.
- *Droughts*: Droughts normally occur during long dry seasons in some regions especially Indonesia's east. Droughts also potentially promote the spread of tropical diseases such as malaria and dengue fever, impact on food crop-producing regions and the power supply from hydroelectric power plants.
- *Forest and land fires:* Forest and land fires are mainly triggered by natural factors and human activities such as land clearing. Low welfare and education standards in communities living in and around forests are among the

contributing factors. The most prominent factor contributing to forest fires is the logging activity conducted by concession holders who clear forests without heeding regulations and environmental considerations. Kalimantan and Sumatra are prone to land fire because the land is composed of peat, which is easily burned following uncontrolled land clearing.

- *Epidemics, disease outbreaks and extraordinary events*: These are hazards caused by the spread of communicable diseases in a certain region. Diseases that are currently of major concern in Indonesia include dengue fever, malaria, bird flu, anthrax, hunger oedema and HIV/AIDS. Poor environmental conditions, bad food and life style and climatic changes are some of the contributing factors.
- *Technological disasters:* The types of technological disaster that happen in Indonesia are transportation accidents (involving ships, aircrafts and land transport such as trains), industrial accidents (gas pipe leaks, poisoning, environmental pollution) and household accidents (short-circuits, fires).
- *Social unrest:* Indonesia's social culture, which is made up of diverse ethnicities, races, community groups, languages and religions, is a rich national asset and at the same time a highly vulnerable feature. Indonesia's proneness to conflict is made worse by gaping economic disparities and low educational levels. A declining sense of nationalism is also a contributing factor as demonstrated by some regions aspiring to break away from the Unitary State of the Republic of Indonesia.

6.2.2　Data against disaster

There has been an upward trend in the number of disaster occurrences in Indonesia during the last decade. Figure 6.1 presents the total number of disasters and people killed per year from 1999 to 2008 in Indonesia based on DIBI (Indonesian Disaster Data and Information). It can be seen that disaster occurrences increased slightly from 101 in 1999 to 190 in 2002. It then rose dramatically to 895 occurrences in 2004 before slightly decreasing in 2005 and then soaring again to 1302 events in 2008. The highest number of fatalities took place in 2004, which was mainly caused by the Aceh Boxing Day tsunami. The 26 December 2004 tsunami has been recorded as one of the most devastating disasters in the modern world. It devastated thousands of communities along the coastline of the Indian Ocean, affected 12 nations, with Indonesia suffering the most. In Indonesia, this disaster left 167 000 people dead or missing, displaced 500 000 people, damaged 110 000 houses and 2000 schools, and 3000 km of road become impassable (Badan Rehabilitasi dan Rekonstruksi (BRR) and International Partners, 2005).

The distribution of the type of disaster in Indonesia between 2000 and 2009 is shown in Table 6.1. The raw data was based on Emergency Events Database (EM-DAT) data which is maintained by the Centre for Research on the Epidemiology of Disasters (CRED). We can see that there are differences between EM-DAT data and DIBI data which is because of the differences

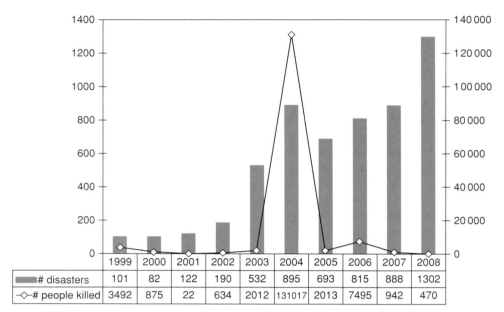

	1999	2000	2001	2002	2003	2004	2005	2006	2007	2008
▬# disasters	101	82	122	190	532	895	693	815	888	1302
–◇–# people killed	3492	875	22	634	2012	131017	2013	7495	942	470

Figure 6.1 Number of disaster occurrences and people killed in Indonesia from 1999 to 2008 (source: after DIBI, 2009)

in the classification method used. From Table 6.1, it can be seen that from the total number of disasters, natural disasters (61.92%) occur more often than technological disasters (38.08%). The most common natural disaster in Indonesia is flooding (24.69%), followed by earthquake (15.06%) and landslide (10.88%). Surprisingly, the highest number of disasters in Indonesia was from transportation accidents (30.54%) that took 3867 lives (2.12%). Although earthquakes make up only 15.06% of the total events, they accounted for 95% of the total killed in Indonesia.

If we look at the top 10 causes of fatalities from natural disasters in Indonesia in the last 30 years, earthquakes are the major threats (Table 6.2). The top six causes of fatalities come from earthquakes and five of them happened after 2004. The severity of an earthquake is not just in terms of fatalities, but in the significant amount of damage to houses and economic losses. The Aceh earthquake in 2004 and the Nias earthquake in 2005 created a need for 120 000 new houses for people and economic loses were US$4.1 billion. The Yogyakarta (Central Java) 6.3 Richter scale earthquake on 27 May 2006 destroyed 157 000 houses and estimated economic losses were US$3.1 billion (BAPPENAS *et al.*, 2006). The two most recent earthquakes in Tasikmalaya (7.0 on the Richter scale), West Java, on 2 September 2009 damaged 65 700 houses and took 81 lives, and in Padang (7.6 on the Richter scale), West Sumatra, on 30 September 2009 killed 1117 people and left 135 000 houses heavily damaged. Figure 6.2 shows a government building damaged by the Padang earthquake.

Table 6.1   Number of disasters in Indonesia by types of disaster from 2000 to 2009 (source: after EM-DAT, 2009)

Disaster category	Disaster subgroup	Disaster type	Number of events	% of total events	Number of people killed	% of total killed
Natural disaster	Geophysical	Earthquake (seismic activity)	36	15.06	173 596	95.00
		Volcano	10	4.18	2	0.00
	Meteorological	Storm	2	0.84	4	0.00
	Hydrological	Flood	59	24.69	2784	1.52
		Mass movement wet (landslide)	26	10.88	1012	0.55
	Climatological	Drought	1	0.42	0	0.00
		Wildfire	4	1.67	0	0.00
	Biological	Epidemic	10	4.18	1190	0.65
		Sub total	*148*	*61.92*	*178 588*	*97.73*
Technological disaster	Industrial accident	Industrial accident	7	2.93	70	0.04
	Miscellaneous accident	Miscellaneous accident	11	4.60	210	0.11
	Transport accident	Transport accident	73	30.54	3867	2.12
		Sub total	*91*	*38.08*	*4147*	*2.27*
		Total	*239*	*100.00*	*182 735*	*100.00*

Figure 6.2   Collapsed building following 30 September 2009 earthquake in Padang (photo by A. Farkah)

Table 6.2   Top 10 natural disasters in Indonesia, sorted by numbers of people killed from 1980 to 2009 (source: EM-DAT, 2009)

No.	Disaster	Date	People killed
1	Earthquake (seismic activity)	26-Dec-04	165 708
2	Earthquake (seismic activity)	27-May-06	5778
3	Earthquake (seismic activity)	12-Dec-92	2500
4	Earthquake (seismic activity)	30-Sep-09	1117
5	Earthquake (seismic activity)	28-Mar-05	915
6	Earthquake (seismic activity)	17-Jul-06	802
7	Epidemic	13-May-98	777
8	Drought	Sep-97	672
9	Epidemic	Jan-98	672
10	Epidemic	01-Jan-04	658

Table 6.3   Earthquake occurrence worldwide and estimated deaths (source: after USGS, 2009)

Year	Earthquake magnitude (Richter Scale)			Estimated deaths
	6–6.9	7–7.9	8–9.9	
1980–1989	980	101	4	58 880
1990–1999	1339	147	6	114 646
2000–2009	1420	128	13	465 100

Table 6.3 presents the trends of earthquakes of a magnitude greater than point 6 on the Richter scale worldwide between 1980 and September 2009. The number of such earthquakes rose significantly during the last three decades, especially of the 8.0–9.9 earthquake magnitude. This group has doubled from six occurrences between 1990 and 1999 to 13 in the last decade. This trend is also true in Indonesia. The number of major earthquakes has increased significantly since the Aceh giant earthquake in 2004. After the 2004 earthquake, 32 major earthquakes have been recorded compared with only 15 earthquakes between 1992 and 2004 (US Geological Survey, 2009). With the increasing number of earthquakes occurring and considering their affect on people and houses, then it has become clear that a good strategy for housing reconstruction has to be developed. One of the options is to carry out community-based housing reconstruction.

6.2.3   Vulnerability

Disaster is a function of three factors: hazardous existence, vulnerability and incapability to overcome negative impacts. The hazardous existence factor is based on the geological, meteorological or ecological characteristics of certain areas, while vulnerability can be classified into physical, social, economic and environmental factors.

The vulnerability of Indonesia is high. Physically, the infrastructure is not well designed to deal with disasters. Although most parts of Indonesia are exposed to earthquakes, the building code for housing is hard to implement. Many houses are not designed to be resistant to earthquakes. The increase in population size and density also contribute to the vulnerability of Indonesia. This is one of the reasons why there are more fatalities caused by similar disasters in recent years compared with 10 or 20 years ago.

As a developing country, government funding available to establish Indonesia as a country resilient to disaster is still limited. As a result, providing adequate infrastructures to cope with disasters has not yet been established. In many tsunami-prone areas, early warning systems have not been installed, emergency shelters are not available, nor is evacuation route information. Ethnic diversity, different beliefs, and deep gaps between wealth and poverty are other factors that contribute to vulnerability. The level of education and knowledge especially about disaster management is very low. This has affected the capacity of people to deal with disasters.

6.3 National policy

6.3.1 Shifted paradigm

The 2004 Boxing Day tsunami in Aceh became a wake-up call for Indonesia on how to face disasters and since then, there has been a lot of discussion on the subject. The new paradigm on disaster management officially began with the approval of Disaster Management Law Number 24 Year 2007 by the House of Representatives. Although it was proposed at the beginning of 2004, before the Aceh tsunami, the Law was enacted in Jakarta on 26 April 2007.

Disaster Management Law Number 24/2007 brings a shifted paradigm in disaster management in Indonesia. Firstly, there is a change from emergency response to a focus on risk management. Secondly, protection is a basic right of all people. Thirdly, disaster management is the responsibility of all stakeholders, not only the government. Finally, disasters are considered as daily occurrences not as an extraordinary issue. Table 6.4 contrasts the old and new paradigm on disaster management in Indonesia after the establishment of Law no 24/2007.

Table 6.4   Shifted paradigm on disaster management in Indonesia (source: after Hadi, 2007)

Old paradigm	New paradigm
Emergency response	Risk management
Protection is a blessing given by the government	Protection is the people's human right
Government responsible	All stakeholders responsible
Handling disasters is an extraordinary issue	Handling disasters is the daily task of administration and development

Moreover, under Law Number 24/2007, the government was encouraged to establish a National Disaster Management Agency (BNPB), a non-departmental body equal to ministry level. On 26 January 2008, BNPB was established by the enactment of Presidential Regulation Number 8 Year 2008. Other regulations that have been produced by the government of Indonesia after the enactment of Law No 24/2007 are:

- Government Regulation Number 21 Year 2008 on Disaster Management Operations.
- Government Regulation Number 22 Year 2008 on the Disaster Management Financing.
- Government Regulation Number 23 Year 2008 on the Participation of International Agencies and Non-Government International Agencies in Disaster Management.
- Head of BNPB Regulation Number 3 Year 2008 about Local Disaster Management Agency (BPBD).
- Minister of Home Affairs Regulation Number 46 Year 2008 about Local Disaster Management Agency Organisation and Works Mechanism.

6.3.2   National Action Plan

The National Action Plan for Disaster Risk Reduction in Indonesia is formulated as a commitment by the Government of Indonesia to the UN Resolution No. 63/1999, which was followed by the Hyogo Framework for Action 2005–2015 and the Beijing Action Plan. National Development Planning Agency (BAPPENAS) and the National Coordination Body for Disaster Management (BAKORNAS PB) identified that there are five key priority areas for disaster risk reduction during 2006–2009 that must be addressed (BAPPENAS and BAKORNAS PB, 2006):

- Ensure that disaster risk reduction is a national and a local priority with a strong institutional basis for implementation.
- Identify, assess and monitor disaster risks and enhance early warning.
- Use knowledge, innovation and education to build a culture of safety and resilience at all levels.
- Reduce underlying risk factors.
- Strengthen disaster preparedness for effective response at all levels.

Each priority area is followed by key activities planned for during 2006–2009.

6.3.3   Problems

There are several problems in implementing disaster management in Indonesia. Some critical areas in implementing it are:

- *Legislation*: The dissemination of disaster management law and its follow up to all stakeholders is a challenge because it is still new. The changing of government systems from centralistic to a de-centralistic system also creates a problem because local government has full authority to set its own institutional organisation. As a result, the unit authorised for disaster management in the region is very much determined by the policy of the respective region.

- *Coordination*: Good coordination among organisations involved in disaster management is still hard to achieve. The ego of each sector is still high. This weakness has led to a problem of data communication, hazard dissemination and spatial planning.
- *Information management*: Availability of comprehensive data and information is still a problem in implementing disaster management in Indonesia. Several institutions are responsible for developing their own information system. This creates many versions of data that overlap each other. Date and information exchange between institutions is also difficult because of the unavailability of regulations and policies on their exchange (Triutomo, 2009).
- *Hazard information*: Hazard information is not integrated and still sectoral because it is undertaken separately by different agencies. As a result, one area could have a different map for a specific hazard.
- *Spatial planning*: A lack of coordination still exists among the organisations concerned with the planning and management of people's settlements in mainstreaming disaster risk reduction elements, including the enforcement of building codes (Triutomo, 2009). The inconsistency in implementing the spatial plan has contributed in increasing vulnerability.
- *Education*: Disaster management as a subject is not part of the school curricula nationally. Only a few schools in disaster prone areas have taught the disaster management topic to their students.

6.4 Community participation in reconstruction

6.4.1 *Community definition*

The word 'community' has a lot of meanings and people define it in different ways. Before 1910 there was little social science literature concerning 'community' and it was really only in 1915 that the first clear sociological definition emerged (Smith, 2001). Hillery (1955) cited in Kumar (2005) states that 94 different definitions of 'community' have been found in the scientific literature. All definitions used some combination of space, people and social interactions.

Willmott (1986), Lee and Newby (1983), and Crow and Allen (1995) cited in Smith (2001) explore the term of community in three different ways, they are:

- *Place*. Territorial or place community can be seen as where people have something in common, and this shared element is understood geographically. Another way of naming this is as 'locality'.

- *Interest*. In interest or 'elective' communities people share a common characteristic other than place. They are linked together by factors such as religious belief, sexual orientation, occupation or ethnic origin.
- *Communion*. In its weakest form we can approach this as a sense of attachment to a place, group or idea (in other words, whether there is a 'spirit of community'). In its strongest form 'communion' entails a profound meeting or encounter – not just with other people, but also with God and creation.

Among theories of psychological sense of community, the definition by McMillan and Chavis (1986) is by far the most influential, and is the starting point for most of the recent research on the psychological sense of community (Wright-House, 2009). They state that community is a feeling that members have of belonging, a feeling that members matter to one another and to the group, and a shared faith that members' needs will be met through their commitment to be together (McMillan and Chavis, 1986). McMillan and Chavis (1986) state that there are four elements of sense of community:

- *Membership*: Membership includes five attributes: boundaries, emotional safety, a sense of belonging and identification, personal investment and a common symbol system. These attributes work together and contribute to a sense of who is part of the community and who is not.
- *Influence:* Influence is a bidirectional concept. Members need to feel that they have some influence in the group, and some influence by the group on its members is needed for group cohesion.
- *Integration and fulfilment of needs*: This is the feeling that members' needs will be met by the resources received through their membership to the groups. Members feel rewarded in some way for their participation in the community.
- *Shared emotional connection*: The 'definitive element for true community', it includes shared history and shared participation (or at least identification with the history).

The other term of community proposed by Abarquez and Murshed (2004) is that it can be used to refer to groupings that are both affected by and can assist in the mitigation of hazards and reduction of vulnerabilities. Hence, in the context of disaster risk management, Abarquez and Murshed (2004) define community as a group that may share one or more things in common such as living in the same environment, similar disaster risk exposure, or having been affected by a disaster. Common problems, concerns and hopes regarding disaster risks may also be shared.

6.4.2 Why community?

According to the United Nations International Strategy for Disaster Reduction (UN-ISDR, 2009) disaster is a serious disruption of the functioning of a community or a society causing widespread human, material, economic or environmental loss that exceeds the ability of the affected community or society to cope using its own resources. Analysing this definition, community is the most important word because if the event, for example a landslide, occurred in the area that was uninhabited, then it cannot be categorised as a disaster. Therefore, every effort made in the recovery or reconstruction process in order to return the community to normal life should benefit them. This must be the first consideration.

On a positive note, a disaster can be shown as an opportunity to build to a better quality than previously. Labadie (2008) states that recovery and reconstruction efforts can help to mitigate possible future effects of disasters by making the community more sustainable and more survivable. Hence, the policies on reconstruction have to be developed towards this goal and taken up by all those involved in reconstruction.

The post-disaster housing reconstruction project should not just attempt to build the house physically; another important objective is to build the social capital of the people. By involving the community in the reconstruction process, it can help the survivors to relieve the trauma, stress, depression and feelings of hopelessness that they have suffered. Community-driven approaches ensure the housing reconstruction fund is directed to the right person, i.e. to the one who really needs it. This approach can reduce marginalisation among beneficiaries because every member of the community can participate in the reconstruction process. It will secure equal rights for everybody.

People know what is best for them; what their needs are, what their problems are and how to solve them. In many countries, there are still many cultural considerations that have to be considered when building a house, and this can vary from one country to another country or even from area to adjacent area. Only the community knows such things. By working and planning together in community-based reconstruction, it can strengthen the community. The sense of belonging and togetherness can increase, and make the beneficiaries more united so they can come together and face any problems that might arise.

6.4.3 Type of participation

Chambers (1983) cited in Kumar (2005) states that it was primarily in the 1980s, with the emergence of 'participatory' methods, that the focus on 'community' started gaining prominence. The popularity of 'community participation' is evident from the proliferation of 'participatory' projects from the 1980s onwards. However, Midgley *et al.* (1986) cited in Kumar (2005) pointed out that even though it was central to the issue of participatory development, the concept of 'community' was poorly defined. Kumar (2005) adds that 'community' participation projects are also often found to be ambiguous as to whether the 'community' is meant to be a means or an end to the development programme.

To explain the level of participation of community we can refer to Arnstein's theory. Arnstein (1969) created eight levels of citizen participation, called

'A Ladder of Citizen Participation'. The ladder consists of eight rungs. The bottom rungs of the ladder are rung 1 (Manipulation) and 2 (Therapy). These two rungs describe levels of non-participation where power holders have power to educate or cure the participants. Rungs 3 (Informing), 4 (Consultation) and 5 (Placation), progress to levels of tokenism where citizens may indeed hear, be heard or give advice. Further up the ladder are levels of citizen power; rungs 6 (Partnership) that enables them to negotiate and 7 (Delegated Power) and 8 (Citizen Control) where citizens obtain the majority of decision-making seats, or full managerial power.

Arnstein's model was later modified by Choguill (1996) to fit into under-developed countries. Choguill (1996) classifies a ladder of community participation as neglect, rejection, manipulation and support. Later on, Davidson *et al.* (2007) combines these two theories to suit community participation in housing reconstruction projects (Figure 6.3). It can be seen that the level of control of the community reduces from the top of the ladder to the bottom. If the level of participation goes to the bottom rung of the ladder, the community has little or no power to control or manage the reconstruction. In this case, they may be consulted about their needs and expectations with no assurance that these concerns will be taken into account, or they may be merely informed about the shape the housing project will take or even manipulated into taking part in the project (Davidson *et al.*, 2007).

On the top rung of the ladder, empowerment and collaboration can offer communities control of the housing reconstruction project. These two levels should be the minimum level where a housing reconstruction programme could be called a 'Community-Based' or 'Community-Driven' approach. In practice, beneficiaries can act as the owner, as the supervisor or even as the contractor of their own housing reconstruction project.

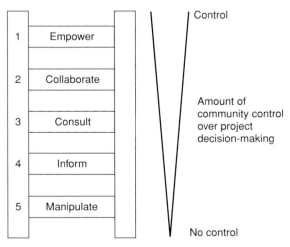

Figure 6.3 Ladder of community participation (Davidson *et al.*, 2007)

6.5   Community-based reconstruction practices

Considering the scale of destruction in Aceh and Nias after the 26 December 2004 tsunami and the 28 March 2005 8.6 Richter scale earthquake, in 2005 the Government of Indonesia established the Rehabilitation and Reconstruction Agency for the Regions and Community Life of Aceh and Nias (BRR). The agency was officially established on 16 April 2005 by Government Regulations in Lieu of Law (Perpu) Number 2 Year 2005 and had a mandate for four years until April 2009. This Perpu was then stipulated to become Law Number 10/2005 on 25 October 2005. BRR was tasked to restore livelihoods and infrastructure and to strengthen communities in Aceh and Nias by directing a coordinated, community-driven reconstruction and development programme (BRR, 2006). For housing reconstruction, BRR created the Deputy for Housing and Settlements, which has three key duties: maintaining coordination with non-governmental organisations (NGOs), data verification, and completing the construction of a portion of houses.

Much can be learned from the devastating tsunami disaster of 2004, especially in the field of housing reconstruction. The massive reconstruction of houses, such as in Aceh and Nias, is not an easy task for governments, donors, international agencies and NGOs involved in the reconstruction task. Providing housing for beneficiaries has been the most problematic and challenging sector of the entire tsunami reconstruction programme.

6.5.1   *Housing reconstruction problems*

The permanent shelter operation in Aceh experienced many problems and delivery has been far lower than the original targets (UNHCR, 2007). From a total of 120 000 houses, BRR had a target to construct 48 000 houses and was responsible for coordinating the construction of 72 000 units built by NGOs and international agencies (BRR, 2007). Moreover, BRR (2007) admitted that the target to construct 120 000 houses at the half way point of its mandate in 2007 has not been achieved. By 31 March 2006, 41 730 houses had been constructed in 18 regions in Aceh. At the end of 2006 the number rose to 57 000 units and by April 2007 the number had reached almost 65 000 units. Finally, BRR (2009) has successfully fulfilled its task by constructing 140 304 houses.

The reconstruction of Aceh and Nias involved more than 100 organisations. Vebry *et al.* (2007) state that many NGOs active in Aceh were originally humanitarian organisations without any relevant experience in housing reconstruction. Lured by huge donations, hundreds of NGOs jumped into the reconstruction process without any supporting background, knowledge and experience in post-disaster housing reconstruction and rehabilitation, and many did so for the first time (Dercon and Kusumawijaya, 2007; Vebry *et al.*, 2007). Dercon and Kusumawijaya (2007) add that many organisations, especially the smaller ones, started building without a clear overall concept.

They worked in the limelight and often failed. In the best of cases, they then dropped out, halted or stopped their programmes. Others postponed their start-up endlessly and in the worst cases built many bad houses and had to acknowledge costly defeat.

BRR (2007) understood that it is not easy to keep all housing beneficiaries fully and equally satisfied. Some agencies implementing housing construction have their own individual construction standards thus leading to a coordination problem. Due to the lack of uniform standards, the housing construction programme is imbalanced either in completion or quality aspects. According to BRR (2006), in general, low contractor capacity and poor supervision has led to poor construction quality.

The delay in housing delivery is caused by many factors: a shortage of human resources, logistical problems, bureaucratic and institutional problems, difficulties in coordinating the multitudes of organisations, land acquisition problems (particularly for the relocation of villages but also for families that have had to move within their own village because their old plot is no longer inhabitable, or no longer exists), and lack of road access (Oxfam, 2006; ACARP, 2007; Vebry *et al*, 2007).

In the ACARP (2007) survey villages, it has been found that the most common complaint about reconstruction has been over delays in housing construction, followed by issues of quality and design, often exacerbated by poor coordination and communication between the housing providers and intended recipients. In a few communities, families have refused to move into their new houses because they believe they were promised superior models, or because they found the design unacceptable.

In 2005, BRR encouraged Universitas Syiah Kuala (UNSYIAH), the Banda Aceh-based State University, to provide third party monitoring and evaluation of housing reconstruction. The survey conducted from 2005 until 2006 monitored settlement recovery of 805 homes, of about 61 organisations, in 161 locations. It used three key indicators to benchmark the success of each project: a construction quality index (0 to 4), satisfaction index (-9 to 9), and accountability index (0 to 10). The accountability index and satisfaction index are based on the beneficiaries' opinion of their benefactor, whereas the construction quality is measured through direct on-site observation with a building inspector, architect and civil engineers, that refer and comply to the Aceh Building Code standard. All results were made public in full. The average result was: construction quality index 2.58; satisfaction index 1.2; accountability index 6.0 (UNSYIAH and UN-HABITAT, 2006).

Looking at the results, it became clear that the Aceh reconstruction has faced serious problem in construction quality, satisfaction and accountability. The most poignant was the satisfaction index. Since the satisfaction index is closely related to community participation, it means that the reconstruction has failed to meet the beneficiaries' needs, which also means less participation from the community.

6.5.2   *Community-based or contractor-based?*

During the Aceh and Nias reconstruction, there were generally two models of housing reconstruction adopted: a contractor-based approach and a community-based approach. Comparing the results of contractor-based and community-based housing after the third party monitoring process by UN-SYIAH, it has been found that the quality, satisfaction and accountability index of community-based housing was better than contractor-based housing (Table 6.5) (Dercon and Kusumawijaya, 2007).

As well as high construction quality, satisfaction and accountability, housing reconstruction using the community-based approach is also faster than the contractor-based approach. Moreover, it was also found that this method faced fewer problems than the contractor-based approach. In addition, MDF (2008) states that the community-based approach also creates a sense of ownership and pride among beneficiaries. Research by Fallahi (1996), cited in Fallahi (2007), has also proven that the lower the level of the participation rate of recipient individuals in the reconstruction process, the lower the level of the satisfaction rate in the resultant relocation and shelter.

More interestingly, according to ACARP (2007) local infrastructure projects that involved communities in planning and implementation have been generally more successful than turnkey small infrastructure projects. Small infrastructure grants to communities, when accompanied by clear guidelines on participatory planning, transparent management and public disclosure of financial information, have proven to be an extremely cost effective means of delivering quality small scale infrastructure not met by other donor or government projects, while significantly strengthening communities' capacity to plan and implement future self-help projects.

Based on their experience in Aceh, Dercon and Kusumawijaya (2007) state that community-based housing reconstruction can work well because it responds quickly to urgent needs and thus can achieve relief at an early stage; it mobilises solidarity among the members of a community and therefore creates social capital; it allows women to be a part of the reconstruction work; it strengthens local institutions; it achieves good planning which leads to high

Table 6.5   Housing reconstruction index in Aceh (source: Dercon and Kusumawijaya, 2007)

Organisations	Construction quality (0 to 4)	Satisfaction score (−9 to 9)	Accountability score (0 to 10)
All organisations in 2006	2.58	1.2	6.0
All community organisation programmes	2.67	2.1	6.7
All contractor-built programmes	2.55	0.8	5.9

quality results; it can limit disaster vulnerability; and it can be done with good monitoring and thus achieve transparent accountability.

Parwoto of the World Bank, cited in Dercon and Kusumawijaya (2007), stresses that the experience from Aceh has shown against all prejudices and misconceptions that Community-Based Development (CBD) can be achieved on a large scale. CBD helps building social capital. Community-based reconstruction experiences also show more accurate targeting of beneficiaries. Doubts about community-based approaches come from a lack of understanding, experience and knowledge about how to organise it, fear of chaos, a shortage of professionals well trained in its implementation, and fear that it will take too long in a post-disaster situation.

In Sri Lanka, for example, a comparative study of Donor Driven vs Owner Driven approaches in housing reconstruction after the 2004 tsunami by Ratnayake and Rameezdeen in 2008, found that dwellers in owner-driven housing programmes are more satisfied than the dwellers in donor-driven housing programmes. The owner-driven housing programme has been prominent in terms of: quality/durability, space availability, flexibility to make any changes in the future, agreeing to change the design as required, land size, location, and overall facilities provided (electricity, water connection and sanitation). Dweller involvement throughout inception, design and construction resulted in greater success in owner-driven housing programmes than those who were under the donor-driven housing programmes (Ratnayake and Rameezdeen, 2008). Another study by Lyons (2009) also concludes that the Owner-Driven Programme (ODP) in Sri Lanka performed better than the Donor-Assisted Programme. The ODP produced more houses, more quickly, of better construction quality, and at lower cost. Space standards were generally better and the designs, layouts, and locations were more acceptable to beneficiaries.

The Bam earthquake in Iran with a magnitude of 6.6 on the Richter scale occurred on 26 December 2003 resulting in 30 000 dead, 20 000 injured and over 60 000 homeless. Almost 80% of Bam was ruined and its historical landmark – a giant medieval fortress complex of towers, domes and walls, all made of mud-brick – was totally destroyed (Fallahi, 2007). The key policy adopted in the Bam reconstruction was where community-active participation in the process of designing, planning and constructing units was strongly encouraged. Householders were able to choose their own plans and layouts and act as supervisors of their own projects, thus paving the way to establish a line of cooperation between designers and contractors. This approach also ensured that government loans resulted in the desired houses being built for the people (Fallahi, 2007). Moreover, Fallahi (2007) states there were two important factors contributing to the success of the Bam reconstruction programme, which were the financial and construction material aid from the Housing Foundation, and the survivors' participation in the process of rebuilding. Active survivor participation in housing leads to operational cost and time reduction, and can reduce the negative psychological impact of the disaster.

6.5.3   Reconstruction model

During the Aceh and Nias housing reconstruction, NGOs adopted different models of procurement methods for housing, one of which is the community-driven approach. The two leading organisations implementing this method are UN-HABITAT and UPLINK (Urban Poor Linkage).

UN-HABITAT: ANSSP

The Aceh Nias Settlements Support Programme (ANSSP) is a joint project between UNDP and its implementing partner UN-HABITAT. This programme was targeted to provide 3600 houses in Banda Aceh, Aceh Besar, Pidie, Simeuleu and Nias starting in 2005. UNDP (2008) reports that between 2005 and 2007 this programme successfully constructed 3587 houses in 23 villages.

ANSSP focused on placing people at the centre of the decision-making process for housing rehabilitation and reconstruction, meaning the process was primarily community-based driven. This did not mean that everything was implemented by the community, but it was more of a reference to the manual needs felt by and defined by the community concerned. Community contracting applied to this, hence the minimum role that the community played was as a construction manager. Basic values developed which was the basis for implementing all the activities of ANSSP: transparency, accountability, equality, honesty, solidarity and participation (UN-HABITAT and UNDP).

This programme supports a process of Community Action Planning (CAP) to ensure beneficiaries are involved in every step of the rebuilding process – from choosing a design, choosing a supplier, taking responsibility for managing finances, and managing all phases of the construction. Facilitators were trained for this purpose and provided model plans for housing. Households were then able to adapt these to meet their individual needs. Communities were also involved in mapping and preparing spatial plans (UNDP, 2008). The general principles of CAP (UN-HABITAT and UNDP) are:

- CAP is a process for action; not a *blueprint* for future development.
- The solution to problems comes from the community and the role of the facilitator is more on formulating the problems in the community.
- CAP is not determined from outside the community but grows/emerges from within the affected community (that suffers as victim).
- Avoid activities such as lecturing or teaching the people; but concentrate more on workshops as a form of discussion with the community.
- Guarantee that all relevant groups participate in the activities, particularly the women and other community groups that are often forgotten.
- Facilitate input from all groups; do not allow one group or leader to dominate the discussion.

- Avoid too many pictures and text; it is better to absorb and remember several points rather than make a long list.
- Use simple language and avoid difficult terms.
- Make the material as simple as possible.

The design of ANSSP consists of six steps (see Figure 6.4) (UN-HABITAT, 2007):

- *Step 1*: The purpose of this step is to establish contact with the community; to recruit and train the field team of facilitators; to engage in household and small group meetings where local problems, capacities and opportunities are discussed. The baseline information is collected at this stage, including the identification of beneficiaries.
- *Step 2*: Cluster groups of eight to 12 households and a representative committee are established. The latter is entrusted to manage the activities on behalf of the clusters. Prior to the formation of the clusters a list of beneficiaries is prepared and verified through community consultation. It is then publicly announced. Moreover, the cluster groups prepare, in consultation with the communities, a Community Action Plan (CAP). Land maps are drawn in a community adjudication process.
- *Steps 3 and 4*: Focus on the preparation and approval of the project proposal; on the disbursement of the block grant; on procurement of materials and services; and on the actual reconstruction of the houses. Witnessed by formal and informal village leaders, a community contract is signed between each cluster and UN-HABITAT. Bank accounts are opened in the name of the cluster by three signatories.

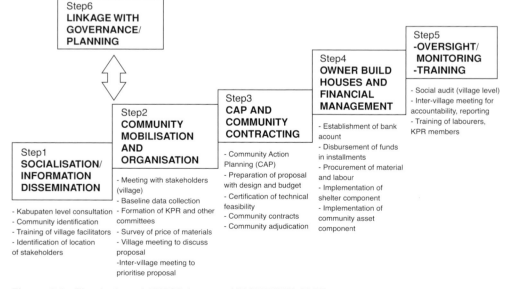

Figure 6.4   The design of ANSSP (source: UN-HABITAT, 2007)

- *Step 5*: Focuses on monitoring, evaluation and project completion, stressing quality assurance.
- *Step 6*: Stresses the need to link this process with the formal governance structure throughout the implementation process.

With beneficiaries deeply involved in the reconstruction work, many communities gained knowledge of the standard features that every house should have, such as a proper foundations and strong roof beams. Many beneficiaries were industrious and saved money by working on the construction of their own homes instead of hiring labourers. They used the savings to buy extra materials or ornamentation for their houses (UNDP, 2008).

JUB/UPLINK

Jaringan Udeep Beusare (JUB) or Udeep Beusaree (means living together) network is a network between 25 coastal villages in Aceh and the Aceh Besar district which was worst hit by the 2004 tsunami (Figure 6.5). In these neighbouring villages, only 47% of the original population survived (Table 6.6) (ACHR, 2005). These villages stretch for 10 km. In managing the reconstruction project, this network was facilitated by Urban Poor Linkage (UPLINK), a network of poor community groups, professionals and NGOs in 14 Indonesian cities. The strategy adopted by Uplink was to empower the communities by organising, giving advocacy and creating networking.

The JUB/UPLINK reconstruction project is not just a physical reconstruction project providing houses and infrastructure, but it also integrates the social dimension. In the construction sector, the objective of this project is to provide 3500 houses and build infrastructure for the community. The

Figure 6.5   JUB village after the 2004 tsunami (photo by M.A.S.M. Omar)

Table 6.6   JUB villages data (source: after ACHR, 2005)

Key features	Number
Number of villages in the network:	25 villages
Pre-tsunami population:	14 144 people (3507 families)
Number of surviving families:	2754 families (78% of original families)
Total surviving population:	6683 people (47% of original population)
Number of female survivors:	2642 people (39% of total survivors)
Number of male survivors:	4041 people (61% of total survivors)
Number of houses needed:	3500 houses

construction cost was US$4200 per house and US$1300 for infrastructure per household. All houses are free for beneficiaries. Non-construction benefits of this project are to create economic renewal, reinforcement of social relations and cultural cohesion, and environmental regeneration.

Members of the community have been involved in all aspects of the project. A complete survey was carried out of the ruined villages, including participatory land mapping and a full account of surviving families. Ownership certificates are being issued based on this work. Further involvement includes land surveying and mapping, housing design, materials purchase and data collection; as well as playing crucial roles in the construction process as material suppliers, labourers, and building inspectors (DIABP, 2008).

There is a strong pro-community policy implemented by UPLINK on managing the housing and infrastructure reconstruction fund so that 60% of the fund stays in the village. This policy aims to accelerate the economy of the community. In order to do this, the resources available in JUB villages are exploited wherever possible. Construction workers, supervisors and suppliers are recruited from local villages. To overcome the soaring cost of building materials, the project tries to maximise the use of local materials that are available in the village. They establish their own production centre, e.g. producing bricks. To provide reinforcement for all houses, this project also built a steel workshop. These workshops employ local workers which then helps the local economy.

The basic principles of housing reconstruction are hard work, integrity, accountability, transparency, and gender equity and social justice (UPLINK, 2008). Some important roles of the beneficiaries in housing reconstruction (UPLINK, 2008) are:

- House owners are implementers. They are builders and inspectors. Construction does not involve contractors.
- Beneficiaries are invited to choose one design from among the five house designs offered by UPLINK.
- Beneficiaries are responsible for the supervision of the house construction assisted by UPLINK building inspectors so as to maintain the construction quality as required for earthquake safe construction.

- Beneficiaries are invited to procure some material items, while UPLINK provides timber, steel, cement, sand, and rock through the JUB Material Bank.
- UPLINK applies a card system for material allocation and distribution.

During the reconstruction process, UPLINK gives technical support (for example digital mapping, planning, drawing designs, cost estimation and construction supervision) to the community. Housing design is very important in the earthquake prone countries so there is a no-compromise requirement of seismic safety. People can make whatever design changes they like to the basic house models being built in the project. However, the one non-negotiable rule in the project is that all the houses must be built to be safe, following the principles of special reinforcement for earthquake resistance, with four ring beams (at foundation, sill, lintel and roof levels), reinforcement at corners, double-thick brick walls, etc. No human being can build a house to withstand a tsunami, but they can definitely build a house that will not fall on anybody and kill them in an earthquake (ACHR, 2006).

Housing reconstruction begins with participatory mapping activities to identify land ownership and borders, then uses the map for village planning and a land tilting supporting document. Participatory village planning based on the agreement of people, including housing, infrastructure (streets, dykes, public facilities and drainage), alternative energy, and mitigation (evacuation routes and building escape routes). Other important activities are the identification of beneficiaries and database creation.

Before construction begins, members of communities have to be trained. Training is done in collaboration with the International Labour Organisation (ILO), for the house owner, construction worker and building inspector. They are provided with an inspection handbook for quality control, training in correct building construction, and steel reinforcement for a good quake designed house (UPLINK, 2008). Figure 6.6 shows the type of housing provided by UPLINK.

In economic development terms, loans are provided for beneficiaries, especially women's groups, to start their businesses. Training has been carried out as part of the project to extend the range of skills of the villagers and to give them greater opportunities to improve their livelihoods. Typical activities include acupuncture, block-making, composting, mushroom cultivation, disaster management, rice farming, earthquake-resistant construction and pedi-cab businesses. Seventy drivers have been given loans to buy their own pedi-cabs. Repayments are made into a collective fund, which can be drawn on for other economic activities and family emergencies (WHA, 2007).

Other innovations in JUB village reconstruction include creating Eco-Village concepts. Methods include using local building materials, recycling, carrying out organic wastewater treatment, and using renewable energy. Organic farming and the planting of special saline-reducing plants is helping the soil return

Figure 6.6   Type of houses built in JUB villages (photo by Y. Kusworo)

to good agricultural condition, following the salt contamination suffered by much of the land.

Natural barriers such as mangrove forests, cypress, coconut, banana, papaya and pine trees have been planted between the villages and the sea to act as buffer layers that will absorb the force of waves. Dykes and ditches will be added. It is recognised that these barriers would not be sufficient to protect against a similar tsunami but that education and pre-preparedness will be crucial to saving lives in the future. Well signposted and wide escape routes are being prepared for each village to evacuation centres in the hills. These hills behind the villages are also being developed as productive agricultural assets, growing income-generating plants including chillies, vegetables and fruit trees. Fishponds are also being established (WHA, 2007).

The JUB/UPLINK community-driven housing reconstruction programme has shown innovation. According to DIABP (2008) this programme has demonstrated how a challenge can be turned into an opportunity to develop new and stronger community links by putting beneficiaries at the heart of the post-disaster reconstruction programme. Community organisation, advocacy and networking were the three main factors in physical and non-physical reconstruction. Compared with other procurement methods, this approach is fast, cheap, transferable and comprehensive due to efficient monitoring and evaluation systems. The programme has been recognised for its unprecedented achievements. It was awarded the 2008 Best Practices in Architecture and Urban Design category by UN-HABITAT and in the same year it also won the Dubai International Award for Best Practices.

Third party monitoring on housing reconstruction for both UN-HABITAT and UPLINK's villages have confirmed that their facilitated houses have achieved high satisfaction, good quality and high accountability by the community (UNSYIAH and UN-HABITAT, 2006). Through their programmes, both organisations have empowered the community to undertake the reconstruction. The basic principles they adopted are the same, which are integrity, solidarity, transparency, accountability and equality.

6.6   Summary

Indonesia is a disaster prone country and is highly vulnerable. Although still facing many problems, the government of Indonesia has tried hard to implement the disaster management principle. This effort has resulted in shifting the paradigm from an emergency approach to a risk management approach and created the national action plan for disaster management.

To be named as a community-based housing reconstruction programme, the level of participation from the community should be on the level of collaborative or empowering. Post-disaster housing reconstruction should not only propose to provide houses for the beneficiaries. It is important that it should also re-empower the social capital of the community. This can be done by implementing the community-based housing reconstruction programme where the community is actively involved and takes control through the whole reconstruction process. This method has proven to be fast, the most cost effective, creates fewer problems and provides high quality housing for the community. As a result, it achieves high satisfaction from the beneficiaries. This method also helps the community to regain their confidence and ease the trauma they have suffered.

References

Abarquez, I. and Murshed, Z. (2004) *Community-Based Disaster Risk Management: A Field Practioners' Handbook.* Pathumthani: ADPC.

ACARP (2007) *The Acehnese Gampong Three Years On: Assessing Local Capacity and Reconstruction Assistance in Post-tsunami Aceh.* Report of the Aceh Community Assistance Research Project (ACARP).

ACHR (2005) *Housing by People in Asia.* Newsletter of the Asian Coalition for Housing Rights. Number 16. August 2005. http://www.achr.net/000ACHRTsunami/Download%20TS/ACHR%2016%20with%20photos.pdf [accessed 25/09/2009].

ACHR (Asian Coalition for Housing Rights) (2006) *Tsunami Update*. http://www.achr.net/000ACHRTsunami/Download%20TS/Tsunami%20Update%20June%20full%2006.pdf [accessed 25/09/2009]

Arnstein, S.R. (1969) A ladder of citizen participation. *JAIP*, **35**, 216–224.

BAPPENAS, The Provincial and Local Governments of D.I. Yogyakarta, the Provincial and Local Governments of Central Java, and international partners. (2006) *Preliminary Damage and Loss Assessment Yogyakarta and Central Java Natural Disaster*. The 15th Meeting of The Consultative Group on Indonesia, Jakarta, June 14, 2006.

BAPPENAS and BAKORNAS PBP (2006) *National Action Plan for Disaster Reduction 2006–2009*. Jakarta: Badan Perencanaan Nasional (BAPPENAS) and Badan Koordinasi Nasional Penanggulangan Bencana (BAKORNAS PB).

BRR (2005) *Aceh and Nias One Year after the Tsunami: the Recovery Effort and Way Forward*. Jakarta: BRR and International Partners.

BRR (2006) *Aceh and Nias, 2 Years after Tsunami: Progress Report*. Banda Aceh: BRR.

BRR (2007) *Strengthening the Narrative of Community Life: Two Year Report Executing Agency for the Rehabilitation and Reconstruction of Aceh-Nias*. Banda Aceh: BRR.

BRR (2009) *Housing: Roofing the Pillars of Hope*. BRR Book Series. Banda Aceh: BRR.

Choguill, M.B.G. (1996) A ladder of community participation for underdeveloped countries. *Habitat International*, **20**, 431–444.

Davidson, C.H., Johnson, C., Lizarralde, G., Dikmen, N. and Sliwinski, A. (2007) Truths and myths about community participation in post-disaster housing projects. *Habitat International*, **31**, 100–115.

Dercon, B. and Kusumawijaya, M. (2007) Two years of settlement recovery in Aceh and Nias: what should the planners have earned? 43rd ISOCARP Congress, Antwerp, Belgium, 19–23 September 2007. Congress Papers.

DIABP (2008) Dubai International Award For Best Practices. http://www.dubaiaward.ae/web/WinnersDetails.aspx?s=36&c=103 [accessed 10/10/2009].

DIBI (2009) Data and Information on Disaster in Indonesia. http://dibi.bnpb.go.id/DesInventar/dashboard.jsp [accessed 10/10/2009].

EM-DAT (2009) *EM-DAT: The OFDA/CRED International Disaster Database*. www.emdat.be. Université catholique de Louvain, Brussels, Belgium. http://www.emdat.be/result-country-profile?disgroup=natural&country=idn&period=1980$2009 [accessed 10/12/2009].

Fallahi, A. (2007) Lessons learned from the housing reconstruction following the Bam earthquake in Iran. *The Australian Journal of Emergency Management*, **22**, 26–35.

Kumar, C. (2005) Revisiting 'community' in community-based natural resource management. *Community Development Journal*, **40**, 275–285.

Hadi, S. (2007) Mainstreaming Disaster Risk Reduction in Indonesia. Climate Change, Disaster Risks and Poverty Reduction: Towards a Common Approach to Reduce Vulnerability. Stockholm, 23–25 October, 2007.

Labadie, J.R. (2008) Auditing of post-disaster recovery and reconstruction activities. *Disaster Prevention and Management*, **17**, 575–586.

Lyons, M. (2009) Building back better: the large-scale impact of small-scale approaches to reconstruction. *World Development*, **37**, 385–398.

McMillan, D.W. and Chavis, D.M. (1986) Sense of community: a definition and theory. *Journal of Community Psychology*, **14**, 6–23.

Multi Donor Fund (MDF) (2008) *Investing in Institutions: Sustaining Reconstruction and Economic Recovery, Four Years After the Tsunami*. Progress Report December 2008. Jakarta: Multi Donor Fund (MDF).

National Information Agency (2004) *Indonesia 2004: An Official Handbook*. Jakarta: Lembaga Informasi Nasional Indonesia.

Oxfam (2006) Oxfam International Tsunami Fund International, Second Year Report, December 2006. http://www.oxfam.ca/sites/default/files/oxfam-international-tsunami-fund-second-year-report.pdf

Ratnayake, R.M.G.D and Rameezdeen, R. (2008) Post Disaster Housing Reconstruction: Comparative Study of Donor Driven vs Owner Driven Approach. *Proceedings of CIB W89 International Conference on Building Education and Research (BEAR)*. Sri Lanka. February 2008.

Smith, M.K. (2001) 'Community' in the *Encyclopedia of Informal Education*. http://www.infed.org/community/community.htm.

Triutomo, S. (2009) *Indonesia: National Progress Peport on the Implementation of the Hyogo Framework for Action*. Jakarta: Badan Nasional Penanggulangan Bencana (BNPB).

UN-HABITAT (2007) 'People's Process' The Path to Recovery. http://www.unhabitat-indonesia.org/docs/files/10-People'-Process-CLEAR.pdf [accessed 16/10/2009].

UN-HABITAT (2008) 2008 Best Practices Database. http://mirror.unhabitat.org/bestpractices/2008/mainview.asp?BPID=2247 [accessed 16/10/2009].

UN-HABITAT (2009). Aceh Nias Settlements Support Project (ANSSP) http://www.unhabitat.org/content.asp?cid=4914&catid=109&typeid=13&subMenuId=0 [accessed 16/10/2009]

UN-HABITAT and UNDP. *ANSSP Guidelines. Volume 1: Orientation and Information*. http://www.unhabitat-indonesia.org/files/book-153.pdf [accessed 16/10/2009].

UN-HABITAT and UNDP. *ANSSP Guidelines. Volume 2: CAP (Community Action Planning & Village Mapping)*. http://www.unhabitat-indonesia.org/files/book-1407.pdf [accessed 16/10/2009].

UNDP (2008) The Third Annual Report on the UNDP Aceh/Nias Emergency Response and Transitional Recovery (ERTR) Programme 2007–2008. Indonesia: United Nations Development Programme.

UNHCR (2007) UNHCR's Response to the Tsunami Emergency in Indonesia and Sri Lanka, December 2004 – November 2006. An independent evaluation by Bobby Lambert 2007. Geneva: UNHCR Policy Development and Evaluation Service.

UNISDR (2009) *2009 UNISDR Terminology on Disaster Risk Reduction*. Geneva: United Nations International Strategy for Disaster Reduction Secretariat.

UNSYIAH and UN-HABITAT (2006) *Post Tsunami Settlement Recovery Monitoring in Aceh*.

UPLINK (2008) Gampoeng Loen Sayang. http://uplink.or.id/v2/downloads/GLS.pdf [accessed 10/10/2009].

USGS (2009) Graph and Earthquake Statistics 1980–2009. http://neic.usgs.gov/neis/eqlists/graphs.html [accessed 12/10/2009].

Vebry, M, Manu, C. and Berman, L. (2007) Community Development Approach in Aceh Reconstruction, Reflecting on Lessons Learned for Yogyakarta – Lesson Learned from The Field, a Practical Guideline in Modern Project Management Style in Post-Disaster Areas. International Seminar on Post-Disaster Reconstruction: Assistance to Local Governments and Communities, Urban and Regional Development Institute. Yogyakarta. 10 July 2007.

WHA (2007) World Habitat Award: Finalist 2007. http://www.worldhabitatawards.org/winners-and-finalists/project-details.cfm?lang=00&theProjectID=742064F6-15C5-F4C0-9966E4B4F5449223 [accessed 10/10/2009].

Wright-House (2009) *Psychological Sense of Community: Theory of McMillan & Chavis 1986.* http://www.wright-house.com/psychology/sense-of-community.html [accessed 6/03/2009].

7 Stakeholder Consultation in the Reconstruction Process

Nuwani Siriwardena and Richard Haigh

7.1 Introduction

A stake is an interest in or a share in an undertaking, while a stakeholder is an individual with a stake. In the context of reconstruction, stakeholders include those individuals or groups that benefit from a reconstruction intervention. Further, stakeholders can be harmed or have their rights affected by the same. Fundamentally, stakeholders are individuals or groups that affect and are affected by any reconstruction activity.

Stakeholders can also affect an activity's functioning, goals, development, and even survival. Stakeholders are beneficial when they help achieve reconstruction goals, but can be antagonistic when they oppose the mission. Stakeholders may have the power to be either a threat or a benefit. They may exert their influence either deliberately or incidentally. Consequently, those responsible for reconstruction projects need to be wary of their stakeholders and their influences.

Diverse sources may trigger stakes. Frequently, stakes can be influenced by economic, cultural or political considerations. For example: a local business owner may be financially advantaged or disadvantaged by a reconstruction project; a local resident may be concerned about the effect a project will have on his property, livelihood or local environment; a politician may be concerned about voter perception of the project, and the degree to which the project is equitable across different groups. Understanding the stake is vital if a stakeholder's interest is to be managed effectively and an appropriate level of engagement identified.

This chapter considers the importance of identifying and engaging with stakeholders during the reconstruction process. It explores the need to identify and manage stakeholders' expectations. The chapter begins with an introduction to the stakeholder concept and describes how stakeholder groups can be classified. The discussion moves on to explore the challenge of managing

Post-Disaster Reconstruction of the Built Environment: Rebuilding for Resilience, First Edition.
Edited by Dilanthi Amaratunga and Richard Haigh.
© 2011 Blackwell Publishing Ltd. Published 2011 by Blackwell Publishing Ltd.

wide-ranging expectations in post-disaster housing reconstruction projects. A range of examples from housing reconstruction programmes around the world highlight that stakeholder expectations are frequently not met. The chapter concludes with a generic process for identifying, engaging and communicating with stakeholders. The process can be tailored for the needs of specific reconstruction projects and programmes.

7.2   Defining stakeholders

Freeman and Reed's (1983) definition of stakeholders as, 'any identifiable group or individual who can affect the achievement of an organisation's objectives, or who is affected by the achievement of an organisation's objectives', is widely cited across the literature. Nickols (2005) adds that a stakeholder is a person or group with an interest in seeing an endeavour succeed and without whose support the endeavour would fail. This implies that stakeholders are individuals, groups of people, organisations and institutions who can influence the strategic decisions of an organisation. Stakeholders can be at any level or position in society, from the international to the national, regional, household or intra-household level. However, several authors have expressed differing viewpoints on the latter definition, due to its breadth and ambiguity making it open to include everyone (see for example: Mitchell *et al.*, 1997; Carroll, 1999). Different classes of stakeholders can be identified by possession of attributes, such as the stakeholder's power to influence the organisation, the legitimacy of the stakeholder's relationship with the organisation, and the urgency of the stakeholder's claim on the organisation. Bunn *et al.* (2002) have highlighted three focal points in the field of stakeholder research. Firstly, the focus on dyadic ties between a stakeholder and the organisation, the organisation and its shareholder, or the organisation and its employees; secondly, the pressure imposed by stakeholders forcing the firm to respond to their voice; and finally, the focus on public policy issues, such as ethics and corporate social responsibility. It is clear that in addition to economic and legal rights or duties, the contract between the organisation and its stakeholder groups – which is not necessarily formalised as it is the case in economic theory – demands that organisations perform social, ethical and environmental responsibilities as well (Podnar and Jancic, 2006). These infer that an organisation has a role and responsibility to deliver the expectations of different categories of stakeholders (Carroll, 1999). Thus, stakeholders are conceptualised as having direct relationships with one another, and relationships emerge depending on the context and necessity (Winn, 2001). It is therefore reasonable to conclude that defining, categorising and identifying the salience and different relationships of stakeholders are socially constructed.

7.3   Stakeholders and post-disaster reconstruction

Alongside human causalities, disasters are usually associated with enormous damage and destruction to the built environment. This impact restricts the development of the economy, annihilating the physical and personal lives of victims, while increasing the likelihood of epidemics. It frequently takes a prolonged period to reinstate the personal, economic, and social lives of the affected due to lack of financial and intellectual resources (Pardasani, 2006). It also opens the way for large scale construction projects: a temporary, one time only and short-term undertaking that is creating the built environment.

Walker *et al.* (2008) define stakeholders of a construction project as individuals or groups who have interest or some aspects of rights or ownership in the project, and can contribute to or may be impacted by, either the work or outcomes of the project. Representatives of different and sometimes discrepant interests are regarded as stakeholders of a construction project (Olander and Landin, 2005). This infers that stakeholder is a collective noun embracing a wider group of people. Newcombe (2003) has accommodated in his definition of construction project stakeholders, 'groups or individuals who have a stake in, or expectation of, the project's performance and include clients, project managers, designers, subcontractors, suppliers, funding bodies, users and the community at large'. Further, the Project Management Institute (2004) identifies stakeholders of a project as those individuals and organisations that are actively involved in the project, or whose interests may be affected as a result of project execution or completion.

It can be deduced that stakeholders of a construction project are any identifiable group or individual who can affect, or is affected by, the achievement of a project's objectives. Capturing their input is a key component of the project development process. Thus, stakeholders of a construction project will vary depending on the nature of the project. A typical housing project undertaken by a contractor will likely be driven by profitability; in contrast, post-disaster housing reconstruction may have much broader social goals. Accordingly, Asgary *et al.* (2006) perceives stakeholders in post-disaster reconstruction as persons, groups, organisations, and systems that have a 'stake' in the reconstruction and that are either likely to be affected by the reconstruction, whose support is needed or who may oppose the reconstruction plans, policies, or projects.

Sundnes and Birnbaum (2002) describe disasters as a mismatch between resources and tasks. This mismatch is felt when it overwhelms the community's ability to cope (Twigg, 2001), but significantly, will also bring about the need to engage with a much wider set of stakeholders. Several parties come to a common platform to bridge the mismatch. For example, the lack of resources typically prompts international assistance flow to the countries affected (Hopkins, 2005). Similarly, international non-governmental organisations (INGOs), non-governmental organisations (NGOs), humanitarian organisations, donor agencies, community-based organisations (CBOs), professional organisations, and the wider public often have a role in the recovery of the affected country. These players may be seen as the typical stakeholders in a disaster, although this will depend on the magnitude of the disaster and the nature of the area affected.

Across the literature, different categories of stakeholder are identified in reconstruction programmes. For example, in the aftermath of the Gujarat earthquake, the Participatory Planning Guide for Planning Disaster Reconstruction (2004) groups stakeholders as follows:

- *Community and citizens' groups*: i.e. groups belonging to or are affiliated with a certain caste, religion, occupation or profession.
- *The government*: Encompassing public and semi-public entities in a wide range of sectors (line agencies) and roles (i.e. elected and administrative officials), at the village, district and state levels, with reconstruction-related responsibilities.
- *Civil society organisations*: Including NGOs, CBOs, civic groups, voluntary associations.
- *Private/corporate sector*: i.e. the business and industrial groups.
- *Professional groups*: Including academic, research, and training organisations, consulting firms, etc.
- *Media*: From newspaper, radio, and television networks.

In a very different context, Rotimi *et al.* (2006) note that those responsible for reconstruction in the aftermath of the Manawatu floods in 2004 and Matata debris flow in 2005, both in New Zealand, had to cooperate with a wide range of stakeholders. In addition to those mentioned in the Gujarat example, typical stakeholders included asset owners (may be private or public and the business community), lifeline agencies, the New Zealand Civil Defence and Emergency Management groups (national, territorial and local government departments, police, fire brigade, relief and welfare agencies, health and safety personnel), insurance companies, and construction and reinstatement organisations.

It is evident that most reconstruction projects will have a large number of stakeholders and this is because, as Bhatt (2003) notes, every disaster has to deal with two sets of stakeholders: one that is active in the area before the disaster (normal development); and, one that becomes prominent after the disaster (response actors), at which time actors usually interact, collide or connect, spending much of their resource managing interpersonal relationships.

It is difficult to identify an exhaustive list of stakeholders that may be associated with a post-disaster reconstruction project. The context and nature of the project will naturally determine them. As demonstrated in the generic process outlined in section 7.5, identifying stakeholders can be a major activity in itself and there are many approaches for doing so. As a starting point, Table 7.1 details an indicative list of stakeholders that are frequently encountered on reconstruction projects following a disaster.

It is evident that reconstruction is a collective effort that may last many years. The priorities of each stakeholder group will vary, depending on the country, role and the resource availability. Engaging with each stakeholder group will be vital if the objectives of the reconstruction project or programme are to be achieved efficiently and effectively. In order to do this, it is necessary to group and prioritise them; as the long list in Table 7.1 demonstrates, treating all stakeholder groups in the same manner would be unrealistic and almost certainly

Table 7.1 Typical stakeholder groups encountered on a post-disaster reconstruction project

Stakeholder group	Example
Individuals	Company owners
Families and households	Long-term local residents
Traditional groups	Clans, religious bodies
Community-based groups	Self-interest organisations of resource users, neighbourhood associations, gender or age-based associations
Local traditional authorities	Village council of elders, a traditional chief
Political authorities recognised by national laws	Elected representatives at the village or district levels
Non-governmental bodies that link different communities	A council of village representatives, a district-level association of fishermen
Local governance structures	Administration, police, the judicial system
Agencies with legal jurisdiction over natural resources	A state park agency
Local governmental services in the area	Education, health, forestry and agriculture
Relevant non-governmental organisations	Local, national or international levels
National interest organisations	Workers' union
Cultural and voluntary associations	Unique national landscapes, an association of tourists
Businesses and commercial enterprises	Local cooperatives to international corporations
Education	Universities and research organisations
Financial	Local banks and credit institutions
Government	National, regional, local
Foreign aid agencies	Staff and consultants of relevant projects and programs
International government bodies	UNICEF, FAO, UNEP

undesirable. It is therefore important to identify a suitable classification for the stakeholders that emerge during a reconstruction project.

7.4 Classifying stakeholders

The reconstruction process gives birth to groups of stakeholders with differing degrees of power, legitimacy and proximity to any resultant projects. A number of such means for classifying stakeholders have emerged (see for example: Freeman and Reed, 1983; Mitchell *et al.*, 1997).

Primary stakeholders are those without whose continuing participation the reconstruction project is unlikely to be able to proceed. Some stakeholders may therefore be considered critical to the project's completion, such as a major donor or regulatory body. Secondary stakeholders are those who can influence or are influenced by the project, but who are not essential to continue with the project. Although secondary, such stakeholders may still be very important to the project. For example, if a project antagonises some local residents, it may fail to meet its goals, or they may be able to influence decision makers.

Stakeholders can be internal or external to the project. Internal stakeholders are those who are members of the project coalition or team, or who provide finance. These might include those responsible for the actual delivery of the project, as well as major donors. External stakeholders are those affected by the project in a significant way. These individuals and groups are frequently diverse, but may include local residents, community groups, regulatory authorities, religious bodies, and national interest organisations.

Stakeholders can also be contrasted between those who are contracted to provide services to the project, such as consultants or suppliers, and those that have no contracted responsibility or formal redress, such as members of the community. Un-contracted stakeholders can have power to disrupt projects through their actions, but are not easily liable for their actions.

In terms of decision making, it is worthwhile to consider stakeholders as being supportive, neutral or anti. The idea is that through effective engagement, stakeholders can be shifted to the supportive side.

In an alternative approach, the concept of 'stakeholder salience', put forward by Mitchell *et al.* (1997), categorises stakeholders by the attributes they possess: power, legitimacy and urgency. These attributes are used to determine how salient the stakeholders are to the project. By using a combination of these attributes Mitchell *et al.* identify seven types of stakeholder.

(1) Dormant stakeholders (power, no legitimacy and no urgency).
(2) Discretionary stakeholders (legitimacy, but no power and no urgency).
(3) Demanding stakeholders (urgency, but no legitimacy and no power).
(4) Dominant stakeholders (power and legitimacy, but no urgency).
(5) Dangerous stakeholders (power and urgency, but no legitimacy).
(6) Dependent stakeholders (legitimacy and urgency, but no power).
(7) Definite stakeholders (power, legitimacy and urgency).

Power is the capacity to induce, persuade or coerce the actions of others and is displayed when one part in a relationship is able to impose its will on the other part. Power may be displayed through force, material or financial resources. For example, a donor who supplies financial resources may have a utilitarian power. Urgency is a determination of how important the project is to the stakeholder and how prepared they are to achieve their own outcomes, whether positive or negative. Stakeholders who possess only the legitimacy attribute are called 'discretionary stakeholders' as it is at the discretion of the decision makers on the project as to whether or not these stakeholders deserve attention.

It is evident from these examples that the number, type and classification of stakeholders can vary significantly depending on the nature of the project under consideration. For the remainder of this chapter, the focus will be on stakeholder engagement in post-disaster housing reconstruction projects, an often reported aspect of recovery programmes where stakeholder expectations have been found to vary significantly, and in many cases, not met.

7.5   Expectation gaps in post-disaster housing reconstruction

Housing reconstruction projects bring stakeholders with differing degrees of power, urgency and legitimacy into the same platform. Managing stakeholder expectations forms the basis for achieving the objectives of time, cost and quality. This will minimise the communication gaps, time lags and conflict. However, in the aftermath of many housing reconstruction efforts, dissatisfaction has been expressed by some stakeholders. Areas of criticism usually include time, cost, quality, coordination with infrastructure and linkage to livelihoods. Managing stakeholder expectations appears to be an important, if often overlooked, task of reconstruction. Housing reconstruction requires the deliberate and coordinated efforts of many stakeholders, if recovery of the affected community is to be delivered effectively and efficiently.

Before considering the challenges in the reconstruction process and some examples, it is important to clarify what is meant by housing reconstruction. The terms shelter and housing are frequently given multiple and ambiguous meanings. For example, the term shelter is often used to refer to everything from an evacuee leaving home to stay in a neighbour's house for a few hours while awaiting the passing of a dangerous threat, to an evacuee staying with relatives for several years in a different part of the country while awaiting for the rebuilding of a house in the local community (Quarantelli, 1995). For the purpose of this chapter, shelter provides accommodation during the transition which is of a temporary nature, while a house is of a permanent and habitable nature. Both are used for residential purposes.

Different approaches to housing reconstruction have emerged depending on the phase of the disaster life cycle, the level of urgency, the underpinning philosophy of funders and other decision makers, the level of resources and expertise available, and the type of disaster. By way of example, some programmes aim to fulfil additional social, political or economic goals, which may influence the approach (Barakat, 2003). The process of housing reconstruction is often divided into four periods: pre-disaster, immediate relief, rehabilitation, and reconstruction. Consequently, Quarantelli (1995) identifies four corresponding types of housing: emergency sheltering, temporary sheltering, temporary housing and permanent housing.

Emergency shelter is usually established after a disaster at the instigation of individuals and households based on chance availability, convenience, proximity and perceived safety. It is transitional to temporary shelter when hazard conditions make permanent housing temporarily uninhabitable. Emergency shelter is spontaneous and depends on the local conditions and the immediacy of the need. Temporary sheltering is designed for the use in the few months following a disaster, which alleviates the problem of immediate need of accommodation. Traditionally, emergency or temporary shelter has taken the form of plastic sheeting, tents or emergency centres set up in communal buildings or relief camps (Barakat, 2003). In these early forms, expectations are usually minimal and focus upon the provision of a shelter to keep away hostile climatic conditions, a place to store salvaged belongings, and a sense of emotional security and a need for privacy. Temporary housing is the place where victims can stay for a longer time before it is safe to return to their original residence and could also become permanent shelter (Bolin, 1994). By this point, the range of expectations can start to diverge. The location of the site, the cost of the shelter, impact on the local urban environment, public safety, the socio-economic impact on displaced families, quality standards, intended period of use, and the provision of proper sanitary facilities can all become points of contention where expectations differ (Johnson, 2002).

For permanent housing, the divergence in what is expected by some stakeholders and what is delivered tends to become even greater. Areas of dispute frequently include the consistency in application of policies and guidelines, timescales for delivery, equity in the distribution of resources and prioritisation, the level of transparency and accountability, socio-cultural compatibility, links to livelihoods, the application of risk reduction in design and execution, the level of supporting infrastructure, aesthetics and interior design, and the quality and durability of the finished house (Davidson *et al.*, 2007).

Inevitably, expectations for permanent housing are significantly higher than for those in the earlier periods following a disaster. Permanent housing reconstruction is usually a long-term, significant, complex, and diverse process. It helps to re-establish lost property and hopefully, provides better and more robust accommodation than existed prior to the disaster (Barakat, 2003). Despite this, evidence suggests that in many housing reconstruction projects, there is a significant gap between the expectations of stakeholders, and what is actually delivered (see for example: HIC, 2005; Davidson *et al.*, 2007; Perry, 2007; Lyons, 2009).

The problems encountered during reconstruction of housing following Hurricane Katrina are well documented. McGee (2008) reports that the task of reconstruction rested upon the state, local and federal governments. There was an abundance of finance but the approach was highly wasteful, bureaucratic and inefficient. This premise was further supported by Writer (2006) who noted that:

'[...] two audits found that up to 900 000 of the 2.5 million applicants who received aid under FEMAs emergency cash assistance program – which included the $2000 debit cards given to evacuees – were based on duplicate or invalid Social Security numbers, or false addresses and names. . . .'

Reconstruction was delayed due to red tape, scrapping of previously announced plans, the introduction of a new set of rules, and employment of obsolete plans. Residents were left alone with a lack of certainty about the exact plan (Raskin *et al.*, 2008).

In New Zealand the regulatory framework had a major impact on the reconstruction efforts after the Manawatu Floods in 2004 and Matata Debris flow in 2005. A barrier was created between the practitioners and policy makers who were involved in reconstruction, due to disharmony resulting from different interpretations of legislation (Le Masurier *et al.*, 2006).

Housing problems in the post-Great Hanshin Earthquake highlighted the socio-economic polarisation apart from the damage which was prevalent in Kobe (Hirayama, 2000). As a result of differential in the condition of housing between the suburbs and the inner-city, housing damage was unequal. For example, dilapidated housing, wooden rental housing, wooden structured terraced housing and the housing stock in the inner-city area were devastated (Hirayama, 2000). Housing damage and the process of housing recovery increased the geographical disparity that had already existed before the earthquake, while more recently developed areas suffered from poor community decision-making and a lack of strong leadership (Shaw and Goda, 2004).

Similarly, there are widespread reports of varying expectations in the aftermath of 2004 tsunami. Differences included expectations in location, design, space, safety, accessibility, changeability of the house, the level of community empowerment, the level of consultation with stakeholders, livelihood reconstruction, cultural and social facets, village planning, and the presence of supporting infrastructure and other facilities. For example, reconstruction and rehabilitation of housing in Aceh and Nias became a major effort, with input from a large number of international and Indonesian organisations. According to Steinberg (2007), beneficiaries were told that construction would be completed by the end of 2006. However, due to a range of factors, this overly ambitious target was not met. Slow physical progress resulted from mounting obstacles, including unbuildable land, difficulties in the selection of beneficiaries, environmental problems on some sites, cost escalation, difficulties in sourcing construction materials, unsuitable specifications, insufficient budgets for supporting infrastructure, and an absence of livelihood reconstitution (Steinberg, 2007; Ahmed, 2008). In extreme cases, as the Aceh Community Assistance Research Project (2007) later reported, families refused to move into their new houses because they believed they were promised superior models or because they found the design unacceptable.

Similar problems were encountered in Sri Lanka. However, as with several other affected countries, the construction of permanent housing followed two very different strategies, with different levels of success reported in respect of stakeholder satisfaction. The owner-driven approach involved the affected households getting actively involved in the design and build of their permanent homes (Mulligan and Shaw, 2007). The government provided a cash grant of US$2500 for a fully damaged house in four instalments, and US$1000 in two instalments for a partly damaged house. NGOs provided additional payments, materials and technical assistance. In contrast, donor-driven housing involved a more traditional construction project approach, whereby agencies or private enterprises undertook the construction work to practical completion, but funded by donor agencies. In such instances, the government was usually responsible for funding and coordinating services and infrastructure for the developments, which would usually be procured through the private sector.

A research study carried out by Ingirige *et al.* (2008), in Galle District of Sri Lanka examined the advantages and disadvantages of the two strategies. Beneficiaries of owner–driven housing were found to be significantly more satisfied than those occupants of donor-driven schemes. Commonly cited benefits included space availability, the ability to influence design and flexibility to future changes. Owner–driven schemes also provided the basis for employment, stronger adherence to socio-cultural norms in design and a stronger sense of emotional attachment. Barenstein's (2005) study in India supported these findings with advantages including cost efficiency, incremental building allowing occupancy before the house is fully finished, and higher occupancy rates. In both studies, owner-driven approaches are not without critics, however. Some problems were encountered, in particular with respect to quality. Despite this, owner-driven schemes appeared to benefit greatly from the active engagement of a major stakeholder – the beneficiary – in the rebuilding process.

These examples, both positive and negative, from disaster-affected communities in many different contexts are indicative of the type of gaps that often develop between what stakeholders expect and what is actually delivered. The reasons for such divergence are complex and multi-faceted but many are linked to weak communication and a failure to engage effectively with all relevant stakeholders. Inappropriate housing can often be attributed to a failure to engage sufficiently with the beneficiaries. Likewise, supporting infrastructure is often delivered by other projects, and communication and coordination with this much wider set of stakeholders, usually including Government, will be necessary to ensure that any housing is appropriately served. Further, inequity and increases in social tension can occur due to recovery programmes not identifying and subsequently ignoring the needs of marginalised communities. Spending time to understand the context in which reconstruction will take place is vital and requires carefully planned stakeholder engagement.

7.6   Developing a stakeholder engagement strategy

The divergent approaches of owner- and donor-driven housing emphasise that the basis from which any stakeholder engagement is undertaken may be very different. Regardless, numerous methods and tools can be used to help any reconstruction project or programme to engage with its stakeholders more effectively. The final part of this chapter provides guidance around the area of stakeholder identification, engagement and communication. The approach is adapted from a range of tools and techniques that have been published in the stakeholder literature including Mendelow (1981), Mayers (2005), Chevalier and Buckles (2008) and Bourne and Walker (2008), as well as methods developed by the authors during stakeholder analysis workshops in post-Tsunami Sri Lanka. It details a process for stakeholder engagement. The process is a high level overview but can be personalised for a specific programme or project. Broadly, the engagement process will:

- Provide guidance for identifying and engaging with stakeholders.
- Ensure that stakeholder engagement activities are integrated and undertaken in a co-ordinated manner.
- Improve the effectiveness of engagement efforts undertaken.

7.6.1 Scope and identify stakeholders

The first stage is to identify a core problem or action where you need to identify the stakeholders. This may be as broad as a long-term risk reduction programme with a range of interventions, a specific housing reconstruction project, or a local activity such as capacity building and training.

The next step is to identify the stakeholders that can influence or be affected by the problem or action. A range of methods can be used to achieve this. Experts may be consulted, such as staff, key agencies (such as non-governmental organisations), local people, or academics that have a lot of knowledge about the situation. Announcements at meetings, in newspapers, on local radio or other media may be used to invite stakeholders to come forward. This will attract those who believe they will gain from communicating their views and are able to do so. Census and population data may provide useful information about the numbers of people by age, gender, religion, residence, and ensure that all groups are represented. Similar information may also be available from directories, organisational charts, surveys, reports or written records issued by local authorities, donor agencies, and government bodies.

When attempting to identify stakeholders it may be helpful to consider (adapted from Chevalier and Buckles, 2008):

- Who are the communities, groups or individuals who may be affected by decisions?
- Who are the main authorities in the area?
- Who has access to the land, area or resources at stake?
- Which communities, groups and individuals are most dependent on the resources?
- Who is responsible for claims, including customary rights and legal jurisdiction, in the territory or area where the resources are located?
- Which communities, groups or individuals are most knowledgeable about, and capable of dealing with, the territories or resources?
- How does use of the resources change depending on the seasons, the geography and the interests of the users?

7.6.2   *Classify and map stakeholders*

In order to facilitate both initial and subsequent engagement with stakeholders, it is important to identify and classify stakeholders. The main objective of the exercise is to get a better understanding of the stakeholder and their stake in the project. This will involve:

- Identifying their mission and stake
- Identifying different categories of stakeholder

It is likely that any initial attempt to identify stakeholders will result in a long list. It is therefore necessary to confirm that each individual or group that has been identified is actually a stakeholder and also what the nature of the stake is. Some simple questions can help to clarify this:

- How is the stakeholder important? What is their 'stake'?
- What does the stakeholder require from the success or failure of the project?

Understanding the stake is vital in order to prioritise stakeholders and design effective engagement strategies for the future. If there is some doubt about the stake, it may be necessary to undertake surveys or similar, to establish interest and expectations.

The stake may be a right, legal or moral, to be treated in a certain way, or to an asset or property. It may be due to being impacted in some way by the project, or its outcomes, or affected by a decision related to the project. It may also be in the form of an important contribution to the project, such as the supply of resources or funding.

Once the stake has been identified, a wide range of methods can be used to classify stakeholders, as detailed in section 7.4. These may consider whether the stakeholder is primary or secondary, direct or indirect, or to what extent the stakeholder has attributes such as power and urgency.

In order to summarise the characteristics of stakeholders, it is often useful to use tables or charts as a means to group or prioritise stakeholders so that appropriate engagement strategies can be identified. Prioritisation can be achieved by means of simple index tables, using consensus to assign values and rank the importance of stakeholders. Alternatively, a power-interest matrix can be used, as proposed by Mendelow (1991), which helps inform the level of engagement required for different stakeholder groups.

Rainbow diagrams are also a popular tool for grouping stakeholders according to specific characteristics (Mayers, 2005; Chevalier and Buckles, 2008). A rainbow diagram is created by drawing a horizontal line with half a circle around it, and two further semicircles inside the chart. The rainbow is further divided into three equal parts: one part to the left, one in the middle, and one to the right. The diagram might be used to consider each stakeholder's ability to influence the project (power), and the extent to which it is affected (value and action). In such an instance, stakeholders that are the most affected by the problem or action would be placed in the small semicircle, the stakeholders moderately affected in the middle, and the stakeholders who are the least affected in the outer. Stakeholders who influence the core problem or action the most can be placed on the left, those who moderately influence in the middle and those who influence the least on the right. The diagram can be adapted by using other characteristics that better describe the main differences between key stakeholders.

7.6.3   Establish stakeholder engagement plans

Classifying and mapping stakeholders provides a method for grouping stakeholders and thereby provides an ideal basis to plan appropriate ways to engage with them, based on their characteristics. Engagement plans typically involve:

- Defining the required level of stakeholder involvement and associated scope of activities.
- Setting expectations about when and how stakeholder groups will be engaged throughout the project.
- Proactively managing activities to sustain stakeholders' level of engagement.
- Building buy-in and commitment, and addressing any potential road blocks across the project.

Based on the characteristics of the stakeholder, it should be possible to consider their level of support for the project and their receptiveness, as discussed in the Stakeholder Circle® approach (see Bourne and Walker, 2008). The nature and frequency of communication and engagement can be tailored to the specific needs of each group. For example, excessive communications aimed at a stakeholder that is neutral and not interested, might be counter-productive and antagonise them. The frequency of any communication would need to be considered carefully. A stakeholder that actively opposes the project but is highly receptive may be better targeted with targeted briefing and consultation. A simple newsletter would be unlikely to persuade them, but may be appropriate for someone that is passively supportive and ambivalent or somewhat receptive. An engagement plan might also consider how to capitalise on those groups that are actively supportive and receptive. How might such groups be used to reach out to other less supportive stakeholders?

In summary, the engagement plan should consider who, what, when and how to engage.

7.6.4   Engage stakeholders

The optimum method and frequency of engagement with stakeholders will vary considerably. Typical methods include:

- Distribution of regular (scheduled) project newsletters.
- Stakeholder briefings.
- Communication of milestones and progress.
- Participation in testing and pilots.

Regardless, a vital final stage is to undertake periodic checks to measure the degree of buy-in and engagement. If the level of support is falling with a particular group, the engagement plan may require updating to address particular concerns.

An essential aspect of engaging effectively with stakeholders is to recognise that the stakeholder community is not static. Individuals and groups that are essential to its success in the planning stage may not be as important once the project is underway. The stakeholder community membership may change over time.

7.7   Summary

Post-disaster reconstruction takes place in a challenging environment. Understanding the wide range of groups that should be involved in the reconstruction, as well as those who might be affected, positively or negatively, is vital in order to achieve goals set out for recovery. Effective stakeholder engagement is not easy, but with appropriate time spent on the identification and mapping of different stakeholder groups, it is possible to identify their characteristics and plan accordingly.

References

Aceh Community Assistance Research Project (ACARP) (2007) *The Acehnese Gampong Three Years On, Assessing Local Capacity and Reconstruction Assistance in Post-Tsunami Aceh.* Report of ACARP.

Ahmed, K.I. (2008) Challenges and opportunities of post-disaster shelter reconstruction: the Asian context. Fourth International i-Rec Conference: Building Resilience: Achieving Effective Post-disaster Reconstruction, 30 April–2 May, Christchurch, New Zealand.

Asgary, A., Badri, A., Rafieian, M. and Hajinejad, A. (2006) Lost and used post-disaster development opportunities in Bam Earthquake and the role of stakeholders. Third International Conference on Post-Disaster Reconstruction: Meeting Stakeholder Interests, 17–19 May, Florence, Italy.

Barakat, S. (2003) *Housing Reconstruction after Conflict and Disaster.* Network Paper Number 43. London: Humanitarian Practice Network, Overseas Development Institute.

Barenstein, J.D. (2005) Housing Reconstruction in Tamil Nadu One Year after the Tsunami. http://www.odi.org.uk/HPG/papers/Housing_Tsunami%20_duyne.pdf [accessed 22/10/2007].

Bhatt, M. (2003) Increasing accountability through external stakeholder engagement: the experience of the Disaster Mitigation Institute (DMI). One World Trust Workshop on 'Increasing Accountability through External Stakeholder Engagement', 23–24 October 2003, London, UK.

Bolin, R. (1994) Post-disaster sheltering and housing: social processes in response and recovery. In Dynes, R. and Tierney, K. (Eds), *Disasters Collective Behavior, and Social Organization.* Newark, DE: University of Delaware Press.

Bourne, L. and Walker, D.H.T. (2008) Project relationship management and the Stakeholder Circle™. *International Journal of Managing Projects in Business*, **1**, 125–130.

Bunn, M.D., Savage, G.T. and Holloway, B.B. (2002) Stakeholder analysis for multi-sector innovations. *Journal of Business and Industrial Marketing*, **17**, 181–203.

Carroll, A.B. (1979) Three-dimensional conceptual model of corporate performance. *The Academy of Management Review*, **4**, 497–505.

Carroll, A.B. (1999) Corporate social responsibility. *Business and Society*, **38**, 268–295.

Chevalier, J. and Buckles, D. (2008) *SAS2: A Guide to Collaborative Inquiry and Social Engagement.* London: Sage Publications.

Davidson, C.H, Johnson, C., Lizarralde, G., Dikmen, N. and Sliwinski, A. (2007) Truths and myths about community participation in post-disaster housing projects. *Habitat International*, **31**, 100–115.

Freeman, E.R. and Reed, D.L. (1983) Stockholders and stakeholders: a new perspective on corporate governance. *California Management Review (pre-1986)*, **25**, 88–106.

HIC (Habitat International Coalition) (2005) Human Rights of Tsunami Survivors in Tamil Nadu and Sri Lanka Being Violated. http://www.hicnet.org/articles.php?pid=1816 [accessed on 13/11/2006].

Hirayama, Y. (2000) Collapse and reconstruction: housing recovery policy in Kobe after the Hanshin Great Earthquake. *Housing Studies*, **15**, 111–128.

Hopkins, M. (2005) Business, CSR and the Tsunami: Time to Re-think Corporations and Development? http://www.mhcinternational.com/articles/csr%20development%20and%20the%20tsunami.htm [accessed 21/02/2008].

Ingirige, B., Haigh,R., Malalgoda, C. and Palliyaguru, R. (2008) Exploring good practice knowledge transfer related to post-tsunami housing (re-)construction in Sri Lanka. *Journal of Construction in Developing Countries*, **13**, 23–42.

Johnson, C. (2002) What's the big deal about temporary housing? Planning considerations for temporary accommodation after disasters: example of the 1999 Turkish earthquakes. Proceedings of the First International Conference on Post-disaster Reconstruction: Improving Post-Disaster Reconstruction in Developing Countries, 23–25 May, Université de Montréal, Canada.

Le Masurier, J., Rotimi, J.O.B. and Wilkinson, S. (2006) A comparison between routine construction and post-disaster reconstruction with case studies from New Zealand. 22nd ARCOM Conference on Current Advances in Construction Management Research, 4–6 September, Birmingham, UK.

Lyons, M. (2009) Building back better: the large-scale impact of small-scale approaches to reconstruction. *World Development*, **37**, 385–398.

Mayers, J. (2005) *Stakeholder Power Analysis*. London: International Institute for Environment and Development.

McGee, R.W. (2008) An economic and ethical analysis of the Katrina disaster. *International Journal of Social Economics*, **35**, 546–557.

Mendelow, A. (1981) Stakeholder mapping: power/interest matrix. In: Proceedings of the 2nd International Conference on Information Systems, December 7–9, Cambridge, MA.

Mitchell, R.K., Agle, B.R. and Wood, D.J. (1997) Toward a theory of stakeholder identification and salience: defining the principle of who and what really counts. *The Academy of Management Review*, **22**, 853–886.

Mulligan, M. and Shaw, J. (2007) What the world can learn from Sri Lanka's post-tsunami experiences. *International Journal of Asia Pacific Studies*, **3**, 65–91.

Newcombe R. (2003) From client to project stakeholders: a stakeholder mapping approach. *Construction Management and Economics.* **21**, 841–848.

Nickols, F.W. (2005) Why a stakeholder approach to evaluating training? *Advances in Developing Human Resources*, **7**, 121–134.

Olander, S. and Landin, A. (2005) Evaluation of stakeholder influence in the implementation of construction projects. *International Journal of Project Management*, **23**, 321–332.

Pardasani, M. (2006) Tsunami reconstruction and redevelopment in the Maldives: a case study of community participation and social action. *Disaster Prevention and Management*, **15**, 79–91.

Participatory Planning Guide for Post-Disaster Reconstruction (2004) Joint report prepared by EPC-Environmental Planning Collaborative, Ahmedabad, India and TCG International, LLC, Washington, DC with the support of USAID/India and the Indo-US Financial Institutions Reform and Expansion (FIRE-D) Project, January 2004.

Perry, M. (2007) Natural disaster management planning: a study of logistics managers responding to the tsunami. *International Journal of Physical Distribution & Logistics Management*, **37**, 409–433.

Podnar, K. and Jancic, Z. (2006) Towards a categorization of stakeholder groups: an empirical verification of a three-level model. *Journal of Marketing Communications*, **12**, 297–308.

Project Management Institute (2004) *A Guide to the Project Management Book of Knowledge (PMBOK)*, 3rd edn. Newtown Square, PA: Project Management Institute.

Quarantelli, E.L. (1995) Patterns of sheltering and housing in US disasters. *Disaster Prevention and Management*, **4**, 43–53.

Raskin, M., Kjar, S.A. and Rahm, R. (2008) What is seen and unseen on the Gulf coast. *International Journal of Social Economics*, **35**, 490–500.

Rotimi, J.O.B., Le Masurier, J. and Wilkinson, S. (2006) The regulatory framework for effective post-disaster reconstruction in New Zealand. Third International Conference on Post-Disaster Reconstruction: Meeting Stakeholder Interests, May 17–18, Florence, Italy.

Shaw, R. and Goda, K. (2004) From disaster to sustainable civil society: the Kobe Experience. *Disasters*, **28**, 16–40.

Steinberg, F. (2007) Housing reconstruction and rehabilitation in Aceh and Nias, Indonesia—rebuilding lives. *Habitat International*, **31**, 150–166.

Sundnes, K.O and Birnbaum, M.L. (2002) *Health Disaster Management—Guidelines for Evaluation and Research in the Utstein Style.* University of Wisconsin: Healthcare Fields Intensive Care and Emergency Medicine.

Twigg, J. (2001) *Corporate Social Responsibility and Disaster Reduction: A Global Overview.* London: Benfield Greig Hazard Research Centre, University College London.

Walker, D.H.T., Bourne, L.M. and Shelley, A. (2008) Influence, stakeholder mapping and visualization. *Construction Management and Economics*, **26**, 645–658.

Winn, M.I. (2001) Building stakeholder theory with a decision modelling methodology. *Business & Society*, **40**, 133–166.

Writer, S. (2006) Audits: millions in Katrina aid wasted. MSCBC Online, February 13. www.msnbc.msn.com/id/11326973 [accessed 22/08/2009].

8 Project Management of Disaster Reconstruction

Udayangani Kulatunga

8.1 Introduction

Recent experience following disasters such as the Indian Ocean tsunami, the Kashmir earthquake and Hurricane Katrina revealed a lack of preparedness for such hazard events, which delayed recovery and reconstruction work, thereby subjecting those affected to increased vulnerability and challenging social and economic conditions. In many instances, those communities affected were living in temporary shelter for more than one year. Both reactive and proactive strategies for before, during and after a disaster are therefore important to ensure the well-being of the affected community and to make disaster management a success. Often the enthusiasm seen just after the disaster does not last long, thus creating a gap between the short-term relief and long-term developments plans. This is mainly due to a lack of strategies for long-term development of the affected community, and their built and human environment. In addition to these strategies, effective project management during post-disaster development is essential for effective and efficient reconstruction activities.

In general terms, project management is the application of knowledge, skills, tools and techniques to accomplish project requirements (Project Management Institute, 2004). Project management generally consists of planning, organising, executing and controlling to allow the successful achievement of specified goals. The main activities of project management include: developing a project plan that defines and confirms the project objectives; identifying tasks, budget, resources and methodologies to achieve the objectives; and, determining timelines for completion. There are a number of success criteria for effective project management (Department for Business, Enterprise and Regulatory Reform, 2007). They include to: deliver the outcomes and benefits required by the organisation, its delivery partners and other stakeholder organisations;

Post-Disaster Reconstruction of the Built Environment: Rebuilding for Resilience, First Edition.
Edited by Dilanthi Amaratunga and Richard Haigh.
© 2011 Blackwell Publishing Ltd. Published 2011 by Blackwell Publishing Ltd.

create and implement deliverables that meet agreed requirements and time targets; stay within financial budgets; involve all the right people; make best use of resources in the organisation and elsewhere; take account of changes in the way the organisation operates; manage any risks that could jeopardise success; and, take into account the needs of staff and other stakeholders who will be impacted by the changes brought about by the project. Although the context may be different to many other project types, if post-disaster reconstruction projects are to be managed successfully, it would appear logical that they should also adhere to such performance criteria.

However, the aftermath of a disaster typically poses a number of challenges that make the project management of reconstruction activities more difficult. These challenges may include:

- *Speed*: The need for rapid reconstruction work to provide shelter for the displaced community.
- *Capacity*: Due to the atypically high demand for reconstruction work in the local affected area, there is usually a lack of local construction industry capacity in terms of professionals, labour and material. Injury or loss of life resulting from the direct effects of the disaster can further exacerbate these shortages.
- *Funds*: After a disaster, the affected community, local authorities and government are often faced with wide ranging demands for emergency works but have only limited funds available.
- *Accountability*: Financial assistance provided by domestic and international donors, including public and private organisations, will typically demand a high degree of accountability for its usage.
- *Multiple actors*: A large number of organisations, many of them who do not usually work together, will usually be engaged in the reconstruction process, thereby making coordination, monitoring and quality controlling difficult. Identifying and managing the different roles and responsibilities of multiple actors is a further challenge.
- *Emergence of new organisations*: New organisations, which have reconstruction goals or responsibilities, often emerge during post-disaster relief and reconstruction phases. The experience and capabilities of such organisations may be unclear, which places further strains on monitoring in order to prevent substandard work.
- *Communication and information*: The involvement of multiple actors, often from different disciplinary, national and cultural backgrounds, makes communication and information management more challenging.

Project management during post-disaster reconstruction needs to ensure that such challenges are properly addressed to rebuild the affected community's built environment to an appropriate standard. In doing so, it should make certain that the desired quality of reconstructed buildings and infrastructure are achieved within appropriate timescales, cost limits and other parameters such as minimal environment impacts, and addressing sustainable construction goals. Planning, organising, executing and controlling these activities, which will involve multiple actors, must also be carried out while limiting the room for corruption and mishandling of resources.

This chapter considers how project management can be implemented in a post-disaster environment. It discusses a number of important project management domains: procurement and contract management; sourcing of labour, material and plant; resource management; quality control; financial management; governance; and, risk reduction.

8.2   Procurement and contract management

A procurement strategy within the construction context can be explained as obtaining the whole spectrum of goods, materials, plant and servicers for design, build and commission of a building that provides best possible value for money over its lifecycle (Cartlidge, 2009). The approach to construction procurement mainly depends on the type of client, and the nature and complexity of the project. The procurement of post-disaster reconstruction work is complex due to a range of factors, including time and resource constraints, and communication and coordination difficulties that emerge due to the nature and number of stakeholders involved.

An effective procurement strategy will ensure rapid progress of reconstruction work without hampering its progress. Well thought out guidelines for the procurement of workforce, material, plant and equipment can help to ensure smooth progress of reconstruction work. In contrast, a failure to identify procedures to procure contractors in Sri Lanka for reconstruction work after the tsunami negatively affected the progress of work (Nissanka *et al.*, 2008). Soon after the 2004 Boxing Day tsunami, the Sri Lankan government imposed a number of rules when selecting contractors for reconstruction work. One rule stipulated that only contractors who are registered with the Institute for Construction Training and Development (ICTAD) could be used. However, before this rule came into practice, some of the donor agencies started reconstruction work using community-based principles, which involved local residents as the workforce. The introduction of this new rule temporarily halted this work, before later being rescinded, which further confused many of the actors.

The procurement strategy must also establish the parameters for selecting contractors to undertake the reconstruction work. These parameters may include whether the selection is based on low cost, quality, the experience of the workforce, or resource capabilities for example. Failure to adopt clear and appropriate parameters for selection may lead to repeat work or not provide value for money. For instance, after the tsunami in Sri Lanka, the government gave authority for the donors to select the contractors for donor-funded reconstruction projects. As a result of selecting contractors based on a low cost parameter, some of the reconstructed houses lacked the quality standards that were required at the outset.

Furthermore, the procurement strategy for reconstruction work needs to strengthen the communication and liaison between the stakeholders. Moreover, post-disaster reconstruction work should ensure proper utilisation and accountability of finance due to the limited resources and use of donor funding. Therefore, the procurement strategy must try to avoid a blame and claim culture, while maintaining the quality of work.

8.2.1 Some procurement routes advisable for post-disaster reconstruction

The use of traditional procurement methods that separate design and construction tends to prolong reconstruction activities due to the time taken for design, tendering, document preparation and contractor selection. Traditional procurement methods are generally not deemed advisable because of this. In contrast, integration of design and construction is encouraged in order to speed up construction work. Further, the use of contractors' experience and knowledge for the design process is evident from integrated procurement methods such as design and build. Design and build procurement approaches may overlap the design phase with on-going construction work, thus lowering overall construction time. Having a single point responsibility for post-disaster reconstruction work is another perceived advantage and may help to ensure an appropriate quality of work. Myburgh *et al.* (2008) suggest that the design and build procurement method and its variations can respond to the client's needs, thus accelerating commencement and completion of the work.

Disaster reconstruction work can involve new construction, as well as repair and refurbishment work. This can involve uncertain quantities that cannot be pre-assessed. In this type of a situation, cost-reimbursable procurement approaches are often deemed preferable. Cost reimbursement methods can take a number of forms:

- Cost plus percentage fee
- Cost plus guaranteed maximum fee
- Cost plus fee with bonus incentives
- Cost plus fee with time penalties

Within these methods, the cost incurred by the contractor is reimbursed by the client with the agreed addition (e.g. percentage fee).

Target-cost contracts are another procurement method that can be used for post-disaster reconstruction. In the target-cost contract methods, the client and the contractor assess the work and come to an agreed cost target. The cost overruns or cost saving are shared by both the parties, thus making this approach a pain/gain sharing method. The target-cost contract method can encourage contractors to work efficiently and promote innovation that was not seen from its predecessor: cost-reimburse methods.

New forms of procurement approaches such as partnering, alliancing and joint ventures are becoming popular for post-disaster reconstruction work. Within these procurement approaches, pain/gain share reward methods such as target-cost contracts can be used. The study by Le Masurier *et al.* (2006) favoured project alliancing for disaster reconstruction work for large and complex projects with a high level of uncertainty. They argue that the size and the duration of the project should be able to justify the investments put in for the alliances. The aftermath of a disaster could result in a lack of skilled labour, building material and equipment due to high demand. Creating joint ventures with international construction organisations to assist the reconstruction work, in terms of skills and technologies, can be considered.

The success of post-disaster reconstruction work largely depends on the availability and accessibility of resources such as labour, materials, plant and equipment, as inadequate resources could result in delaying the reconstruction work. The following section looks into sourcing of labour, material and plant required for reconstruction.

8.3   Sourcing of labour, material and equipment

One of the challenges for post-disaster reconstruction is to make sure the required manpower and construction material are in place as and when they are needed. Limited resources for reconstruction work can result in delays and increased costs. Further, extensive competition to obtain limited resources may provide room for corruption and mishandling of resources. During the reconstruction of Banda Aceh, Indonesia following the tsunami, limited timber availability resulted in the affected community living in temporary shelter one year after the disaster (Zuo *et al.*, 2009). This occurred despite Indonesia being rich with forests. Timber supply in Banda Aceh was frequently delayed by up to 10 weeks. Such delays badly affected the on-going reconstruction activities. Illegal logging, bribes and illegal payments by truck drivers to corrupt police and state authorities were some of the reasons identified for the lack of sufficient timber. Such illegal payments and limited order amounts increased the cost of timber significantly, resulting in reconstruction costs that were three times higher than the actual construction cost. It was later identified that proper supply chain management when sourcing timber, and good communication links between the suppliers and reconstruction project teams regarding the timelines for timber requirements, could have avoided unnecessary delays (Zuo *et al.*, 2009). Further, a tight governing structure could have been imposed for the authorities to avoid corruption.

Suppliers have a direct impact on the cost, time and quality of work provided to the buying organisation. Thus proper supply chain management is important to ensure that disaster reconstruction work is delivered to the required standards. As Carr and Pearson (1999) argue, the buyers and suppliers need to share sensitive information in order to jointly find solutions to issues related to limited resource availability after a disaster. Therefore, effective two-way communication between the supplier and buyer strengthens reconstruction work.

When procuring the materials and components, multiple suppliers can be used even when supplying the same material or component. The main advantage of this approach is to reduce the risk posed by company failures or poor performance of suppliers. This reduced risk comes at the expense of increased administrative and transactional costs (Dyer, 2000). These costs are frequently compounded by the lack of local and government authorities to manage the larger number of suppliers. This was evident during tsunami reconstruction work in Sri Lanka and Indonesia, and led to delays in reconstruction work. The alternative is to use a single source to supply construction materials and components. Myburgh *et al.* (2008) suggest that the selection of a single source to supply materials, which is based on performance rather than bidding for multiple suppliers, could also have the benefit of increased quality. Further, a single point of responsibility for the supply of materials can lead to long-term relationships between the suppliers and buyers, which are more likely to be based on trust and cooperation. A long-term relationship between a supplier and a buyer can provide benefits due to the trust and relationship developed over time, sharing willingness to share risks, and the reward of continuous work for suppliers. As a consequence, supply chain management that is based on a long-term relationship between suppliers and buyers is becoming popular in modern procurement practices and may yield benefits for post-disaster reconstruction when sourcing labour, material, plant and equipment. In order to maintain the quality and performance of the supplier base, pre-qualifying of suppliers based on certain criteria can also be used (Zuo *et al.*, 2009).

During disaster reconstruction work, community-based sourcing of resources is usually encouraged. Domestic and international donors must therefore be encouraged to look into the availability of material, labour, plant and equipment within the community, prior to looking for external sources. The use of community-based resources may help to strengthen the capacity and commitment of the affected community, increase the sense of ownership for reconstruction activities, and reduce the social tension that sometimes emerges during post-disaster management activities (Schilderman, 2004). For example, when designing earthquake resistant houses in Alta Mayo, Peru, the use of locally sourced material for reconstruction was given a higher priority by the designers. The designers wanted to assist the community by using local material such as timber, earth and aggregate that was common in the community. This created long-term sustainability of the houses rather than relying on foreign technology that could require external aid (Lowe, 1997).

Having discussed the sourcing of labour, material and equipment, the following section looks into the management of these resources.

8.4   Resource management

The capacity of the affected community's local construction industry is usually insufficient to meet the exceptional demand for reconstruction that is created after a major disaster (Seville and Metcalfe, 2005; Rotimi *et al.*, 2005). Singh and Wilkinson (2008) state that mobilising resources for post-disaster reconstruction is especially difficult at the initial stage but normalises as time passes by. For instance, post-disaster reconstruction of Wellington State highway in New Zealand after an earthquake was constrained due to a lack of construction materials: aggregates, cement and concrete. The supply of aggregate was affected due to limited production, damages to quarries and transportation networks, while cement and concrete supply were affected due to a lack of production and limited suppliers.

Effective resource management in the form of scheduling and pre-planning is therefore important to guarantee the availability and accessibility of resources. If the required material cannot be supplied from regular sources, identification of alternative sources is important to ensure rapid progress in post-disaster reconstruction work. For example, concrete that was available as a refuse after the earthquake in Wellington State was suggested as an alternative material for aggregates (Singh and Wilkinson, 2008). Pre-identification of alternative material sources is particularly important at the initial stage of post-disaster planning, or even during the preparedness phase prior to a disaster occurring. The transportation of construction materials from other close-by locations, use of alternative materials, use of alternative transportation mechanisms, importation of construction material, and reducing the exportation of construction materials to other countries, are some of the methods that can be used to ensure the availability of construction materials after a disaster. In order to minimise the immense pressure on the construction materials and components suppliers, post-disaster reconstruction must be done in such a way as to avoid the use of materials and components that are in limited supply.

During post-disaster reconstruction, it is important to mobilise resources to the required areas as and when needed, to facilitate timely reconstruction. Therefore, infrastructure facilities such as roads need to be properly maintained and functional. Seville and Metcalfe (2005) assert that the governing body for such infrastructure facilities need to have a framework to identify, evaluate and manage risks to infrastructure, and plan maintenance and repair work accordingly.

8.5   Quality control

Quality control during post-disaster reconstruction is challenging due to the involvement of a large number of stakeholders, including those who would not usually be involved in traditional construction projects. A lack of quality control measures, a failure to communicate the required standards, and an inability to coordinate multiple stakeholders can result in producing buildings with poor standards. Defects and failures in buildings usually lead to work being repeated and the associated additional cost and time. After the tsunami in Sri Lanka, two main reconstruction programmes were initiated: donor-driven and owner-driven. Under the owner-driven housing programme, the government provided grants from development banks and bilateral donors to affected homeowners to reconstruct their houses. In contrast, within the donor-driven housing programme, domestic or international donors built or assisted in housing reconstruction. Under this programme, the contractors were selected by the donors, who sometimes gave priority to low cost, without appropriate consideration of qualifications. The subsequent poor construction work and lack of appropriate quality control for reconstruction often resulted in buildings of inferior quality. Further, the dwellers were not satisfied with the strength, arrangement of structure, quality of materials used, and improper land fillings and cuttings used by many of the donor-driven housing programmes. Some of the houses constructed were subsequently demolished as a result (Nissanka *et al.*, 2008; Ratnayake and Rameezdeen, 2008).

Quality control should start with the identification of required quality standards during the design and planning stages. These may include factors such as material specifications, codes and building regulations, workmanship and construction techniques. The establishment of such quality standards during the preliminary stage will facilitate appropriate identification of materials, plant, and the construction techniques required for actual construction. Thereafter, the quality control process during the construction phase consists of ensuring the quality standards are adhered to. However, there is a tendency for the quality standards and identified technologies for reconstruction work to be changed due to unforeseen situations during the construction stage. The established quality standards and changes need to be documented properly, thus making document control an important part of the quality control system. The final aspects of quality control are inspection, testing and final checkouts of the constructed facility.

The quality requirements need to be clear and visible to all the stakeholders involved in the work, including the owners of the reconstructed facility. This will help to satisfy the aspirations of all stakeholders involved in the reconstruction activities, and in particular, the occupiers. Appointment of a quality assurance person is a good practice to ensure the required quality standards are properly implemented. The attitude to quality control, however, should be shared by all the parties involved in the construction process, rather than trying to manipulate the quality and required standards for the benefits of individual or organisational gains. Quality control will ultimately lead to increased productivity due to lack of re-work, introduction of new materials and technologies, and avoidance of long-term problems with the reconstructed facility.

The reconstruction activities that took place in Alto Mayo, Peru, after the earthquake in 1990 are a good example of how appropriate quality standards should be identified and introduced (Lowe, 1997). An earthquake measuring 5.8 on the Richter scale struck this region destroying over 3000 houses and damaging a further 5000. A reconstruction programme was initiated with the participation of the community. It identified the failures within existing houses and the required improvements in order to construct buildings more resistant to earthquakes. It was decided to upgrade a local technology called 'quincha' as the houses built using this technology were relatively flexible during earthquakes. During the preliminary stage, a number of improvements were identified and documented. Some of the key improvements included concrete footings to give adequate load-bearing capacity and anchorage, a concrete wall base to protect the walls from humidity, good connections between the structural timber elements, lightweight roofing of corrugated iron sheets or micro-concrete roofing tiles to reduce the risk of roofs falling on inhabitants, and roof eaves of sufficient width to protect walls against heavy rains. Another earthquake struck in Alto Mayo during the following year and the improved quincha houses were able to survive the earthquake.

8.6   Financing

Financial management of disaster reconstruction includes sourcing, proper distribution of finance, and periodic review of the finance to make sure the financers are properly used for its intended task. Financial management of post-disaster reconstruction is challenging due to the involvement of a large number of domestic and international donor agencies. This section discusses financing methods and good practices.

The funding agents need to bear a certain amount of risk when making investment decisions for post-disaster reconstruction work. Governments are often identified as the most efficient institutions to bear the investment risk and lend money for disaster reconstruction because they are able to transfer the risk to the citizens via taxation (Freeman, 2004). However, following a disaster, this depends on the financial capacity of the government that has been affected. For instance, financial instability and weak taxation mechanisms can hinder governments from investing money in disaster reconstruction work. When a disaster occurs in a developing country, the scale of the disaster usually prevents governments from providing sufficient financial assistance (Mechler and Weichselgartner, 2003). Fengler *et al.* (2008) report that the impacts of disasters in developing countries are disproportionately greater in terms of GDP and government revenue. When the tsunami waves hit the south Asian countries in 2004, this scenario was evident.

When the government cannot bear the whole cost of reconstruction, the financing comes from international banks, domestic credit, and domestic and international donor agencies. Efficient usage of these financial contributions are a key concern. Post-disaster financial management is extremely difficult due to the urgency of reconstruction work. Rigid financial systems and regular government budget allocations are inadequate to respond quickly to support the rapid reconstruction work. The affected country's budgeting cycles and reconstruction work rarely match and therefore the exclusive use of the affected country's budgeting system to finance the reconstruction work has practical limitations (Fengler *et al.*, 2008). Systems that go beyond the regular budget system are therefore favoured to provide financial assistance in a timely manner.

However, unconventional methods and off-budget mechanisms for the financial support provided by donor agencies is prone to a lack of accountability and can provide room for corruption. For example, inefficient monitoring of funds was reported during post-disaster reconstruction in Sri Lanka (Nissanka *et al.*, 2008). A number of non-governmental organisations (NGOs) were allowed to manage funds themselves, which led to suspicions of corruption. This has ultimately reduced the actual amount made available for reconstruction. Further, the actual amount of funds received by house owners also diminished due to an absence of government monitoring. As emphasised by the reconstruction and development agency in Sri Lanka, some house owners were given too much money while others were not given enough. Therefore, proper financial management is important to ensure the reconstruction budget is used for the purpose it was intended, whilst also ensuring that the reconstruction work that was promised is actually delivered. Further, it is important to maintain accountability to the affected community and to the investors via making proper arrangements for financial management and accountability (Fengler *et al.*, 2008). The financial management systems should create a credible environment for the donors to invest their money, ensure the required support is received by the intended beneficiaries, strengthen the government's fiduciary standards, and also encourage community support towards reconstruction activities. Further, proper financial management agreements should lead to the implementation of policies, the strategic allocation of resources to prioritise reconstruction, and efficient and effective reconstruction work. In addition to ensuring accountability for the finance, the international donor agencies target the strengthening of domestic resources in developing countries to upgrade the effectiveness of the aid. To ensure the accountability of financial investments, independent monitoring mechanisms and complaint procedures need to be in place in addition to government controlling and monitoring. Poor financial management systems could negatively affect project approvals and their effectiveness. Regular reporting from the institutions, proactive data collection and analysis protocols, and creating modalities for management and evaluation of the financiers, are also important for proper financial management. The financial management system needs to: provide details to benchmark the progress against the expected reconstruction; evaluate funds allocation and spending from all the financial sources; evaluate economic and social impacts; be transparent; identify recording and reporting mechanisms; and, identify methods of external scrutiny and audits. An effective financial management system will also help to establish a consolidated budget with capital and recurrent expenditure while consolidating donor agency's activities (Sarraf, 2005). The establishment of monitoring and controlling systems to assess the proper progress of reconstruction work, and establishment of triggers to provide warning signals to take corrective measures, is also important (Fengler *et al.*, 2008).

8.7   Governance

The nature of the policies and strategies of governance structure has a major influence on: resource sourcing and managing; building standards, techniques, and regulations; and, financial management. Proper governance of post-disaster reconstruction leads to optimal use of public and private resources, and hence helps to ensure value for money.

The disaster-affected community needs to have the required coping capacity within its people and organisations to use effectively the existing resources and thereby withstand the damages caused by the disaster (UN, 2002). The International Strategy for Disaster Reduction (2005) highlights the need to integrate the measures to reduce disaster risk reduction into the main elements of governance structures, such as policies and programmes for sustainable development. The term 'govern' refers to rule or control with authority, and the conduct of policy to influence the correct course of action. Therefore, governance is the mechanism that controls the activities of an entity towards achieving its objectives. Governance structures have a major influence in terms of strengthening capabilities, and creating capacities in the community and rebuilding the community with minimal economic and social losses. In contrast, a lack of proper governance structures can reduce the capacity of the affected community and significantly prolong recovering from the disaster, as for example experienced in the Andaman and Nicobar islands after the tsunami. Similarly, after the tsunami in Sri Lanka, legislation was imposed to create a buffer zone of 100 metres from the sea. The zone was later increased to 200 metres. Changes to the buffer zone delayed the commencement of reconstruction programmes by more than six months. As a result, some donors left the country with their unused grants and moved to other tsunami-affected countries. Although a more carefully considered buffer zone was later agreed, Sri Lanka has already missed out on a considerable number of grants.

The four dimensions of governance include predictability, transparency, participation and accountability (Asian Development Bank, 2004). The ultimate goal of post-disaster reconstruction is to match or improve on the previous living standards of the affected community. In achieving this target, the fundamental principles of governance can be incorporated within the policies and practices that need to be adopted during post-disaster reconstruction. The involvement of multiple stakeholders for financing, reconstruction and management can lead to complex situations. Thus, proper predictability, transparency, participation and accountability are vital.

8.7.1 *Participation*

The participation of the affected community within the policy-making process, and the design and implementation of reconstruction projects, provides an opportunity for the beneficiaries to contribute to reconstruction work. Participatory activities are usually politically and socially encouraged as such participation is deemed more likely to lead to the identification of the exact needs of the affected community, rather than trying to implement and enforce exogenous policies and practices. Further, community participation empowers the beneficiaries to take a leading role, thus increasing their commitment towards the success of the disaster reconstruction and mitigation activities. Such participation of the affected community eases social tensions that can emerge during reconstruction work. Further, the policy making should support and maintain the dignity and self-reliance of the affected community. The participation of the affected community has the advantage of being more likely to produce governance policies and practices that are responsive to the community. Therefore, governance structures should encourage reconstruction at the local level and allow the affected community to identify, plan and respond to their own requirements. The formation of reconstruction committees from the affected community and appointing representatives from the affected community at a national level can lead to more sustainable reconstruction work. Schilderman (2004) detailed the need for community participation and involvement when making policies and regulations for disaster reconstruction. On 30 September 1993, the Indian state of Maharastra was struck by an earthquake measuring 6.3 on the Richter scale. The houses were not built to withstand earthquakes, as this region had not experienced many in its recent history. The houses were built with local skills and material, such as stones for walls, and timber beams and columns, which subsequently failed during the earthquake. The Indian government initiated a reconstruction programme with the support of the World Bank and NGOs. The approach to reconstruction was mainly top-down and didn't allow for user participation due to a preference for more complicated and costly earthquake-resistant building technologies instead of local materials and techniques. Local contractors were not given an opportunity to participate in the reconstruction work; large contractors from outside the community were used instead. As a result of the increased cost the community ended up with houses that were too small to cater for their needs. Many elected to build extensions to their new houses or live in their weakened old houses, thus leaving them in homes at high risk to further earthquakes. Such a situation may have been avoidable if appropriate consultation had been undertaken at the outset.

8.7.2 *Predictability*

Predictability refers to the existence of laws, regulations, and policies to regulate society, as well as their fair and consistent application. A legal system comprising of well-defined rights and duties, with proper mechanisms for enforcement provides an appropriate level of confidence for investors to make their investment decisions. In addition to the proper legal systems, the consistency of public policies is also important, as government policies influence the investment climate, including factors such as exchange rates and the price of commodities.

8.7.3   Accountability

Accountability for individual and organisational actions is important for successful disaster reconstruction work. Accountability towards beneficiaries is called downward accountability; whereas the accountability towards higher levels, such as government, is called upward accountability. Those who are engaged in policy making and decision taking, including everyone from village leaders, local governments, national governments, to domestic and international organisations, needs to be downwardly and upwardly accountable for their activities. The governance structure should clearly identify the roles and responsibilities of parties involved in policy making and decision making, and ensure that the required standards of performance and quality are adhered to (Turner and Hulme, 1997).

8.7.4   Transparency

Transparency is the mechanism that shows the accountability of the policy and decision makers. During a disaster, the tendency for corruption and waste of resources is high due to the demand for rapid reconstruction and relief work. Transparency therefore, should minimise or prevent room for corruption and waste of resources. Furthermore, transparency should ensure the equitable distribution of resources and proper financial assistance for the affected community, regardless of their social and economic status, caste, religion and political views. Reconstruction activities need to ensure the community and the built environment can withstand future disaster events with no or minimal impact. In order to do this, the reconstruction work needs to adhere to certain performance standards, building codes, and land-use planning, and also use improved construction technologies, materials and components. The governance structure should facilitate transparency to ensure these standards and mechanisms are properly implemented during reconstruction activities.

8.7.5   Framework for effective construction

In order to carry out the reconstruction work in a collective and coordinated manner, establishment of an institutional framework for reconstruction work is suggested (Environmental Planning Collaborative and TCG International, 2004). An institutional framework can integrate all organisations, including domestic and international, and public and private; and, include sectors that need reconstruction work such as shelter, livelihood, education, health, and infrastructure. The framework needs to clearly identify the roles and responsibilities of different organisations, the mechanisms for coordination and cooperation between the organisations, and the line of authority between the organisations. However, the institutional arrangements for reconstruction work will vary depending on the scale of the disaster and country context (Fengler *et al.*, 2008). For example, after the tsunami, damage in the Maldives was high as a proportion of the country's economy. Reconstruction activities in the Maldives were led by an experienced central government but supported by the appointment of a centralised coordination board called the National Disaster Management Centre. Under the coordination of this central board, the government created several task forces to organise and implement the relief and reconstruction activities. In contrast, India, which suffered comparatively less damage elected to assign the sub-national governments to govern relief and reconstruction efforts.

Elsewhere and in an alternative approach, after an earthquake struck Colombia in 1999, the government developed an innovative governance model to manage reconstruction work. The reconstruction agency, FORCE, was established with a decentralised management structure to coordinate and monitor reconstruction of the coffee region. Project implementation was carried out by NGOs in 32 zones through a competitive selection process. This mechanism had been successful in terms of public participation, social control and transparency of work. The projects were monitored by a consortium of universities whilst the funds were administered by a fiduciary agency (World Bank, 2003). However, FORCE had several institutional challenges. The exclusion of local government from project identification and implementation resulted in a lack of cooperation. Further, diverse institutional capacities of zone managers resulted in differentiated implementation of reconstruction work. Coordination and information collection problems were encountered due to the decentralised governance structure.

8.8   Disaster risk reduction

Implementation of appropriate disaster risk reduction measures is another important element for successful project management of post-disaster reconstruction work. White (2004) identifies disaster risk reduction as the measures to control disaster losses by minimising the hazard, reducing exposure and susceptibility, and enhancing coping and adaptive capacities. The emphasis for disaster risk reduction should continue even after the disaster, thus building resilience for future disastrous events. The Hyogo Framework for Action 2005–2015 (International Strategy for Disaster Reduction, 2005) highlights the importance of mainstreaming disaster risk reduction measures within urban planning and reconstruction of building and infrastructure projects.

The United Nations categorise two forms of activities for disaster risk reduction: structural and non-structural (International Strategy for Disaster Reduction, 2005). The structural mitigation includes: the removal or upgrade of unsafe buildings; reconstruction of buildings that comply with effective codes, standards, and regulations; and, use of construction materials and techniques that could minimise the damage of disasters. The non-structural mitigation includes: reconstruction in safer sites that are less vulnerable to disasters through land-use planning and control; maintaining protective features of the nature to act as barriers for disasters; and, ensuring health and safety at construction sites.

In addition to the structural and non-structural mitigation measures, consideration needs to be given to community-based approaches that can reduce the risk of disasters whilst improving community preparedness. They include increasing community awareness about disasters, training and education about the disaster-proof reconstruction techniques, coordinating the community activities towards disaster risk reduction, identifying refuge points, establishing essential services such as medicine and food, and increasing the effectiveness of information dissemination.

The approaches to disaster risk reduction during reconstruction work can be divided into two based on the level of involvement of the authorities and community. The top-down approach comprises use of technology-based approaches provided by the authorities towards the affected community. This approach focuses on monitoring techniques such as hazard mapping, implementation of buffer zones and physical mitigation measures such as flood barriers. Often, the top-down approach does not take into consideration social factors, such as cultural beliefs, livelihood patterns, and land ownership of the affected community. Hence, governments and NGOs may be in confrontation with the same people that they want to help. The bottom-up approach consists of developing policies and techniques with consultation of the community. Community-based disaster risk reduction measures ensure that the policies and technologies are catering for the requirements of the community, and provides awareness to the community about the risks that they could encounter and how to protect them in the future.

Cultural dimensions and livelihood patterns of the affected community also need to be considered. Recovery and reconstruction work should be carried out without sacrificing cultural elements or neglecting the livelihood patterns of the affected community for the mere sake of technical efficiency and effectiveness (RICS, 2006). Often, communities and their location are closely linked with an occupation. If displaced people are relocated, they may find their livelihoods disrupted, which creates long-term economic difficulties. In a similar vein, evaluation plans must take account of local cultural values. For example, during the volcanic eruption of Mount Merapi, Indonesia in 2006, the community refused to evacuate their village until they received warnings according to their traditions (Lavigne *et al.*, 2008).

It is evident that any attempts to introduce disaster risk reduction measures must strike an appropriate balance between respect for and engagement of the affected community, and appropriate risk reduction strategies and technologies that meet recognised standards.

8.9   Summary

This chapter has considered aspects of project management in post-disaster reconstruction. Best practices that can be adopted during both short-term relief and longer term reconstruction projects have been considered. These include procurement, sourcing of labour, material and plant, resource and financial management, quality control, and governance. Construction approaches that are built on trust and long-term relationships, that provides win-win situations to stakeholders, and encourage pain-gain sharing methods are suggested for procuring buildings, services and material. Governance structures and policy making that comprise predictability, transparency, participation and accountability are favoured for successful reconstruction activities. Adhering to these characteristics will help to ensure proper utilisation of investment while delivering high quality facilities for the beneficiaries. However, it was noted that depending on the magnitude of the disaster, economic status and coping capacity of the country, the most appropriate procurement strategies, resource and financial management methodologies and governance mechanisms may vary considerably.

A common theme running across many of these issues has been the need to promote community-based reconstruction. This includes: sourcing locally based labour, material and plant; getting the active involvement of the local community in policy making and decision taking; using indigenous technologies for reconstruction; using technologies that encourage local resources; and considering the culture and livelihood patterns of the community. By ensuring that project management strategies adhere to the principles of community participation, the affected community should be more sustainable and better able to deal with disaster hazards in the future.

References

Asian Development Bank (2004) *Governance: Sound Development Management*. http://www.adb.org/Documents/Policies/Governance/default.asp?p=policies   [accessed November 2009].

Carr, A.S. and Pearson, J.N. (1999) Strategically managed buyer-seller relationships and performance outcomes. *Journal of Operations Management*, **17**, 497–519.

Cartlidge, D. (2009) *Quantity Surveyor's Pocket Book.* London: Butterworth-Heinemann.

Department for Business, Enterprise and Regulatory Reform (2007) *Guidelines for Managing Projects*. http://www.berr.gov.uk/files/file40647.pdf [accessed November 2009].

Dyer, J.H. (2000) *Collaborative Advantage: Winning through Extended Enterprise Supplier Networks*. New York: Oxford University Press.

Environmental Planning Collaborative and TCG International (2004) *Participatory Planning Guide for Post-Disaster Reconstruction*. http://www.tcgillc.com/tcgidocs/TCGI%20Disaster%20Guide.pdf [accessed November 2009].

Fengler, W., Ihsan, A. and Kaiser, K. (2008) *Managing Post-Disaster Reconstruction Finance*. Policy Research Working Paper. Washington, DC: The World Bank.

Freeman P.K. (2004) Allocation of post-disaster reconstruction financing top housing. *Building Research and Information*, **32**, 427–437.

International Strategy for Disaster Reduction (2005) *Hyogo Framework for Action 2005–2015*. World Conference on Disaster Reduction, 18–22 January 2005, Kobe, Hyogo, Japan.

Lavigne, F., De Cocter, B., Juvin, N., Flohic, F., Gallard, J., Texier, P., Morin, J. and Sartohadi, J. (2008) People's behaviour in the face of volcanic hazards: perspectives from Javanese communities, Indonesia. *Journal of Volcanology and Geothermal Research*, **172**, 273–287.

Le Masurier, J., Wilkinson, S. and Shestakova, Y. (2006) An Analysis of the Alliancing Procurement Method for Reconstruction Following an Earthquake. Proceedings of the 8th US National Conference on Earthquake Engineering, 18–22 April, California, USA.

Lowe, L. (1997) *Earthquake Resistant Housing in Peru*. Rugby: Intermediate Technology Development Group.

Mechler, R. and Weichselgartner, J. (2003) *Disaster Loss Financing in Germany: The Case of the Elbe River Floods of 2002*. Interim Report, Laxenburg, International Institute for Applied Systems Analysis.

Myburgh, D, Wilkinson, S and Seville, E. (2008) *Post-Disaster Reconstruction Research in New Zealand: An Industry Update*. Resilient Organisations. www.resorgs.org.nz [accessed November 2009].

Nissanka, N.M.N.W.K, Karunasena, G and Rameezdeen, R (2008) Study of factors affecting post disaster housing reconstruction. In: Keraminiyage, K., Jayasena, S., Amaratunga, D. and Haigh, R. (Eds), *Post Disaster Recovery Challenges in Sri Lanka*. The University of Salford, UK: CIB.

Project Management Institute (2004) *A Guide to the Project Management Body of Knowledge*. 3rd Edition. Newtown Square, PA: Project Management Institute.

Ratnayake, R.M.G.D. and Rameezdeen, R. (2008) Post disaster housing reconstruction: comparative study of donor driven vs owner driven approach. In: Keraminiyage, K, Jayasena, S, Amaratunga, D and Haigh, R (Eds), *Post Disaster Recovery Challenges in Sri Lanka*. The University of Salford, UK: CIB.

RICS (2006) *Mind the Gap! Post-Disaster Reconstruction and the Transition From Humanitarian Relief*. www.rics.org [accessed November 2009].

Rotimi, J.O.B., Le Masurier, J, and Wilkinson, S. (2005) *The Regulatory Framework for Effective Post-disaster Reconstruction, in New Zealand*. Resilient Organisations. www.resorgs.org.nz [accessed November 2009].

Sarraf, F. (2005) *Integration of Recurrent and Capital 'Development't Budgets: Issues, Problems, Country Experiences, and the Way Forward*. Paper prepared for Public

Expenditure and Financial Accountability (PEFA) Program. Washington, DC: World Bank.

Schilderman, T. (2004) Adapting traditional shelter for disaster mitigation and reconstruction: experiences with community-based approaches. *Building Research and Information*, **32**, 414–426.

Seville, E. and Metcalfe, J. (2005) *Developing a Hazard Risk Assessment Framework for the New Zealand State Highway Network*. Land Transport New Zealand Research Report 276, Canterbury, New Zealand.

Singh, B. and Wilkinson, S. (2008) *Post-Disaster Resource Availability Following a Wellington Earthquake: Aggregates, Concrete and Cement*. Resilient Organisations. www.resorgs.org.nz [accessed November 2009].

Turner, M. and Hulme, D. (1997) *Governance, Administration and Development – Making the State Work*. Hampshire/New York: Kumarian Press.

UN (2002) *Living With Risk – A Global Review of Disaster Reduction Initiatives*. Geneva: United Nations.

White, P., Pelling, M., Sen K., Seddon, K., Russell, S. and Few, R. (2004) *Disaster Risk Reduction: A Development Concern*. London: Department for International Development.

World Bank (2003) *Implementation Completion Report on a Loan in the Amount of US$ 225 Million to the Government of Colombia for the Earthquake Recovery Project*. Washington, DC: World Bank.

Zuo, K., Potangaroa, R., Wilkinson, S. and Rotimi, J.O.B. (2009) A project management prospective in achieving a sustainable supply chain for timber procurement in Banda Aceh, Indonesia. *International Journal of Managing Projects in Business*, **2**, 386–400.

9 Legislation for Effective Post-Disaster Reconstruction: Cases from New Zealand

James Olabode Rotimi, Suzanne Wilkinson
and Dean Myburgh

9.1 Introduction

New Zealand is susceptible to natural hazards such as cyclones, volcanoes, earthquakes and tsunamis; therefore it needs to be prepared for mitigation and management of these hazard events. In New Zealand hazard management is covered by the Ministry of Civil Defence and Emergency Management (MCDEM) under their 4Rs programme of Reduction, Readiness, Response and Recovery. This chapter focuses on the regulatory policies that directly impact on the reconstruction of physical assets with an emphasis on the Building Act 2004 and Resource Management Act 1991. Reference is made to the Civil Defence Emergency Management (CDEM) Act 2002, which is the key document that prescribes the activities of disaster management agencies in New Zealand. These legislative documents provide the overarching framework for post-disaster reconstruction.

This chapter argues that legislation must be made robust enough to facilitate the implementation of recovery and reconstruction programmes. Two New Zealand flood case studies illustrate where legislation can hinder reconstruction. As a major aspect of pre-planning for disasters, prior considerations for feasible legislative frameworks before a disaster occurs are pertinent. Thus a community mitigates its disaster risks and is poised to implement recovery programmes within these enabling frameworks.

9.2 Recovery and reconstruction

For disaster management activities to be successfully implemented, pre-planning in one form or another needs to be carried out (Cousins, 2004).

Post-Disaster Reconstruction of the Built Environment: Rebuilding for Resilience, First Edition.
Edited by Dilanthi Amaratunga and Richard Haigh.
© 2011 Blackwell Publishing Ltd. Published 2011 by Blackwell Publishing Ltd.

One aspect of pre-planning is the constitution of viable policies and procedural arrangements that will facilitate recovery after disasters. Studies conducted have shown that legislation drives the implementation of recovery policies, especially where special powers, rights or responsibilities need to be defined (ACTIONAID Nepal, 2004). Spence (2004) suggests that legislation and regulatory frameworks have to be appropriate, so that the framework provides a suitable environment for the interaction and interrelationship of disaster management stakeholders during response and subsequent recovery activities.

The formulation of public policies for reduction, readiness, response and recovery should be the rational starting point for disaster management (Comerio, 2004). These public policies have to be coupled with a strong commitment by national governments for their successful implementation. Rolfe and Britton (1995) are of the opinion that the pace of reconstruction is severely impacted by political and cultural conflicts over recovery plans, thus the successful achievement of disaster management goals will depend on the political environment.

The organisation and coordination of recovery is usually complex because a wide range of activities occur simultaneously with an equally wide range of needs that have to be met. Experiences from past disaster management arrangements (even in advanced economies), are indicative of the continuous struggle to meet the recovery needs of all stakeholders. Some advanced countries are being caught off-guard in spite of their previously acclaimed disaster management policies (Kouzmin et al., 1995; Schneider, 1995; Smith and Dowell, 2000; Chan et al., 2006; Mitchell, 2006).

Legislation and regulatory provisions significantly influence the organisation for long-term recovery after natural disasters. Particular attention needs to be paid to changes that should be put in place to facilitate reconstruction programmes. Without developed frameworks, reconstruction and re-development programmes may be carried out on an ad-hoc basis with little regard for the needs of an affected community. A robust regulatory framework would enable the achievement of community resilience. In other words, the regulatory framework should be enabling of rebuilding programmes for damaged physical facilities and consequently contribute to the overall well-being of the affected community.

The frequency with which different disaster emergencies have been declared is proof of New Zealand's disposition to natural disasters. Table 9.1 presents an outline of recent disaster events (excluding the Christchurch earthquakes of September 2010 and February 2011) that have necessitated the declaration of states of emergencies around the affected areas in New Zealand. There is a high disposition to rainfall-related hazards (floods and landslips), but these events have been largely confined to rural areas and are of low-magnitude. Relative to other world disasters, these local events had low scope of impact in terms of their physical and societal dislocations. These low-magnitude events do not negate the real risks that New Zealand communities are faced with and neither does it reduce the importance of disaster risk management strategies. More recently New Zealand experienced two major earthquakes in the Christchurch city area. The first (in September 2010) was a 7.1 magnitude earthquake which rocked Christchurch city, but with a low scope of impact. The second was a more severe 6.3 magnitude earthquake (at a shallower depth but with the epicentre closer to Christchurch) on February 22, 2011 with multiple fatalities. There were 146 confirmed casualties and 10 000 houses that will need to be rebuilt (as at the time of reporting).

New Zealand has had its share of significantly large natural disasters in its history. The Ministry for Culture and Heritage gives a dateline of major natural disasters that happened in New Zealand history (Ministry for Culture and Heritage, 2009). Some examples of these are: the Taupo landslide of 7 May 1846 resulting in about 60 deaths including Ngati Tuwharetoa leader Mananui Te Heuheu Tukino II; the 1855 Wairarapa earthquake that altered the landscape of the Wellington region with five to nine recorded deaths in Wellington, Manawatu and Wairarapa; and the deadliest earthquake in the Hawke's Bay (Napier) region of 1931. The official death toll as a result of the Napier earthquake was put at about 256.

Large scale natural disasters have been few and far between and it can be concluded that New Zealand is relatively inexperienced in the management of catastrophes and that such events may require more extensive preparatory work than the usual. Hopkins *et al.* (1999) hold a similar view; they conclude that large-scale disasters would pose considerable economic, physical

Table 9.1   Emergency declarations in New Zealand (2004 –2007). Reproduced by permission of Ministry of Civil Defence & Emergency Management (source: adapted from www.civildefence.govt.nz)

Date declared	Date terminated	Geographical area/region affected	Nature of emergency
2007			
21 Dec	22 Dec	Gisborne	Earthquake
30 July	31 July	Milton	Flooding
10 July	13 July	Far North DC	Flooding
05 July	07 July	Taranaki District	Tornado
2006			
07 July	08 July	Rangitikei District	Flooding
2005			
17 May	30 May	Whakatane District	Flooding, Landslips
18 May	20 May	Tauranga District	Flooding, Landslips
2004			
17 July	30 July	Whakatane District	Flooding, Landslips
17 July	23 July	Opotiki District	Flooding, Landslips
17 Feb	25 Feb	Manawatu-Wanganui	Flooding
17 Feb	18 Feb	Marlborough District	Flooding
17 Feb	27 Feb	South Taranaki District	Flooding
16 Feb	17 Feb	Manawatu District	Flooding
16 Feb	17 Feb	Rangitikei District	Flooding

and social challenges that could make the task of recovery and reconstruction extensive.

In terms of legislation and regulation there is evidence to suggest that there is little provision in legislation to facilitate large scale reconstruction programmes in New Zealand. Feast (1995) for example, identified several issues in relation to planning and construction legislation that would impede reconstruction of Wellington, following a major earthquake. Feast's study suggested that much of the legislation (in particular the Resource Management Act and Building Act) that existed during the period were neither drafted to cope with an emergency situation nor developed to operate under the conditions that will inevitably prevail in the aftermath of a severe seismic event. Commenting on the Resource Management Act (RMA), Feast (1995) explains that its consultation procedure may be precluded by the problems of meeting the reconstruction requirements of a devastated city within a reasonable period.

Since Feast's analysis, the Building Act (BA) and Civil Defence Emergency Management (CDEM) Act have been revised, with amendments made to the RMA (Governmental review and realignment of the RMA is ongoing). However these legislative documents still have considerable obstacles to post-disaster reconstruction. An example of the scale of the problem being experienced under RMA provisions was summed up by the Minister for Environment in his first reading to Parliament about the need to reform the RMA. An extract of his comments made on 19 February 2009 follows:

> '. . . since the RMA became law there has been growing criticism about the slow and costly plan preparation and consenting processes. Decision-making processes under the RMA must become more efficient. The amendments in this Bill will provide timely and welcome support to other government measures to stimulate the economy' (Smith, 2009).

Recovery under procedural burdens may become an endless process of partially fulfilled expectations. This will most certainly be exacerbated by poor planning and unsustainable policies. Such poor planning activities may result in increases in vulnerability of disaster-affected individuals or groups and could lead to a recurring disaster-poverty cycle, as was experienced in the Latur 1993 earthquake (Jigyasu, 2004) and the Gujarat 2001 earthquake (Shaw *et al.*, 2003) where reconstruction objectives were largely unmet. Failed recovery after the Latur and Gujarat earthquakes could be considered extreme and unlikely in the New Zealand context, but it is evident that every post-disaster management programme requires deliberate and sustained approaches that are built upon well grounded policies and strategic frameworks (Coghlan, 2004; Comerio, 2004; Quarantelli, 2007).

Becker *et al.* (2008) reported that New Zealand's recovery arrangements have been approached haphazardly with little forethought to long-term consequences. Response arrangements during the snowstorms in the Canterbury region in 2006 were ineffective and are indicative that more ground needs to be covered in the realm of emergency management. During the Canterbury snowstorms the affected areas experienced electricity outages of upwards of three weeks after the event (Hendrikx, 2006). A catastrophe of the magnitude of the New Orleans disaster that followed Hurricane Katrina in 2005 or the Indian Ocean tsunami in 2006 would pose significant challenges for reconstruction works in New Zealand, and rebuilding programmes would find it hard to cope with its legislative framework.

If reconstruction is to proceed at the speed most often desired for an early recovery from disasters, improvements to implementation guidelines and changes to legislative provisions are needed. These improvements may take the form of reviews, repeals and/or waivers to the subsisting legislation and regulatory frameworks.

There is continuous tension between strictly applying reconstruction regulations which aim at preventing a recurrence of previous community vulnerabilities; and allowing an affected community to move back to their former habitation quickly. Clearly, the quicker communities return as many of their homes to habitability as possible, the better it will be for restoring a sense of normality (recovery). According to Kennedy *et al.* (2008) disaster management activities should aim to 'build back safer'. For example, McDonald (2004) explains that the redevelopment programme in Napier, New Zealand still exposes the community to the same level of destruction it experienced during the earthquake in 1931. The city was largely rebuilt using previous planning parameters (the current study therefore does not recommend speedy reconstruction at the expense of quality of delivery but proposes enabling legislation that will provide robust processes for early recovery after a major disaster). Decisions on the implementation of reconstruction regulations have to be made with trade-offs between idealistic goals and expediency.

9.3   Legislative and regulatory considerations post disaster

Legislation and regulatory requirements can have significant influence on the rate of recovery after a disaster event. The overall desire is for legislation to enhance the recovery and reconstruction process so that it improves (or presents opportunities for improving) improves the functioning of an affected community and risks from future events can be reduced while the community becomes more resilient.

Opportunities for increased resiliency do not remain for long after disasters (Cousins, 2004). The desire to return to normalcy builds quickly after disasters, and with a good flow of external resources, the opportunities to introduce mitigating measures become limited over time (Berke and Campanella, 2006). Menoni (2001) notes that market forces apply pressure to reconstruct infrastructure as quickly as possible. For example, the restoration of transportation networks to long distances and commercial and office buildings could hamper genuine efforts to implement lessons learnt from a disaster and consequently increase pre-earthquake vulnerabilities. Speed of reconstruction is important, otherwise victims might begin to rebuild in their own way and at locations that controlling agencies are unable to prevent (Olshansky, 2005).

Pressures to rebuild key lifelines quickly are borne by national and local administration with the implication of reduced quality of delivery. This approach has led to even more disasters and the increased vulnerability of a poorly planned and designed built environment to future disasters (Shaw *et al.*, 2003; Jigyasu, 2004). For example, buildings reconstructed in the same vulnerable locations create increased and additional risks (Wamsler, 2004). The clamour to rebuild quickly also amplifies the social, economic and environmental weaknesses that result in large-scale disasters (Ingram *et al.*, 2006). Extra quality and embedded forethought can help reconstructed built assets and community to be more resilient, but there is inevitably a trade-off between time, cost and quality, which recovering communities have to make (Olshansky, 2005).

In New Zealand there is an apparent emphasis on readiness and response activities, with little consideration given to planning for sustained recovery activities (Angus, 2005). Where recovery is considered, it seems to be for the short term, as is evident in emergency awareness campaigns that encourage communities to prepare for up to three or more days after an event (MCDEM, No date-b). Recent emergency events clearly show that longer-term recovery plans would be required beyond seven days for the complexities associated with the rebuilding of damaged built assets. Routine construction processes have been observed to be modified on an ad hoc basis during the recovery phases in previous hazard events in New Zealand (Le Masurier *et al.*, 2006; Becker and Saunders, 2007). Whilst such an approach can work reasonably well for small-scale emergencies, the effectiveness of reconstruction could be improved by modifying the recovery framework in advance of a disaster. Recent earthquake events in Christchurch have forced a rethink of approaches to recovery and reconstruction and have prompted the NZ Government to appoint a Minister for Earthquake Recovery to oversee the fast-tracking of special legislation and other measures to facilitate reconstruction programmes. Though significant progress has been made in reviewing disaster-related legislation (notably since the development of the new CDEM Act), there remain opportunities for improvement (Rotimi, 2006). Within the context of past experiences in New Zealand, there is an imperative to have revised systems in place before a larger scale disaster occurs. Larger scale disasters present different set of challenges of which New Zealand response organisations have no experience. Pre-planning for reconstruction should therefore avoid any disaster event becoming protracted. In the words of the Chairman of the Earthquake Commission in New Zealand, Neville Young, 'natural disasters are by definition unpredictable and it is much more difficult to plan response under the stress of post-disaster trauma than in the calm before the storm' (Earthquake Commission, 2005). New Zealand is now faced with the challenge of reconstructing Christchurch over the next few years and applying the lessons learnt in the pre-planning that may be required for other disasters of similar/greater magnitude.

It is useful to note that research on the impact of regulations on building rehabilitation or on how procedural barriers discourage physical development and rehabilitation is sparse but developing (May, 2004; Burby *et al.*, 2006). Much of what exists in housing and disaster management literature is anecdotal, but suggests that there is a relationship between building/environmental regulations and rehabilitation works (Martin, 2005). Some of these anecdotes do not provide enough empirical data for further research and Schill (2005) suggests that the lack of empirical data makes it difficult to influence public policy. However poor the empirical data on these relationships are, it has not diminished the fact that development regulations could become burdensome in rehabilitation and reconstruction projects and are worthy of consideration (Gavidia and Crivellari, 2006; Marano and Fraser, 2006).

Martin (2005) describes burdensome regulations as those which incorporate excessive rules and regulations and red tape (statutory procedures) that add unnecessarily to the cost of housing. Though Martin's study refers to the effects of building codes on housing, the same parallel can be drawn for reconstruction projects also. Therefore burdensome regulations impact negatively on recovery such that physical facilities are unable to be rebuilt at the speed desired by the community and property owners.

Listokin and Hattis (2004) provide useful analysis on two kinds of barriers that building codes could pose to rehabilitation works. These are the 'hard' and 'soft' barriers to rehabilitation. The hard barriers are impediments to rehabilitation as a result of over-regulation, which would not add appreciably to building value or public safety (Burby *et al.*, 2006) and could discourage housing development or rehabilitation because they are added burdens (May, 2004). For instance, building and environmental regulations that do not reduce the vulnerability of built assets to a hazard event are unnecessary. Also to insist on expensive structural solutions in a highly hazardous zone, where a simple alternative will be to restrict development in that zone, is another example of regulation that could fall under the Listokin and Hattis (2004) hard category.

Soft impediments, on the other hand, are administrative requirements that require extra time, money and effort to accomplish rehabilitation and reconstruction projects (Listokin and Hattis, 2004). These are red tapes (bureaucratic procedures) that could delay new construction and the rehabilitation/reconstruction of physical facilities (May, 2004; Marano and Fraser, 2006). Such soft impediments are the focus of this chapter.

Bureaucratic procedures must be supportive of emergency management under different emergency scenarios whether routine or chaotic. However, research suggests that bureaucracies have been less supportive of the expediency that is desired in disaster response and recovery (Rosenthal and Kouzmin, 1997; Olshansky, 2005). Bureaucracies derive their strength and weaknesses from a modus-operandi that is time consuming; the typologies are 'procedure-bound' and are unable to foster creativity, improvisation, and the adaptability needed in disaster situations (Harrald, 2006). May (2004) suggests three sources of regulatory process barriers which are in line with this chapter's current focus on legislation in New Zealand. These process barriers are outlined below:

- *Regulatory approvals*: These are delays associated with consent processes and approvals that arise from cumbersome decision-making processes and the duplication of regulations. These types of delays are inherent in building and environmental legislation.
- *Regulatory enforcement strategies and practices*: These are overly rigid practices that foster an unsupportive regulatory environment for the development and rehabilitation of the housing stock. In post-disaster situations rigid enforcement strategies discourage genuine recovery efforts as would be shown later.
- *Patchwork of administrative arrangements*: This could result from duplication of administrative structures (as in layers or hierarchies of control) and gaps in regulatory decision processes. May (2004) explains that this patchwork frustrates regulatory implementation and adds to complexities in regulatory processes.

Process barriers could also result from *administrative conflicts* in and among disaster agencies (Listokin and Hattis, 2004). Rivalry may result from existing organisational silos which could cause breakdown in communication, cooperation and coordination between disaster stakeholders (Fenwick *et al.*, 2009). The absence of a coordinating agency could also cause confusion as were experienced in some of the local disasters reviewed in this chapter. Hence, a broad range of cooperative effort is needed for the success of post-disaster reconstruction activities. Stakeholder organisations must coalesce to plan for resource utilisation in the restoration of physical assets. Coordination is therefore central to multi-organisational response and recovery programmes (Comfort *et al.*, 1999; McEntire, 2002).

Another useful dimension to the problems with burdensome regulations is provided by Listokin and Hattis (2004). It is that regulatory procedures could become too rigid forcing implementers to 'go by the book' even though variations may be warranted. This places implementers in a state of continuous fear of liability should things go wrong. Some latitude of control and discretion is often required to aid decision making as long as such decisions are pragmatic. Commenting on the rebuilding programme after the flooding incident in New Orleans, Stackhouse (2006, page 36) says 'removing democratic processes from the rebuilding process has the advantage of expediting decision making by allowing politically dangerous but practical outcomes'. This statement suggests that greater freedom in decision making by officials of coordinating agencies could increase the speed of rebuilding programmes after significant disaster events.

From the foregoing analysis, it is evident that legislation and regulations pertaining to post-disaster reconstruction could hinder the achievement of reconstruction objectives. Speed is of the essence in disaster reconstruction while pre-planning activities help to improve the speed and quality of reconstruction delivery (Harrald, 2006).

9.4   Improving recovery through legislation

Legislation should give legal backing to disaster management policies (Interworks 1998; ACTIONAID Nepal, 2004). However, it is believed that coordinating authorities in New Zealand would be unable to cope with a high volume of demand for their services in the event of a significant hazard event (Hopkins, 1995). Resource availability is an issue because there is a high potential for shortfalls in resources (Hopkins, 1995; Lanigan, 1995; Page, 2004; AELG, 2005; Singh, 2007). Overall community recovery would therefore be exacerbated by inadequate resources with the implication of a sustained recovery period beyond that anticipated. Evidence from literature on the recovery from the Bay of Plenty storm in New Zealand 2005 provides valuable lessons on the complexity of issues that could impact disaster recovery efforts. Middleton's (2008) report of the housing situation after the flood is presented in Table 9.2. At 300 days after the event, 35 households still required permanent re-housing out of a total of 300 compulsory evacuations. At the same time nine households were still occupying temporary accommodation. Middleton (2008) suggests that the situation was traceable to the inadequacy of personnel in carrying out building safety evaluations and the mandatory requirements for processing building and environmental consents. Processing of consents appeared to have been undertaken under a business-as-usual approach that are prescribed by existing legislation (Rotimi *et al.*, 2008).

It is apparent that there was a gap between the process of identifying homes that were suitable or unsuitable to continue to be lived in by residents on one hand; and of helping the affected to recover from the event so that they got back to their normal life, on the other hand.

Recently the New Zealand Society of Earthquake Engineers (NZSEE) drafted new guidelines on building safety evaluations during a declared state of emergency (New Zealand Society of Earthquake Engineers, 2008). The NZSEEs

Table 9.2   Temporary accommodation requirements after the Bay of Plenty storm. Reproduced by permission of David Middleton (2008)

Period in temporary accommodation	Number of households permanently re-housed	Number of households in temporary accommodation
Up to 60 days	0	293
60–150 days	71	222
150–200 days	140	82
200–300 days	38	44
Over 300 days	35	9

Details not available after 16 March 2006 (303 days after the event)

belief is that rapid evaluations undertaken during the emergency declaration period could help to start the process of recovery. Such guidelines show the need to harness human resources towards successful recovery from disasters. With such revisions, it is hoped that the problem of damage evaluations and assessments have been adequately addressed. However, the NZSEE document fails to suggest how externally sourced personnel could be integrated into the re-building process (New Zealand Society of Earthquake Engineers, 2008).

After damage assessments and evaluations, building and environmental legislation should not present impediments to actualising reconstruction and rebuilding programmes. Yet some post-disaster recovery literature has indicated that the implementation of certain aspects of development legislation could hinder the realisation of reconstruction objectives (Meese et al., 2005; Marano and Fraser, 2006) and drag residential rehabilitation (Burby et al., 2006) after disasters.

Reference was made in a report commissioned by Building Research, New Zealand to the conflicts that may exist in the interpretation and implementation of the Resource Management Act (RMA) 1991 and Building Act (BA) 2004 (Ministry of Work and Housing [MWH], 2004). Such conflicts may cause impediments to post-disaster reconstruction processes. For example, the report identified two potential sources of conflict from these legislative documents that may impact on reconstruction projects. The first type of conflict relates to the processing of consents under the BA and the RMA. It is noted that both Acts are coordinated by different agencies, the BA by the Department of Building and Housing and the RMA by the Ministry for Environment. The MWH report suggests that coordinating the implementation of the two documents in a disaster situation may result in legal complications. Both documents may need to be streamlined with one another to avoid such legal complications. The MWH report also identified potential sources of conflict around the interpretation of the substantive contents of both Acts.

Legislation defines powers, rights or responsibilities and promotes the interaction and interrelationship of disaster management stakeholders during initial response and subsequent recovery activities. Every stakeholder needs to understand their individual and collective responsibilities which have been prescribed in recovery plans. The apparent division between those who, in practice, take responsibility for reconstruction and those who set policy and legislation also create barriers that need to be overcome (Le Masurier et al., 2008). The problems in ineffective coordination may be traced to the bureaucratic tendencies of public officials (Schneider, 1995). An evaluative report (Schneider, 2005) on governmental response to Hurricane Katrina suggests that the problems encountered during response activities in the event were the result of poor administrative elements in the emergency management system.

9.5   Impediments to post-disaster reconstruction: the New Zealand Building Act (BA) 2004

The Building Act provides for the regulation of building works, the establishment of a licensing regime for building practitioners, and the setting of performance standards for buildings for purposes outlined in s.3 of the Act. The Act prescribes the requirements of the national building code which requires buildings and other associated features to meet certain performance standards such as durability, fire safety, sanitation services and facilities, moisture control, energy efficiency and access (s.16 and s.17). The Act is administered at the national level by the Department of Building and Housing (DBH) and at the local level by Building Consent Authorities (BCA) and Territorial Authorities through a building consent process (s.12). The responsibilities of BCAs under the Act are complemented by Licensed Building Practitioners who are expected to have undergone an accreditation and certification process to enable them to act in the capacity of consent and code compliance officers.

Building consent processing involves individual property owners, their designers/builders and the Building Consent Authorities. An application for consent is required for all building work in connection with the construction, alteration, demolition or removal of a building (s.40) with some minor exceptions. Consent is granted when the BCA is satisfied that the proposed works are in accordance with the building codes and associated regulations. Under normal circumstances, the building consent process is expected to last 20 days (s.48), though the reality is far from this. The Act requires a strict inspection of work progress during construction at 'hold points' corresponding to progress milestones. Each defined stage must be inspected and certified before subsequent stages can be started. Inspection provides some certainty about code compliance and construction quality, and ensures that constructed works are in accordance with the original work specified in the approved consents. At completion of all works a Code of Compliance Certificate (CCC) is issued (s.91–s.95A).

The BA is laudable as a risk-mitigating document for proposed development and redevelopment projects. It would allow for improved construction technology and facilities that could reduce vulnerabilities after a hazard event, for example. But strict implementation of some of the BA provisions could create problems during reconstruction projects.

Relevant sections of the BA appear to give some clarity on the building and alteration works that will or will not require an application for consent. However, in the event that building or alteration work is carried out in situations of urgency (as may be the case after a disaster), there are no special provisions in the BA for territorial authorities to issue retrospective consents for work that has been undertaken without the necessary building consent. It behoves the property owner therefore to apply for consent before any remedial work is carried out on a damaged property; and consequently a code of compliance certificate (CCC) is issued at completion. A property owner (vendor) may have limited opportunities to sell without a CCC. Property law requires that the owner of a property for sale must guarantee full disclosure of the availability or non-availability of the necessary permits and where appropriate a CCC (The Real Estate Institute of New Zealand, cl. 6[8]). It is not clear what implications such disclosure will have on sale value in terms of how a buyer will view a Certificate of Acceptance in relation to a CCC during sale transactions.

The BAs provision for works that do not have to comply with building codes is likely to generate implementation problems during reconstruction programmes. There is a special waiver under the BA to allow building consents to be granted subject to a waiver or modification of the building code (s.67–s.70). The determination of the appropriate circumstances when this section can be applied has been left to the discretion of BCAs. BCAs are required to prepare policies and guidelines on how this discretion can be exercised, but this is not being done across all councils (DBH, 2005). In somewhat similar requirements, BCAs are to prepare guidelines for collaborating with other councils and disaster agencies for resource sharing and deployments in a disaster situation. Such guidelines could simplify logistics and operational activities during the likely demands for external services, and when there is a spike in building and development consent applications.

Processing of building consents at the early stages of reconstruction and recovery are a potential bottleneck (AELG, 2005). Access to normal resource levels is unlikely and so there will be shortages of qualified persons and material resources to handle impact assessments and consent processing. More flexible approaches to the standard consent process might be necessary to expedite the consenting process and help cope with high volumes of consent applications after a major disaster. As a consequence, the potential complexities during response and recovery after disasters, procedural delays and other bureaucratic processes may impact a property owners' ability to proceed with reconstruction before an approved building consent lapses. The validity period of issued building consents is pertinent to reconstruction. Section 52 of the BA prescribes a 12 month period for the validity of an issued building consent, except a special extension is granted by the BCA. This provision may compound post-disaster recovery, where reconstruction is not started before an issued consent lapses. A repeat application will have to be made by the property owner. An appropriate extension of this period to reflect the realities of post-disaster reconstruction and with due consideration to the magnitude of devastation experienced after an event, may have to be made by respective councils. A valuable lesson for New Zealand was the reported amendment to planning regulations after Hurricane Ivan in the Cayman Islands. The amendment included the extension of the period of validity to development approvals from 12 to 18 months from their dates of issue. It was also reported that the fees charged for consent applications were reduced by 50% to alleviate the effect of the disaster on the community (Cayman Legislative Assembly, 2004).

These kinds of amendments could ensure that the recovery process is not made more onerous by planning requirements.

Another key aspect of the BA that may impact on reconstruction activities post disaster is the limitations and restrictions applicable to buildings on land subject to natural hazards (s.71–s.74). The Act requires that Territorial Authorities must refuse to grant building consents on land subjected to natural hazards unless the land can be protected from the hazard, and where waivers are granted, it requires that notices be placed on the land to indicate the risk of natural hazards it is exposed to. If this provision is strictly implemented, then some categories of house owners may not qualify for insurance claims where there is an identified risk to their land and facilities. For example, risk from ground subsidence is not covered by the EQC, although the rule was bent in Waihi in 2001 due to public outcry (Earthquake Commission, 2005). There are ongoing adjustments to the New Zealand hazard landscape, which means that previously risk-free buildings may be re-classified as risk-prone after the hazard review exercises. Thus new notices could prevent owners of such properties from being compensated in future disasters.

9.6   New Zealand Case Study 1: Manawatu-Wanganui Floods, 2004

The flooding incident occurred in the Manawatu-Wanganui region in 2004. The affected region is located in the lower half of the North Island of New Zealand. The flooded area covered 10 districts and includes Palmerston North and Wanganui Township as prominent cities. The flooding of the Manawatu-Wanganui region was caused by heavy rain and gale force winds which occurred from 15 to 23 February 2004. The event necessitated a region-wide civil defence emergency declaration on 17 February that lasted till midnight of 25 February. At the height of the event over 2300 people were reported to have been evacuated from their homes.

9.6.1   *Damage assessment*

The flood incident triggered the largest emergency management activity in New Zealand in 20 years (Reid *et al.*, 2004). Many rivers breached their banks and considerable areas of farmland were inundated by silt and floodwaters. Damage to infrastructure was significant with damage to roads, bridges and railways recorded. In addition, there were telecommunications, power, gas and water supply outages to thousands of homes. The magnitude of the event stretched the response and recovery capabilities of the local authority and emergency management agencies involved (Reid *et al.*, 2004).

Damage assessments carried out immediately after the flooding event gave recovery estimates as NZ$160–180 million for the rural sector; and NZ$120 million for roads and council infrastructure (Van der Zon, 2005).

In addition, costs of NZ$29.5 million and NZ$3.5 million were estimated to stop future flooding of the lower Manawatu and Rangitikei rivers that run through the region respectively. Approximately 500 houses were damaged, four bridges destroyed and 21 bridges seriously damaged. Roads and rail closure including power and phone outages were widespread. Livestock losses were estimated at 1300 during the event (MCDEM, No date-a).

9.6.2   Reconstruction and recovery work

Reconstruction work on damaged utilities commenced immediately after the flooding incident. Various utility providers, consultants and contractors worked 24-hour days to repair damaged roads and bridges, and to restore disrupted services (Reid *et al.*, 2004; Le Masurier *et al.*, 2006). The focus was on rebuilding quickly; hence existing contractual relationships were exploited. Reconstruction was carried out through collaboration between CDEM agencies, local authorities, utility companies and insurance companies. Recovery was coordinated through the regional council's CDEM group arrangements as provided for in the CDEM Act. For the other territorial authorities the event was managed using the previous Civil Defence Act 1983 arrangements (Wilkinson *et al.*, 2007).

The CDEM Act 2002 provided a structure appropriate for dealing with events such as the floods and did not hinder authorities from dealing with the event (Reid *et al.*, 2004). Hence the evolved structure involving the Interdepartmental Officials Domestic and External Security Coordination (ODESC), the Domestic and External Security Coordination (DES), the National Crises Management Centre (NCMC) and the Local Emergency Operations Centres (EOC) was considered suitable, flexible and robust in the event (Reid *et al.*, 2004). Road authorities did not diverge from normal legislation and regulations and building consents were sought and granted as usual.

However, the implementation of environmental control requirements became a source of frustration. According to AELG (2005) much time was lost by utility companies trying to develop an understanding with the regional council about emergency actions that could cover all situations under the RMA, rather than require a formal process for each activity. The Infrastructure Recovery Task Group leadership and the Regional Council had to outline the procedures to be followed in the form of a guidance note (AELG, 2005). Van der Zon (2005) highlighted the problem that arose with the deposition of slip materials for example. The regional council required all slip materials to be deposited at designated landfill sites. These landfill sites were located far away from the disaster zone which would mean long journeys to deposit slip. Subsequently the Regional Council had to allow a more pragmatic approach which meant that slip materials could be moved and deposited locally.

There were reported delays caused by the sourcing of reconstruction funds especially for road infrastructure. Transfund, the road-funding authority that had direct access to government funds, did not become involved as early as required, thus the prioritisation process for infrastructure works was hampered (Van der Zon, 2005). Much more was expected from Transfund to secure certainty over funding of emergency road contracts in the early stages of recovery.

The 2004 flooding incident exposed disaster management problems that could arise from the management of emergencies across jurisdictional boundaries. There were issues connected with the management of recovery around the Whangaehu valley that is shared between the Wanganui and Rangitikei Districts. Reid *et al.* (2004) believes recovery management could pose jurisdictional conflicts as to who should take responsibility for what in such situations. Reid *et al.*'s report identified the need for advance negotiation and a memorandum of understanding amongst Districts Councils to determine which district has primacy in the event of a civil emergency. Without some form of clarity about responsibilities, borderline lands or properties could be mismanaged, overlooked or at worst, emergency activities could be duplicated.

Reid *et al.* (2004) criticised communication and information flow between agencies during response and recovery. It was reported that the Local Councils were slow in some cases to realise the importance of being proactive in seeking information on the range of activities that took place during the event. The local authorities did not have the opportunity to prepare detailed plans and standard operating procedures (SOPs) for information gathering and dissemination before the incident. For example, requests for information from the NCMC did not match the Local Council's requirements (Reid *et al.*, 2004). The expectation was for Local Councils to be the receptacle of information by identifying the exact nature of the flooding incident and then to identify who can assist and how these people could be contacted and directed.

9.7   New Zealand Case Study 2: Matata (Bay of Plenty) Floods, 2005

Matata is a small farm community under the jurisdiction of the Whakatane District located in the North Island of New Zealand. The town is 24 kilometres to the north-west of Whakatane. On 18 May 2005, a band of intense rain (308 millimetres of rainfall within 20 hours) fell in the catchments behind the Matata Township. The rain triggered floods and several large debris avalanches and landslips. Debris flow reached State Highway 2 and railways around Pikowai to Awakaponga with boulders the size of cars strewn all around.

The Matata/Tauranga area had experienced significant flooding in the previous year, but this particular incident was more localised, concentrated and unparalleled in its magnitude. The engineering solutions consultants contracted for rehabilitation works at Matata confirm that the flooding incident had a chance of between 0.5% and 0.2% of happening in any year (Tonkin and Taylor Ltd, 2005). The flooding incident necessitated a civil defence emergency declaration on 18 May 2005 and this remained in place until the end of May.

9.7.1 *Damage assessment*

Total government valuation including land value and capital value of properties affected along the flood path of the hazard was initially estimated to be about NZ$10 million for unsafe buildings and NZ$3 million for buildings subject to restricted use. The MCDEM situation report (Recovery Report Number 6) on the initial damage evaluation at Matata and environs gives a breakdown of physical damages in Table 9.3 and Table 9.4.

Response and subsequent reconstruction activities commenced immediately after the flooding incident. It was reported in the 'WDC Recovery News' a Newsletter published by the Whakatane District, that a week after the incident there were already collaborative activities between the Department of Building and Housing (DBH), and the Whakatane and Tauranga District Councils. The collaboration involved the assessment of flood-damaged properties, processing of urgent building consents, and the provision of guidance on rebuilding procedures. A Recovery Coordinator was appointed by the MCDEM to act as interface between the central government and Whakatane District Council (WDC). The Coordinator was required to produce a recovery plan and to

Table 9.3   Affected houses in 2005 Matata flood path. Reproduced by permission of Whakatane District Council (source: WDC Recovery Report)

No.	Description	No. of properties affected (Matata/Awakaponga)
1	Unsafe houses	28
2	Unsafe – houses washed away	3
3	Unsafe land	14
4	Restricted use – houses	16
5	Restricted use – land	1

Table 9.4   Other housing situation reports. Reproduced by permission of Whakatane District Council (source: WDC Recovery Report)

Nature of work required	Matata/ Awakaponga	Edgecumbe/ Otakiri
Houses requiring removal of wall linings – unable to be occupied (total less unsafe/restricted houses)	24	9
Houses requiring removal of wall linings – pre-line approved	6	0
Houses requiring removal of wall linings – post-line approved	0	0
Houses requiring removal of wall linings – CCC approved	0	0
Septic tanks and drains to be cleaned	85	1
Septic tanks and drains cleaned	84	0

determine the quantum of the government's assistance package required by the community, in conjunction with the WDC Recovery Manager.

9.7.2 *Reconstruction and recovery work*

Several task forces were set up in line with CDEM guidelines. A Hazard Task force was appointed whose original scope of work included identifying what action plans and processes need to be put in place to address the short-term and long-term risks still facing Matata as a result of the event. The Hazard Task Force worked with the Infrastructure Task Force responsible for clearing flood debris, restoring roads and restoring potable water to the affected areas (Wilkinson *et al.*, 2007). Both ONTRACK and Transit NZ owned a significant part of the infrastructure in the affected area and were required to work collaboratively with the Hazards and Infrastructure Task Groups to identify long-term solutions. Other task forces that were set up included the Rural Task Team and Task Force Green (Le Masurier *et al.*, 2006).

Reconstruction works commenced with road clearance for rocks, stones and debris resulting from the flood. Wilkinson *et al.* (2007) explain that there was no tendering for the works undertaken during this initial response period. However, a fast-tracked tendering arrangement was implemented on subsequent reconstruction projects (4–6 weeks after the incident). Fortunately, there were existing relationships with contractors executing civil work projects in the vicinity, thus mobilisation of the required resources did not pose a great challenge. However, because of the priority attached to the reconstruction projects, progress on existing developmental projects was disrupted.

The rate of progress achieved at the initial response was commendable. However, reconstruction activities slowed down considerably afterwards. Private property owners were seriously affected and some were unable to rebuild because they were plainly at risk from similar events in the future. Several flood mitigation project options were proposed coupled with planning controls to reduce risks and to protect lives and property. The WDC Recovery News (Issue 7), 2005 reported that the following redevelopment controls were recommended on sites where damage had occurred until the extent of future hazard zones was confirmed:

- Limit redevelopment works through s.72 of the Building Act.
- Keep a record of hazard information on land information memoranda (LIM) and project information memoranda (PIM).
- Undertake all redevelopment works in accordance with the Public Works Act, which requires a limitation of works, so far as practicable, to hazards that have already developed.
- Variation of District Plans to reflect the improved level of hazard information in Matata Township.

A revealing insight into reconstruction and recovery after the flooding incident is found in Spee's (2008) study of the psychological and social impacts of the event on the Matata community. Spee (2008, page 18) generated a list of stressors that relate to the reconstruction problems experienced by individuals and the community following the event. These stressors include:

- The inability to return to homes until months later, 15 months in one case.
- Two years after the event people were still waiting to receive resource consent to rebuild on their sections.
- Moving four times in one year.
- Constantly making plans which needed to be adjusted due to resource consent timeframes being moved.
- A state of limbo as people waited to learn of their property's fate (i.e. whether the land was considered safe to rebuild).

Spee (2008) concludes that the longer term recovery events had caused more stress and frustration than the initial periods after the disaster. It would seem that the greatest impact on holistic recovery was the inadequacies experienced during the rebuilding efforts. Individuals and property owners were in a state of limbo for too long after the event. Fifty families were still in temporary accommodation five months after the incident (Rowan, 2005) and a formal disaster recovery plan only came into effect 18 months after the event.

Construction of flood mitigation structures that were approved by the council could not commence because environmental resource consents for such works were still being processed as at June 2008 (Becker *et al.*, 2008). Without the mitigation measures in place, property owners were unable to get insurance cover and without insurance payments no rebuilding could take place. There was widespread misunderstanding on compensation claims and settlement with the Earthquake Commission (EQC) making compromises to enable residents to receive compensation for their flooded properties (Dixon, 2005). In any case building consent processing in the at-risk areas of Matata was suspended until March 2007. Middleton's (2008) analyses of the housing situation (see Table 9.2) after the event re-affirms the impact that (re)development and legislative requirements had on the whole-of-Matata recovery.

9.8   Lessons for post-disaster legislation changes

9.8.1   *Powers to coordinate reconstruction activities*

It was observed in both flooding incidents that response and reconstruction works were managed through the collaborative effort of CDEM groups, local authorities, utility companies and insurance companies during recovery in both cases. The CDEM did not play a lead role in directing activities in spite of the legal provisions under the CDEM Act that enabled it to do so. Emergency powers were never exercised during the declared state of emergencies (AELG, 2005). Utilities (with their respective contractors) were allowed to determine their own reconstruction priorities without specific directives from the CDEM groups. The following modes of interaction between CDEM groups and lifeline utilities are possible (AELG, 2005). The different modes are dependent on the scale and type of hazard events that are being managed:

- Utilities determine their own restoration priorities with CDEM gathering information and monitoring performance.
- CDEM and lifelines work together to identify priorities and to implement performance through agreements.
- CDEM determine priorities and then request utilities to perform in line with the set priorities.
- CDEM direct specific actions calling on the powers in the CDEM Act.

The first and second modes were the operating situation in both flooding incidents and could pose barriers to effective reconstruction activities. None of the entities involved in the management and coordination of reconstruction in previous disaster events had any specific remit to work outside their own interests (Resilient Organisations, 2006). Therefore, for larger scale disasters it might not be out of place for more proactive action from CDEM groups in the form of the third and fourth modes of coordination.

9.8.2 Processing of building and environmental consents

In both flooding incidents, there was a fast track approach adopted at the initial response stages. Rapid procurement routes were employed to engage contractors for emergency road network clearance and debris removal by key utilities. This approach was possible because there were already established relationships with these contracting organisations. However, reconstruction activities were slowed down at the construction stage when building and environmental compliance requirements began to take effect. There appears not to have been a waiver to normal stipulated processes except in the situation reported in the Manawatu-Wanganui flood where the Recovery Task Group leader and the Regional Councils prepared guidance notes outlining more expedient procedures to be followed for debris clearance and disposal. Delays caused by statutory compliance procedures were in addition to delays already caused by property assessment and evaluations after the event. Such delays were also apparent in the Gisborne earthquake in 2007, necessitating a review of building safety evaluation procedures to speed up recovery in future disasters (New Zealand Society of Earthquake Engineers, 2008).

There was anecdotal evidence to suggest widespread duplication of resources in both flooding incidents. Damage assessments were required by different agencies like the EQC, private insurers and the local councils and it was not uncommon for assessment exercises to be repeated on the same properties by these agencies (Reid *et al.*, 2004). This situation had a 'knock-on' effect, thus delaying the actual implementation of the various reconstruction projects.

In summary, both flooding incidents emphasise the importance of pre-planning for large scale response and recovery programmes in New Zealand. By undertaking some prior planning, responding agencies are better equipped to implement recovery plans within reasonable time frames forestalling the sort of frustrations experienced in the two events reviewed. A good supportive framework should guide effective coordination of resources after an event, allowing for more effective and efficient recovery activities.

9.9 Summary

This chapter has shown that poor legislative provisions can have a detrimental effect on post-disaster reconstruction activities. Two case studies demonstrate the impact that building and environmental legislation could have on reconstruction programmes. Also the inherent problems surrounding the implementation of the Building Act in the event of a significant disaster in New Zealand are presented. For vulnerable countries like New Zealand, there needs to be more conscious preparation for mitigation and management of hazard events, and this includes reducing the burdensome impact legislation can have on post-disaster activities. The efficiency of post-disaster reconstruction activities is impacted as a result of delays caused by poor planning and implementation, restrictive legislation and regulatory provisions. If communities are to recover from a natural hazard, then legislative effects need to be better understood and managed. The recent earthquakes in Christchurch are a reminder that more flexible approaches (also within special legislative provisions) are required to address the reconstruction and recovery challenges presented by large natural disasters.

References

ACTIONAID Nepal (2004) *Disaster Management in Nepal: Analysis of Laws and Policies.* Nepal: ACTIONAID, Nepal.

AELG (Auckland Engineering Lifelines Group) (2005) *Resources Available for Response and Recovery of Lifeline Utilities* (Technical Publication No. 282): Auckland Regional Council Technical Publication. Retrieved from www.aelg.org.nz/

Angus, L. (2005) New Zealand's Response to the 1994 Yokohama Strategy and Plan of Action for a Safer World. Paper presented at the World Conference of Disaster Reduction, Kobe-Hyogo, Japan.

Becker, J. and Saunders, W. (2007) *Enhancing sustainability through pre-event recovery planning.* www.qualityplanning.org.nz/ [accessed 13/10/2008].

Becker, J., Saunders, W., Hopkins, L., Wright, K. and Kerr, J. (2008) *Pre-event Recovery Planning for Land-use in New Zealand: An Updated Methodology* (No. 2008/11). Institute of Geological and Nuclear Sciences Limited. www.gns.cri.nz/

Berke, P. R. and Campanella, T. J. (2006) Planning for postdisaster resiliency. *The ANNALS of the American Academy of Political and Social Science*, **604**, 192–207.

Burby, R. J., Salvesen, D. and Creed, M. (2006) Encouraging residential rehabilitation with building codes: New Jersey experience. *Journal of the American Planning Association*, **72**, 183–196.

Cayman Legislative Assembly (2004) The Development and Planning (Amendment) Bill. www.caymanprepared.ky/

Chan, Y., Alagappan, K., Gandhi, A., Donovan, C., Tewari, M. and Zaets, S. B. (2006) Disaster management following the Chi-Chi earthquake in Taiwan. *Prehospital Disaster Medicine*, **21**, 196–202.

Coghlan, A. (2004) Recovery Management in Australia: A Community-Based Approach. Paper presented at the NZ Recovery Symposium, Napier, New Zealand.

Comerio, M.C. (2004) Public policy for reducing earthquake risks: a US perspective. *Building Research and Information*, **32**, 403–413.

Comfort, L., Wisner, B., Cutter, S., Pulwarty, R., Hewitt, K., Oliver-Smith, A., Wiener, J., Fordham, M., Peacock, W. and Krimgold, F. (1999) Reframing disaster policy: the global evolution of vulnerable communities. *Environmental Hazards*, **1**, 39–44.

Cousins, T. (2004) A Holistic Framework for Recovery: What Happens When and Works Best. Paper presented at the NZ Recovery Symposium, Napier, New Zealand.

DBH (Department of Building and Housing) (2005) *Building Officials' Guide to the Building Act 2004*. www.building.dbh.govt.nz

Dixon, L. (2005, 28 July) The insurance issue explained. *Recovery News*, 2.

Earthquake Commission. (2005) *Chairman's Report*. www.eqc.govt.nz/ [accessed 18/08/2007].

Feast, J. (1995) Current Planning and Construction Law: The Practical Consequences for Rebuilding Wellington after the Quake. Paper presented at Wellington After the 'Quake: The Challenges of Rebuilding Cities, Wellington, New Zealand.

Fenwick, T., Seville, E. and Brunsdon, D. (2009) Reducing the Impact of Organisational Silos on Resilience. Resilient Organisations Research Project (No. 2009/01). University of Canterbury, New Zealand.

Gavidia, J. and Crivellari, A. (2006) Legislation as vulnerability factor. *Open House International*, **31**, 84–89.

Granot, H. (1997) Emergency inter-organizational relationships. *Disaster Prevention and Management*, **6**, 305–312.

Harrald, J. R. (2006) Agility and discipline: critical success factors for disaster response. *The ANNALS of the American Academy of Political and Social Science*, **604**, 256–272.

Hendrikx, J. (2006) *Preliminary Analysis of the June 12 2006 Canterbury Snow Storm* (No. CHC2006-088). National Institute for Water and Atmospheric Research Ltd. www.niwa.cri.nz/

Hopkins, D. (1995) Assessment of Resources Required for Reinstatement. Paper presented at Wellington after the Quake: The Challenge of Rebuilding Cities, Wellington, New Zealand.

Hopkins, D., Lanigan, T. and Shephard, B. (1999) The Great Wellington Quake: A Challenge to the Construction Industry. Paper presented at Wellington after the Quake: The Challenge of Rebuilding Cities, Wellington, New Zealand.

Ingram, J. C., Franco, G., Rio, C. R.-d. and Khazai, B. (2006) Post-disaster recovery dilemmas: challenges in balancing short-term and long-term needs for vulnerability reduction. *Environmental Science and Policy*, **9**, 607–613.

Interworks (1998) *Model for a National Disaster Management Structure, Preparedness Plan, and Supporting Legislation.* www.iaemeuropa.terapad.com/ [03/09/2008].

Jigyasu, R. (2004) Sustainable Post-disaster Reconstruction through Integrated Risk Management. Paper presented at the Second International Conference on Post-disaster Reconstruction, Coventry University, UK.

Kennedy, J., Ashmore, J., Babister, E. and Kelman, I. (2008) The meaning of 'build back better': evidence from post-Tsunami Aceh and Sri Lanka. *Journal of Contingencies and Crisis Management,* **16**, 24–36.

Kouzmin, A., Jarman, A.M.G. and Rosenthal, U. (1995) Inter-organizational policy processes in disaster management. *Disaster Prevention and Management,* **4**, 20–37.

Lanigan, T. (1995) Physical Reconstruction: Availability of Material, Labour and Plant from within New Zealand and the Role of the Private Sector. Paper presented at Wellington After the Quake: The Challenge of Rebuilding Cities, Wellington, New Zealand.

Le Masurier, J., Rotimi, J.O.B. and Wilkinson, S. (2006) A Comparison Between Routine Construction and Post-Disaster Reconstruction with Case Studies from New Zealand. Paper presented at the 22nd ARCOM Conference on Current Advances in Construction Management Research, Birmingham, UK.

Le Masurier, J., Wilkinson, S., Zuo, K. and Rotimi, J. (2008) Building resilience by focusing on legal and contractual frameworks for disaster reconstruction. In: Bosher, L. (Ed.), *Hazards and the Built Environment: Attaining Built-in Resilience.* London: Taylor and Francis.

Listokin, D. and Hattis, D. (2004) Building Codes and Housing. Paper presented at the Workshop on Regulatory Barriers to Affordable Housing, US Department of Housing and Urban Development Washington, DC. www.2004nationalconference.com

Marano, N. and Fraser, A.A. (2006) *Speeding Reconstruction by Cutting Red Tape.* www.heritage.org/ [accessed 20/10/2006].

Martín, C. (2005) Response to 'Building Codes and Housing' by David Listokin and David B. Hattis. *Cityscape: A Journal of Policy Development and Research,* **8**, 253–259.

May, P.J. (2004) Regulatory Implementation: Examining Barriers from Regulatory Processes. Paper presented at the Workshop on Regulatory Barriers to Affordable Housing, US Department of Housing and Urban Development, Washington, DC. www.2004nationalconference.com [accessed 22/04/2004].

MCDEM (No date-a) *Civil Defence Declarations since 1 January 1963.* www.civildefence .govt.nz/ [accessed 02/12/2007].

MCDEM (No date-b) *Get Ready Get Thru.* Retrieved from http://www.mcdem.govt.nz/

McDonald, C. (2004) The promise of destruction. *The Australian Journal of Emergency Management,* **19**, 51–55.

McEntire, D.A. (2002) Coordinating multi-organisational responses to disaster: lessons from the March 28, 2000, Fort Worth tornado. *Disaster Prevention and Management,* **11**, 369–379.

Meese III, E., Butler, S.M. and Holmes, K.R. (2005) *From Tragedy to Triumph: Principled Solutions to Rebuilding Lives and Communities.* www.heritage.org/ [accessed 20/10/2005].

Menoni, S. (2001) Chains of damages and failures in a metropolitan environment: some observations on the Kobe earthquake in 1995. *Journal of Hazardous Materials,* **86**, 101–119.

Middleton, D. (2008) Habitability of Homes after a Disaster. Paper presented at the 4th International i-REC Conference on Building Resilience: Achieving Effective Post-disaster Reconstruction, 30 April–2 May, Christchurch, New Zealand.

Ministry for Culture and Heritage (2009) *New Zealand Disasters Timeline.* www .nzhistory.net.nz/ [accessed 13/06/2009].

Mitchell, J.K. (2006) The primacy of partnership: scoping a new national disaster recovery policy. *The ANNALS of the American Academy of Political and Social Science,* **604**, 228–255.

MWH (2004) *Conflict Between the Resource Management Act 1991 and the Building Act 2004 – An Issues Paper* (No. 801/008787). MWH New Zealand Limited.

New Zealand Society of Earthquake Engineers (2008) *Building Safety Evaluation During a Declared State of Emergency: Draft Guidelines for Territorial Authorities.* From www.nzsee.org.nz/ [accessed 20/01/2008]

Olshansky, R.B. (2005) How Do Communities Recover from Disaster? A Review of Current Knowledge and an Agenda for Future Research. Paper presented at the 46th Annual Conference of the Association of Collegiate Schools of Planning, Kansas City. From www.kerrn.org/pdf/

Page, I. (2004) Reconstruction Capability of the New Zealand Construction Industry. Paper presented at the NZ Recovery Symposium, Napier, New Zealand.

Quarantelli, E.L. (1988) Disaster crisis management: a summary of research findings. *Journal of Management Studies,* **25**, 373–385.

Quarantelli, E.L. (2007) Disaster crisis management: a summary of research findings. *Journal of Management Studies,* **25**, 373–385.

Reid, P., Brunsdon, D., Fitzharris, P. and Oughton, D. (2004) *Review of the February 2004 Flood Event.* www.civildefence.govt.nz/ [accessed 28/11/2005].

Resilient Organisations (2006) *Barriers to Post-Disaster Reconstruction.* Te Papa, Wellington: Resilient Organisations. www.resorg.org.nz/

Rolfe, J. and Britton, N.R. (1995) Organisation, Government and Legislation: Who Coordinates Recovery? Paper presented at Wellington after the Quake: The Challenge of Rebuilding Cities, Wellington, New Zealand.

Rosenthal, U. and Kouzmin, A. (1997) Crises and crisis management: toward comprehensive government decision making. *Journal of Public Administration Research and Theory,* **7**, 277.

Rotimi, J.O., Wilkinson, S., Myburgh, D. and Zuo, K. (2008) The Building Act and Reconstruction Programmes in New Zealand: matters arising. Paper presented at Building Abroad: Procurement of Construction and Reconstruction Projects in the International Context, 23–25 October, Montreal. www.grif.umontreal.ca/

Rotimi, J.O.B. (2006) Achieving Reconstruction Objectives Within Appropriate Regulatory Frameworks. Paper presented at the 31st AUBEA Conference on Building in Value, Sydney, Australia.

Rowan, J. (2005) Katrina? What about Matata? *The New Zealand Herald,* 1 October 2005.

Schill, M.H. (2005) Regulations and housing development: what we know. *Cityscape: A Journal of Policy Development and Research,* **8**, 5–19.

Schneider, S.K. (1995) *Flirting with Disaster: Public Management in Crisis Situations.* Armonk, NY: M.E. Sharpe.

Schneider, S.K. (2005) Administrative breakdowns in the governmental response to Hurricane Katrina. *Public Administration Review,* **65**, 515–516.

Shaw, R., Gupta, M. and Sarma, A. (2003) Community recovery and its sustainabilty: lessons from Gujarat Earthquake of India. *The Australian Journal of Emergency Management*, **18**, 28–34.

Singh, B. (2007) *Availability of Resources for State Highway Reconstruction: A Wellington Earthquake Scenario.* Unpublished Master of Engineering, University of Auckland, Auckland.

Smith, N. (2009) *First reading of the RMA Reform Bill.* www.beehive.govt.nz/ [accessed 02/03/2009]

Smith, W. and Dowell, J. (2000) A case study of co-ordinative decision-making in disaster management. *Ergonomics*, **43**, 1153–1166.

Spee, K. (2008) *Community Recovery after the 2005 Matata Disaster: Long-term Psychological and Social Impacts* (No. 2008/12). Institute of Geological and Nuclear Sciences Limited. www.disasters.massey.ac.nz/

Spence, R. (2004) Risk and regulation: can improved government action reduce the impacts of natural disasters? *Building Research and Information*, **32**, 391–402.

Stackhouse, A. (2006) Where to begin: a framework for rebuilding New Orleans [Melissa Vanlandingham]. *Policy Matters*, **3**, 34–39.

The Real Estate Institute of New Zealand. *Auckland District Law Society Agreement for Sale and Purchase of Real Estate*, 6th Edn. www.wellington-realestate.co.nz/

Tonkin and Taylor Ltd (2005) *Matata Debris Flows Hazard and Risk Investigations (Regulatory Review).* Prepared for Whakatane District Council.

Van der Zon, J. (2005) *Post-disaster Reconstruction in New Zealand.* Unpublished Research Report, University of Canterbury, Christchurch.

Wamsler, C. (2004) Managing urban risk: perceptions of housing and planning as a tool for reducing disaster risk. *Global Built Environment Review*, **4**, 11–28.

Wilkinson, S., Zuo, K., Le Masurier, J. and Van Der Zon, J. (2007) A Tale of Two Floods: Reconstruction After Flood Damage In New Zealand. Paper presented at the CIB World Building Congress on Construction for Development, Cape Town, South Africa.

10 Conflict, Post Conflict and Post-Conflict Reconstruction: Exploring the Associated Challenges

Krisanthi Seneviratne and Dilanthi Amaratunga

10.1 Introduction

Conflicts are predominant in many countries in the world. However, conflicts mostly exist in the developing world rather than in developed countries. Apart from a few exceptions conflicts can last for decades and have a huge impact on people, society, property and the economy. War-torn countries are characterised by death and injury to many of people, massive displacement of people as refugees and internally displaced persons (IDPs), widespread destruction of properties, poor institutional capacity, vulnerability to disease and crime, reduced security, lack of access to productive facilities and erosion of social capital.

Post-conflict reconstruction refers to the reconstruction of the enabling conditions for a functioning peacetime society. Therefore, reconstruction after war plays an important role in establishing the country's development and peace. However, it is claimed that post-conflict reconstruction is poorly understood and there remains a great deal to be learned about post-conflict reconstruction. Also reconstruction of countries emerging from conflicts is considered as one of the major challenges for development agencies in the coming decades as their nature, scale and proliferation of conflicts are undermining development in a wide range of countries and diverting international attention and scarce resources from pressing development problems. Therefore, post-conflict reconstruction has emerged as a new research area. A good understanding about conflict, post conflict and post-conflict reconstruction and associated post-conflict reconstruction challenges will serve as a basis for these studies.

Post-Disaster Reconstruction of the Built Environment: Rebuilding for Resilience, First Edition.
Edited by Dilanthi Amaratunga and Richard Haigh.
© 2011 Blackwell Publishing Ltd. Published 2011 by Blackwell Publishing Ltd.

Therefore, this chapter aims to explore the associated post-conflict reconstruction challenges through identifying the links and interdependencies between conflict, post conflict and post-conflict reconstruction, based on a comprehensive literature survey and review (reports, journal articles and conference papers) carried out on conflicts and post-conflict reconstruction. The chapter starts with an introduction to conflicts identifying their causes and impact. The next section describes the conflict circle followed by an introduction to the post-conflict environment. The last section is focused on post-conflict reconstruction through identifying key activities, challenges and dilemmas. A summary of the chapter is provided at the end.

10.2   Conflicts

Conflicts are inherent in all societies. While differences in interests and opinions between groups and countries are accepted as natural, how such differences are expressed and managed determines if the conflict will manifest itself in political or violent ways (Rugumamu and Gbla, 2003). Conflicts tend to be predominantly political when groups within a society pursue their objectives in accordance with the laws and established norms of the society. In other cases groups turn to violence to pursue their interests. To ensure that policy interventions do not instigate, exacerbate or revive situations of violent conflict, there should be a better understanding of what affects the level and dynamics of conflict. Further, it is accepted that well-designed and implemented policies can help reduce it (Rugumamu and Gbla, 2003).

Most recent conflicts have been intrastate conflicts (World Bank, 1998; Zenkevicius, 2007). In 2000, the State Failure Task Force (François and Sud, 2006) discussed various forms of internal conflicts as revolutionary wars, ethnic wars, adverse regime changes and genocide as described below:

- *Revolutionary wars*: Episodes of sustained violent conflict between governments and politically organised challengers that seek to overthrow the central government, to replace its leaders, or to seize power in one region.
- *Ethnic wars*: Episodes of sustained violent conflict in which national, ethnic, religious or other communal minorities challenge governments to seek major changes in their status.
- *Adverse regime changes*: Major abrupt shifts in patterns of governance, including state collapse, periods of severe elite or regime instability, and shifts away from democracy towards authoritarian rule.
- *Genocides and politicides*: Sustained policies by states or their agents, or, in civil wars, by either of the contending authorities, that results in the deaths of a substantial proportion of a communal or political group.

A civil war is defined as 'an internal violent conflict between a sitting government and a rebel organisation where at least 1000 people are killed in combat-related violence with at least 5% of the casualties incurred on each side' (Nkurunziza, 2008). Therefore, civil wars will also be a form of internal conflict. Civil war is considered as the most devastating type of conflict, because it has a much more destructive effect on economic activities, social capital and institutions than do multi-state wars (Elbadawi, 2008).

Many of the conflicts are armed conflicts. The World Health Organisation (Krause and Jutersonke, 2005) defines armed conflict as 'the instrumental use of violence by people who identify themselves as members of a group against another group or set of individuals, in order to achieve political, economic or social objectives'. According to Cuny and Tanner (1995), the nature of armed conflicts has changed dramatically since the end of the Second World War. First the majority of conflicts have involved Southern nations, though with considerable economical and political encouragement from Northern countries. It is widely acknowledged that wars exist in many developing countries (Cuny and Tanner, 1995; El-Masri and Kellett, 2001; Anand, 2005; Fearon *et al.*, 2009). As an example, in 2001 there were 34 armed conflicts involving 29 countries and 26 of those were in developing countries (Gleditsch *et al.*, 2002). Similarly, in 2002 there were 31 armed conflicts involving 24 countries, 22 of which were developing countries (Eriksson *et al.*, 2003). The World Bank (1998) describes conflict as a cause and effect of impoverishment and reports 15 of the world's 20 poorest countries have experienced significant periods of conflict since the 1980s. Developed countries also experience conflict, but it occurs more often in poor countries. Afghanistan, Bosnia, Ethiopia, Iraq, Kosovo, Lebanon, Sudan, Sri Lanka and Somalia are some examples (El-Masri and Kellett, 2001). The most violent conflicts affect poor countries (FAO, 2005) and Anand (2005) states that the main victims of conflict are often the poor. Furthermore, a poor country that has not yet experienced war has a 14% risk of conflict in any given 5-year period (Patrick, 2006).

Notably, civilians and civilian life have become the primary military targets of the warring parties. Targeting and punishing civilians is part of the strategy adopted by governments seeking to quash insurrection by depriving them of their base, or by insurgents attempting to survive. Nkurunziza (2008) indicates that the norm of modern warfare is violence against civilians who represent up to 84% of all war-related casualties (Azam, 2006 cited in Nkurunziza, 2008). These conflicts have also lasted over a long period of time with few exceptions. More than 30 armed conflicts were in progress during the 1990s, of which 24 had lasted more than a decade, with more than 5.5 million civilians dead, and more than 25 million people forcibly uprooted (Hewitt 1997 cited in El-Masri and Kellett, 2001). In Lebanon the civil war lasted for 17 years, in Afghanistan for two decades, in Guatemala for over three decades and in Sudan for four decades (FAO, 2005). In addition, a woeful number of wars have recurred and most argue that it is highly likely that warfare will resume after the end of a conflict (World Bank, 1998; Patrick, 2006).

10.2.1   *Causes of conflicts*

Root causes of conflicts are typically complex and perceived goal and interest incompatibility is perhaps the most basic cause of social conflict.

Social, economic and political discrimination against groups in society, whether by a minority or a majority of the population, is found to be a significant factor in many conflicts (FAO, 2005). Many recent conflicts have been in defence of ethnic or religious identity (World Bank, 1998). Cultural differences and particularly language are additional sources of separateness. Some conflicts are linked to competition for land and other natural resources (World Bank, 1998; FAO, 2005). Population growth without an increase in productivity can increase the pressure on natural resources. The resulting environmental degradation may cause greater competition for the remaining natural resources.

According to the Food and Agriculture Organisation (FAO, 2005), grievances held by many in a society seldom lead to violence without mobilisation by leaders. Although the divisions of ethnicity and religion do not often result in violence by themselves, such divisions may be exploited by leaders who are ideologically inspired to address poverty and political injustice, or inspired by greed and the ability to extract wealth out of the conflict. Other than that, conflicts generate large scale violence if structural conditions are present, such as authoritarian rule and/or lack of political rights, state weaknesses and lack of human and institutional capacity to manage (Rugumamu and Gbla, 2003). The risk of an outbreak of violence increases when these conditions exist concurrently with other problems such as manipulation of ethnicity or other differences in religion, culture and language (Rugumamu and Gbla, 2003).

For example, the high incidence of civil war in Africa is due to Africa's high dependence on natural resource exports which are relatively easy to loot to finance rebellions, low per capita income and low levels of education and poor governance (Elbadawi and Sambanis, 2000 cited in Nkurunziza, 2008). Ethnic diversity (Collier and Hoeffler, 1998 cited in Nkurunziza, 2008) and distributional inequalities are other causes of civil wars in Africa. Rugumamu and Gbla (2003) indicate that it is not always possible to distinguish the cause of a conflict from its effects as these tend to blend when a conflict emerges. A deeper understanding of the causes, consequences and means of resolution of any conflict necessitates recognition of the complex dynamics and motivations of local actors (Kalyvas, 2003 cited in Krause and Jutersonke, 2005).

10.2.2   *Impact of conflicts*

Conflicts have a number of direct and indirect impacts. The more direct effects of war are fatalities and population displacements (Cuny and Tanner, 1995; Rugumamu and Gbla, 2003). As a result of conflicts, people flee, either across borders (as refugees) or as IDPs. The Second World War left a million people homeless in France (Wakeman, 1999). The World Bank (1998) reports that about 35 million people are displaced as a result of conflict. During the war in Rwanda 1 million people were killed and the conflict created 3 million refugees and 4 million IDPs. UNHCR (2003 cited in Nagai *et al.*, 2007) reported that in Sri Lanka over 60 000 people lost their lives and 200 000 had fled abroad. Recent reports suggest that 250 000 people have been displaced and held in camps in Northern Sri Lanka (Senanayake, 2009). In addition, 200 000 people have been displaced in the Eastern Province (International Crisis Group, 2009). Over 3 million people are estimated to have been killed in the conflict in the Congo (FAO, 2005). At the end of the war, there were 95 000 ex-combatants, 100 000 IDPs and nearly 400 000 refugees in Sudan (Kibreab, 2002). The border conflict between Eritrea and Ethiopia generated 700 000 IDPs (Kibreab, 1999 cited in Kibreab, 2002). In Bosnia more than half of the population were displaced during and in the aftermath of the war (Stefansson, 2006). Most displaced families lose their assets, livelihood and accumulated wealth (Cain, 2007). This deprives countries of the most important assets needed to sustain their development efforts (Nkurunziza, 2008).

Widespread destruction of property is typical during conflicts. During war and immediately after conflict, rebels tend to target physical infrastructure as part of their strategy to disrupt the logistical flow of the enemy and to put the sitting government in difficulty (Barakath, 2002; Rugumamu and Gbla, 2003; Nkurunziza, 2008). Physical structures, such as housing, schools, health facilities, roads, bridges, dams, railways, airports, ports, electricity grids, commercial enterprises and telecommunication facilities are often damaged in conflicts (Cuny and Tanner, 1995; Rugumamu and Gbla, 2003; FAO, 2005; Grant, 2005; Nkurunziza, 2008). Even if a facility has not been destroyed it is likely to have been looted of much of its equipment and materials (Barakath, 2002). The breakdown of preventive medical services has long-term effects on the well-being of a population. Children who cannot receive an education will lose all prospects of a brighter future. The breakdown of services, such as clean water, will contribute to increased morbidity and misery (Cuny and Tanner, 1995). A third of the province's housing stock was destroyed by the conflicts in Kosovo, while war in Sierra Leone saw the destruction of an estimated three million homes, leaving over a million people displaced (Barakath, 2003). In the case of France, over three-quarters of the country was affected by the Second World War. Over two million buildings, a quarter of the housing stock, were destroyed or damaged (be Croizé 1991 cited in Wakeman, 1999). Millions of people lived in temporary shelters and run-down apartments without access to basic services (Wakeman, 1999).

The wars in Bosnia and Herzegovina, Croatia and Serbia and Montenegro destroyed or badly damaged about a million dwelling units (Wegelin, 2005). In each country housing production levels plummeted during the war years. A major contributing factor for this decline was that the former public sector housing delivery mechanism through state enterprises hardly existed due to the war. Apart from this, little attention was paid to housing maintenance, particularly of publicly owned housing stock (Wegelin, 2005). Further, a steady decline in average household size and increased urbanisation accelerated as a result of war. These factors have imposed considerable pressure on the housing stock in Bosnia and Herzegovina, Croatia and Serbia and Montenegro (Wegelin, 2005). An additional problem is impoverishment. The combination of war and the demise of public housing production has created a situation in which between 25% and 30% of the population are not able to afford housing (Wegelin, 2005). As a result, a substantial investment in reconstruction is necessary.

Economic ruin is another impact of conflict. For instance, the civil war in Lebanon caused an estimated US$25 billion of damage, while around US$10 billion was spent on peacekeeping and relief in Bosnia and Herzegovina (World Bank, 1998). Donors are injecting large amounts of aid into post-conflict countries (Fearon *et al.*, 2009). According to the World Bank (1998), in 1995, peacekeeping operations alone cost the international community US$3 billion. Wars in the developing world cause harm to welfare systems (Fearon *et al.,* 2009) and constrain the alleviation of poverty (World Bank, 1998). Armed conflicts have produced massive poverty in many of the least prosperous areas of the world (Cuny and Tanner, 1995). Conflict breaks down the underpinning of the economy, disrupts markets and distribution networks and destroys banking and credit systems (World Bank, 1998). Countries at war divert their resources from production activity to destruction by limiting development (Rugumamu and Gbla, 2003; Nkurunziza, 2008) and this can retard progress towards Millennium Development Goals (Anand, 2005). Furthermore, war disturbs the productive base of a community. Physical destruction of infrastructure and services affects agricultural and industrial production. In addition, the indirect effects of reduced security also prevents people from working. For example, mines are laid precisely for the purpose of breaking down normal socio-economic patterns (Cuny and Tanner, 1995). Displacement leads to the disappearance of local markets. This is considered to be one of the impediments for a return to normality. Furthermore, the climate of insecurity caused by conflicts leads to the contraction of internal and external investment.

Displacement as a result of war breaks up communities and erodes social capital (Cain, 2007). Conflicts alter gender roles (World Bank, 1998). Rugumamu and Gbla (2003) indicate that severe conflict destroys more than buildings and primarily targets organisations and individuals who administer the rules and thereby wipe out the most positive forms of social capital. This socio-economic disintegration of a population leaves them more vulnerable to disasters, as their resilience is poor. The climate of violence that exists during conflicts also leads to wide scale human rights abuses, ranging from the targeting of political dissenters to the harassment of the disempowered, particularly women and minorities (Cuny and Tanner, 1995).

Conflicts often weaken institutions at all levels and they exhibit little capacity to carry out their traditional functions (Cain, 2007). Civil wars are not only a challenge for countries in which they occur, but also for the entire region, as surrounding countries must accommodate large numbers of refugees and face additional consequences such as a growing defence budget, spread of diseases, drug production, trafficking, terrorism and the hindrance to potential investment in those countries (Rugumamu and Gbla, 2003; François and Sud, 2006).

10.3   Conflict circle

Conflicts are not sudden events but should be seen as historical structures that are transformed over time (Hettne and Soderbraum, 2005). Accordingly, they suggest a conflict circle that consists of six crucial elements: provention, prevention, intervention, peace settlement, conflict resolution and post-conflict reconstruction as described below. This is viewed as a simplified way of understanding conflict dynamics. The conflict circle could be relatively short if the resolution takes place before it turns violent or prolonged, if early conflict prevention fails (Hettne and Soderbraum, 2005).

- *Provention*: This refers to the early prevention of conflict and precedes the conflict even in its latent form. The term provention describes the combination of the promotion of conditions conducive to peace and the prevention of conditions conducive to violence (Burton, 1990 cited in Hettne and Soderbraum, 2005). The normative position taken here is to try to prevent a conflict even before it emerges by dealing with the structural root causes.
- *Prevention*: Prevention calls for early warning systems, fact-finding missions and confidence building measures (Boutros-Ghali, 1992 cited in Hettne and Soderbraum, 2005). Accordingly, this includes the activities taken to prevent disputes from arising between parties, to prevent exiting disputes from escalating into conflicts and to limit the spread when they occur. The first category of activities explained here seems to coincide with what above was called provention while the latter implies that the conflict has already turned violent. Therefore, conflict prevention is confined to the period after it has become manifest but before it has turned violent.
- *Intervention*: Intervention mostly means military intervention in order to put an end to a violent conflict. Whether it is termed humanitarian intervention or not is due to the nature of crisis. Distinctions are made among different modes of military intervention in acute security crisis. These modes range from unilateral, bilateral, plurilateral, regional to multilateral intervention.
- *Peace settlement*: Peace settlement is the formal ending of a conflict and may include principles of conflict resolution to be applied, or simply be confined to conditions of the ceasefire. It is essential that the terms of the peace agreement address the root causes.
- *Conflict resolution*: Conflict resolution takes place before an event turns violent. However, there are post-conflict resolutions as an end of conflict.

The cessation of conflict goes through political restructuring between the contending groups. Basically, there are three principal forms of conflict resolution in divided societies. Firstly, constitutional change, modifying the skewed ethnic power structure and establishing a power sharing within a particular state formation. Secondly, the dismemberment of the state is an option that remains when the preferred solution of constitutional reform has failed. Finally, a completely reversed process is the integration of neighbouring states into a regional formation, a process, providing solutions to ethnic tensions simply by downplaying the role of borders so central to the old Westphalian order based on national sovereignty (Hettne and Soderbraum, 2005).

- *Post-conflict reconstruction*: This is completely different from the physical rebuilding of the war-torn society and not confined to the issues of physical reconstruction (Barakath, 2002; Grant, 2005; Hettne and Soderbraum, 2005). Scope and depth of post-conflict reconstruction must extend beyond the simple rebuilding of physical infrastructure. Intangible social needs must be met, ranging from healing societal scars from the civil war to eradicating corruption in the public and co-operate arenas (Grant, 2005). According to Barakath (2002) physical reconstruction is required but not at the expense of spending on human skills development. Accordingly, post-conflict reconstruction requires more diverse, interconnected strategies: support for livelihoods, small communities, demobilised soldiers, women, people with disabilities, cultural heritage and structures of governance are all vital to the assistance and reintegration of different war-affected groups. It is the process of normalisation. It is claimed that post-conflict reconstruction is poorly understood, underfinanced and generally neglected. This phase will be discussed in detail in the following sections.

Though provention is somewhat unrealistic, prevention attracts more interest as it can reduce both material cost and human suffering. Despite this, it is claimed that prevention comes too late, because the mechanisms for early management of emerging conflicts are at best embryonic (Hettne and Soderbraum, 2005).

10.4   Post conflict

As the name indicates, the concept of post conflict refers to the period following the end of a conflict in a given country. However, the start and end of a post-conflict period is difficult to define, as there is seldom a clearly defined boundary (FAO, 2005; Nkurunziza, 2008). The transition from war to peace does not follow a straightforward path and does not occur in all parts of the country at the same time. Two major events have been identified that determine the beginning of a post-conflict period (Nkurunziza, 2008). The first is the immediate period following a landmark victory by either of the war-ring parties. This could be the fall of the capital city or seat of political power following a long protracted war. The second event is the date of signature of a comprehensive agreement between the warring parties. Even after the signature of a peace agreement, low-intensity hostilities might continue for months or years at varying levels of intensity, but it reduces them dramatically. Social ten-sions may even increase during the post-conflict period. The FAO (2005) guide has defined the start of the post-conflict period as when main hostilities have ceased to the point that international aid can begin. Theoretically, the post-conflict period ends when the specific attributes inherited from the conflict cease to have influence or when the country returns to normality. As in reality it is impossible to say exactly when a country returns to normality from its post-conflict state, the post-conflict period is arbitrarily defined as the 10 years following the end of a conflict (Nkurunziza, 2008). According to the FAO (2005), the post-conflict period is considered to end at the point when basic legal frameworks and land administration institutions exist and there is rea-sonable capacity within the country for the development of policies. Focus on the post-conflict period varies from emergency relief, development of policies to implementation of policies as shown in the Figure 10.1.

In practice, these activities are likely to overlap and do not occur in a straightforward process. The development and implementation of policy will depend on the extent to which peace can be maintained. Emergency activities focus on establishing basic governance and proving humanitarian services. Activities for the development of policy focus on the planning of necessary administrative and physical infrastructures and these occur when emergency activities are being concluded and when more concrete plans can be made for the future. Activities for the implementation of policy tend to be concentrated later in the post-conflict period when there is a return to a reasonable degree of political and social stability (FAO, 2005).

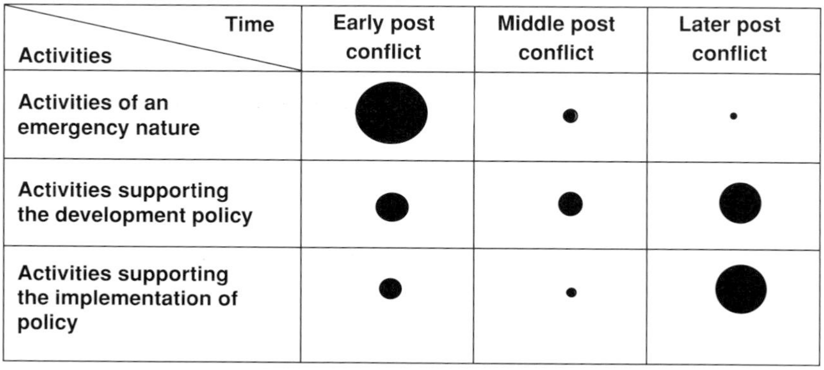

Activities \ Time	Early post conflict	Middle post conflict	Later post conflict
Activities of an emergency nature	⬤	●	·
Activities supporting the development policy	●	●	⬤
Activities supporting the implementation of policy	●	·	⬤

Figure 10.1   Nature of activities in post-conflict period

10.5   Post-conflict reconstruction

Post-conflict reconstruction is seen as a long term, comprehensive project that requires unremitting effort by the governing regime, ex-rivals and the international community (Misra, 2004). Post-conflict reconstruction remains a priority in parts of Africa, the Middle East and South Asia (Fengler *et al.*, 2008).

Post-conflict reconstruction does not refer only to the reconstruction of 'physical infrastructure', nor does it necessarily signify a rebuilding of the socio-economic framework that existed before the onset of conflict (World Bank, 1998). If conflicts continue for many years, transformation of a society and a return to the past may not be possible or desirable. What is needed is the reconstruction of the enabling conditions for a functioning peacetime society. The following section provides details of the key activities of post-conflict reconstruction. Both the extent of conflict worldwide and the impact of modern intrastate conflicts necessitate new strategies and modalities for post-conflict reconstruction.

10.5.1   *Activities of post-conflict reconstruction*

Reconstruction entails the full range of integrated activities and processes that have to be initiated in order to reactivate the development process that has been disrupted by the conflict. The World Bank (1998) identified the following priority activities that need to take place during post-conflict reconstruction:

- Jump starting the economy through investment in key production sectors and supporting the conditions for resumption of trade, savings and domestic and foreign investment, including macroeconomic stabilisation, rehabilitation of financial institutions, and restoration of appropriate legal and regulatory frameworks.
- Reconstructing the framework of governance: strengthening government institutions, including the capacity for resource mobilisation and fiscal management; the restoration of law and order and the organisation of a civil society.
- Repairing important physical infrastructure including key transport, communications and utility networks.
- Rebuilding and maintaining key social infrastructure, i.e. education and health, including the financing and recurrent costs.
- Targeted assistance to those affected by war i.e. reintegration of displaced populations, demobilisation and reintegration of ex-combatants; revitalisation of the local communities most disrupted by conflict through support (such as credit lines) to subsistence agriculture, microenterprises, and the like; and support for vulnerable groups (such as female-headed households).

- Landmine action programmes, including mine surveys, and de-mining of key infrastructures as part of comprehensive development strategies for returning populations living in mine-polluted areas to normal life.
- Financial normalisation: planning calculation of arrears, rescheduling of debt, and the longer-term path to financial normalisation.

According to Rugumamu and Gbla (2003), post-conflict reconstruction typically involves the repair and reconstruction of physical and economic infrastructure and rebuilding weakened institutions. Apart from that, critical interventions undertaken during the post-conflict reconstruction include reviving the economy, reconstructing the framework for democratic governance, rebuilding and maintaining key social infrastructure and planning for financial normalisation. Among these physical reconstruction is arguably accepted as the most visible indicator of economic reconstruction (Nkurunziza, 2008). Furthermore, physical reconstruction and transitional initiatives are important for building trust and sustaining confidence in a war-weary population and for private investors both domestic and foreign. Therefore, it is necessary to undertake physical reconstruction and transitional initiatives before the larger economic reform programme begins.

10.5.2   *Post-natural disaster vs post-conflict reconstruction*

There are some similarities and differences between post-natural disaster reconstruction and post-conflict reconstruction. Both typically involve the repair and reconstruction of physical and economic infrastructure. However, in post-disaster situations the government system typically functions regularly while in a post-conflict reconstruction the government system is often weakened by prolonged struggle. Therefore, post-conflict reconstruction entails rebuilding weakened institutions. Unlike post-natural disaster reconstruction, post-conflict reconstruction assistance often operates amid social tensions and suspicions between key actors within the country, which can and does influence relations among the involved international parties (Rugumamu and Gbla, 2003). Civil war alters both the level and structure of economic activity in ways that persist beyond the war (Rugumamu and Gbla, 2003). While post-disaster reconstruction has a more linear path, in post-conflict reconstruction there is a significant likelihood of reverting to conflict. Finally, natural disasters are typically unforeseen and post-conflict reconstruction is often signalled by a peace agreement and offers some lead-in time (Fengler *et al.*, 2008).

10.5.3   *Post-conflict reconstruction challenges and dilemmas*

The impact of violent conflict poses various post-conflict reconstruction challenges and poverty is one of them. Prolonged conflicts have inflicted severe wounds in war-affected countries making them the least developed countries in the world. Most countries at this stage have devastated or at least severely distorted economies (Castillo, 2001). For example, in 1995, Afghanistan was ranked 170 out of 174 in the United Nations Development Programme (UNDP) Human Development Index, making it one of the least developed countries in the world (Barakath, 2002). When countries are poor, the scale of the challenges to be addressed is high. As conflict disturbs the production and distribution of food supplies, severe food shortages typically exist in affected countries. Food insecurity is exacerbated due to the difficulties in getting access to land, seeds, agricultural implements, livestock and capital. Hunger can reduce peoples' ability to work and weakens them physically (FAO, 2005).

High numbers of refugees and IDPs also add to the enormity of reconstruction (Barakath, 2002). Conflict results in a large number of displaced people. Refugees are people who flee their homes for the safety of another country and IDPs are people who flee from violence but remain in their own country. While refugees have the protection of international laws, IDPs are subject to the laws of the country. For many who return after a conflict the biggest concern is the availability of a sustainable livelihood. Employment opportunities are typically rare in rural areas, but land may not be available for agricultural purposes because of mines and basic services may not be accessible (FAO, 2005).

Furthermore, conflict affects men and women differently. Men are more likely to have been recruited, either voluntarily or forcibly, by one of the opposing forces and may have been killed or captured (FAO, 2005). As a result women face the increased responsibility of looking after children, the elderly and themselves. Conflict may result in significant changes in people's values and expectations. For instance, conflicts shift power from men to women through the empowerment of women, and from one generation to another as young people with more formal education than their elders are recruited by NGOs. Further exposure to urban life may reshape behaviour and attitudes as amenities such as schools, clinics, electricity and a greater variety of goods are experienced. A reluctance to return to rural areas lacking these amenities is evident and people who do return bring these values and perceptions with them (FAO, 2005).

Reduced security is a prevalent feature during post-conflict reconstruction. There is unlikely to be a smooth transition from war to peace (Goodhand, 2002). This impedes the restoration of dignity, trust and faith during reconstruction. The widespread availability of cheap weapons is one of the legacies of such struggles. Many survivors of conflict may be exposed to murder and rape, and experience displacement, separation from family and friends, loss of employment, mental illnesses and inability to cope effectively with the task of rebuilding their livelihood. The primary challenge of post-conflict transition is to prevent the recurrence of hostilities or to make the transition irreversible. This entails the complex political task of addressing the root causes of the conflict. Stabilisation of the economy through investment, employment generation and low inflation can facilitate the complex process of reintegrating former combatants, returnees and displaced persons into society and into productive activities. Reintegration is a condition for the consolidation of peace

(Castillo, 2001). Of course, an unsettled military and political environment can further exacerbate the situation.

Most conflict exists in developing countries burdened by debt that tend to depend heavily on the international community for official aid flows, mostly in the form of grants (Castillo, 2001; FAO, 2005). Obtaining international support for long-term reconstruction and development is another challenge faced by these countries. The levels of international political support and commitment will not continue forever and with other topical issues, political agendas and administration changes, funding pledges may be forgotten, either for political or budgetary reasons. On the other hand the immediate post-conflict environment is not usually conducive to investment by economic agents and it takes time to allocate savings to productive investment (Nkurunziza, 2008). Most post-conflict projects in Africa and other parts of the world continue to draw upon a limited pool of foreign aid (Grant, 2005).

Weakened government capacity and finding skilled personnel is another issue of reconstruction. The power of government to operate in parts of the country may have been limited or blocked during a conflict. Other issues which affect reconstruction are lack of transparency, poor governance, corrupt legal and judicial systems, absence of property rights, lack of independent central banks, weak tax and customs administration and public expenditure management, etc. (Castillo, 2001). A lack of accounting procedures may result in widespread mismanagement of money and create opportunities for corruption (FAO, 2005). Shaping the governance system through establishing the rule of law by strengthening the justice system and law enforcement agencies along with administrative capacity building is another crucial process of reconstruction. At the end of a conflict, the government at central and local level may have little or no capacity to manage the process of reconstruction. Most government agencies will lack experienced staff and buildings and equipment may have been destroyed. Planning and management of major infrastructure projects are found to be problematic even under stable conditions and the limited capacity found in post-conflict settings magnifies these problems. The inability of these countries to make managed and effective decisions in the reconstruction process leads to less effective absorption of post-war funding. However, large sums of investment will be required at a later stage as capacity and development priorities are established (Barakath, 2002).

In post-conflict reconstruction there is no right or perfect formula. The best that the actors in the situation can hope to do is steer a course that does the least harm, by learning from past efforts. Reconstruction must necessarily be a continuous and prolonged process of negotiation between diverse internal interest groups and other political actors (Barakath, 2002). If reconstruction is to be successful and effective, it requires long-term political commitment to the process from both international and local actors (Barakath and Chard, 2002).

Post-conflict situations cannot be clearly defined. According to Barakath (2002), there are two schools of thought concerning the timing of reconstruction and development activities. The first is that peace is a precondition for reconstruction and development. The second is that, through the initiation of reconstruction and development activities at an appropriate time during the conflict, the seeds of long-term recovery will be sown. Evidence from worldwide experience supports the second theory (Barakath, 2002). Therefore rebuilding should be approached as a national project, accessible to public discussion and participation, set to heal inter-community wounds and restore dignity, trust and faith in the system. The process of reconstruction should be carefully planned and integrated into wider aspects involving all parties to the conflict in order to make it more successful. On the contrary, it is found that most reconstruction efforts are initiated as part of a rushed, knee-jerk reaction by external actors. Due to fast-track processes, consultation of partners leaves incomplete those systems that later lead to unsuccessful reconstruction (El-Masri and Kellett, 2001; Barakath, 2002; Barakath *et al.*, 2004). As an example, in Afghanistan, most agencies and organisations submitted plans, budgets and proposals based on desk studies and inadequate assessments due to the fear of being marginalised by the assumed and imposed urgency of the process (Barakath, 2002).

An integrated and holistic approach is needed instead of a fragmented contradictory approach. For example, in Afghanistan, food assistance (relief aid) and food security (agricultural and rural economic recovery) are still regarded as separate sectors despite the long history of misdirected food aid seriously damaging post-war agricultural recovery (Barakath, 2002). Also, the plans of these two sectors similarly make no connection with other sectors, such as mine clearance and employment, which have significant repercussions for the recovery of agricultural land and livelihoods.

Simply repairing damaged infrastructure facilities seldom results in sustainable solutions to a country's infrastructure needs. Achieving a sustainable infrastructure will assist in the attainment of a durable peace only if it contributes to the re-establishment of livelihoods and the connection of disparate ethnic and religious groups. Conversely, the degradation of such infrastructure will limit the potential for local communities to become effective and so reinforce existing divisions. Reconstruction at village and community level and the community's direct influence over these is lacking. Centrally controlled, top-down approaches to reconstruction after war have failed in many parts of the world (El-Masri and Kellett, 2001; Barakath *et al.*, 2004). However, there is a risk of decentralising, in the context of an insecure environment, as it cannot be assumed that there will be high levels of acceptance and adherence to government codes and guidelines (Barakath, 2002). The consequences of conflict can radically alter the demographic, political and economical structure of a country. These changes, brought about by the conflict, need to be identified and incorporated in any rehabilitation strategy (El-Masri and Kellett, 2001; Barakath, 2002).

Even though international interventions are necessary in reconstruction, their existence creates problems when policy is driven by outsiders without national ownership (FAO, 2005). According to Barakath (2002), the first step in the initiation of any reconstruction and development process is recognition of people's resilience and abilities to survive the hardship of conflict by employing various coping mechanisms. These may not be clear to outsiders and as such cannot easily be understood and valued. Instead of pledged money with 'blueprint' solutions by outsiders, countries should be given time to learn lessons from the past and to make decisions for themselves within time frames that are suitable for them. Competing regional and international interests influence the process and outcome of reconstruction (Barakath, 2002).

10.6   Summary

This chapter has provided an introduction to conflict, post conflict and post-conflict reconstruction and has identified the associated post-conflict reconstruction challenges. Differences in interests and opinions between groups and countries lead to conflicts and how such differences are expressed and managed determines if the conflicts will manifest themselves primarily in political or violent ways. Most recent conflicts have been intrastate violent conflicts that last over a longer period, target civilians and predominantly exist in developing countries. Conflicts are linked with competition for natural resources, land, identity defence, external factors, poverty, lack of education and poor governance. Key in the prevention of conflicts is understanding their root causes and consequences. Loss of life, displacement of people, destruction of physical structures, economic ruin, disruption of society, weakened government institutions and regional effects are the major direct and indirect impacts of prolonged violent conflicts. The conflict circle identifies the phases of conflict management as provention, prevention, intervention, peace settlement, resolution and post-conflict reconstruction. These efforts aim to reduce or avoid the potential harm from conflicts and consequent violence.

Post conflict refers to the period following the end of a conflict. However, the terms post conflict and post-conflict reconstruction are basically associated with violent conflicts or wars. Post-conflict reconstruction is an emerging area of concern as it remains a priority in most parts of the world including Africa, the Middle Eastern and South Africa. Post-conflict reconstruction is an umbrella term which covers repair and reconstruction of physical and economic infrastructure, rebuilding and maintaining key social infrastructure, rebuilding weakened institutions, reviving the economy, reconstructing the framework for democratic governance, and planning for financial normalisation. Physical reconstruction is arguably accepted as the most visible indicator of economic reconstruction, which can restore trust, dignity and confidence in a war-weary population and in private investors both domestic and foreign. However, the direct and indirect impact of violent conflicts account for various post-conflict challenges including poverty, reduced security, refugees and IDPs, weakened government capacity, changed expectations and values, excessive debt, etc. Therefore strategies must be developed to overcome these challenges and make post-conflict reconstruction more successful.

References

Anand, P.B. (2005) *Getting Infrastructure Priorities Right in Post-Conflict Reconstruction.* Helsinki: UNU-WIDER.

Barakath, S. (2002) Setting the scene for Afghanistan's reconstruction: the challenges and critical dilemmas. *Third World Quarterly*, **23**, 801–816.

Barakath, S. (2003) *Housing Reconstruction after Conflict and Disaster.* London: Humanitarian Practice Network at Overseas Development Institute.

Barakath, S. and Chard, M. (2002) Theories, rhetoric and practice: recovering the capacities of war-torn countries. *Third World Quarterly*, **23**, 817–835.

Barakath, S., Elkhalout, G. and Jacoby, T. (2004) The reconstruction of housing in Palestine 1993–2000: a case study from the Gaza strip. *Housing Studies*, **19**, 175–192.

Cain, A. (2007) Housing microfinance in post-conflict Angola: overcoming socio-economic exclusion through land tenure and access to credit. *Environment and Urbanisation*, **19**, 361–390.

Castillo, G.D. (2001) Post-conflict reconstruction and the challenges to international organisations: the case of El Salvador. *World Development*, **29**, 1967–1985.

Cuny, F.C. and Tanner, V. (1995) Working with communities to reduce levels of conflict: 'spot reconstruction'. *Disaster Prevention and Management*, **4**, 12–20.

El-Masri, S. and Kellett, P. (2001) Post war reconstruction participatory approach to rebuilding villages of Lebanon: a case study of Al-Burjain. *Habitat International*, **25**, 535–557.

Elbadawi, I.A. (2008) Post conflict transitions: an overview. *The World Bank Economic Review*, **22**, 1–7.

Eriksson, M., Wallensteen, P. and Sollenberg, M. (2003) Armed conflicts, 1989–2002. *Journal of Peace Research*, **40**, 593–607.

Fearon, J.D., Humphreys, M. and Weinstein, J.M. (2009) Can development aid contribute to social cohesion after civil war? Evidence from a field experiment in post conflict Liberia. *American Economic Review: Papers and Proceedings*, **99**, 287–291.

Fengler, W., Ihsan, A. and Kaiser, K. (2008) *Managing Post-Disaster Reconstruction Finance.* Washington, DC: World Bank.

Food and Agriculture Organisation (FAO) (2005) *Access to Rural Land and Land Administration after Violent Conflicts.* Rome: FAO.

François, M. and Sud, I. (2006) Promoting stability and development in fragile and failed states. *Development Policy Review*, **24**, 141–160.

Gleditsch, N.P., Wallensteen, P., Eriksson, M., Sollenberg, M. & Strand, H. (2002) Armed conflict 1946-2001: a new dataset. *Journal of Peace Research*, **39**, 615–637.

Goodhand, J. (2002) Aiding violence or building peace? The role of international aid in Afghanistan. *Third World Quarterly*, **23**, 837–859.

Grant, J.A. (2005) Diamonds, foreign aid and the uncertain prospects for post conflict reconstruction in Sierra Leone. *The Round Table*, **94**, 443–457.

Hettne, B. and Soderbraum, F. (2005) Intervening in complex humanitarian emergencies: the role of regional cooperation. *The European Journal of Development Research*, **17**, 449–461.

International Crisis Group (2009) *Development Assistance and Conflict in Sri Lanka: Lessons from the Eastern Province.* Brussels: International Crisis Group.

Kibreab, G. (2002) When refugees come home: the relationship between stayees and returnees in post-conflict Eritrea. *Journal of Contemporary Studies*, **20**, 53–80.

Krause, K. and Jutersonke, O. (2005) Peace, security and development in post conflict environments. *Security Dialogue*, **36**, 447–462.

Misra, A. (2004) Rain on a parched land: reconstructing a post-conflict Sri Lanka. *International Peacekeeping*, **11**, 271–288.

Nagai, M., Abraham, S., Okamoto, M., Kita, E. and Aoyama, A. (2007) Reconstruction of health service systems in the post-conflict northern province in Sri Lanka. *Health Policy*, **83**, 84–93.

Nkurunziza, J.D. (2008) Civil War and Post Conflict Reconstruction in Africa. United Nations Conference on Trade and Development, Geneva, Switzerland,

Patrick, S. (2006) Weak states and global threats: facts or fiction? *Washington Quarterly*, **29**, 27–53.

Rugumamu, S. and Gbla, O. (2003) *Studies in Reconstruction and Capacity Building in Post-Conflict Countries in Africa*. Harare: The African Capacity Building Foundation.

Senanayake, D.R. (2009) *Sri Lanka's Post Conflict Transition: Reconstruction, Reconciliation and Aid Effectiveness*. Singapore: Institute of South Asian Studies.

Stefansson, A.H. (2006) Homes in the making: property restitution, refugee return and senses of belonging in a post-war Bosnian town. *International Migration*, **44**, 115–137.

Wakeman, R. (1999) Reconstruction and the self-help housing movement: the French movement. *Housing Studies*, **14**, 355–366.

Wegelin, E.A. (2005) Recent housing resettlement and reconstruction in south-eastern Europe. *Global Urban Development Magazine*, **1**(1).

World Bank (1998) *Post-Conflict Reconstruction*. Washington, DC: World Bank.

Zenkevicius, G. (2007) Post-conflict reconstruction: rebuilding Afghanistan – is that post-conflict reconstruction? *Baltic Security & Defence Review*, **9**, 28–56.

11 Private Construction Sector Engagement in Post-Disaster Reconstruction

Richard Sutton and Richard Haigh

11.1 Introduction

This chapter considers the need to engage the private construction sector in order to successfully deliver reconstruction projects after a major disaster. By most definitions, a *disaster* represents a situation in which an affected community is unable to cope and recover from a hazard event without external assistance. Typically, the resources required to deliver effective reconstruction are considerable. While academics and practitioners alike generally acknowledge that the capacity for reconstruction should be endogenous and locally led, it is unlikely that an affected community would be able to manage and deliver major building and reconstruction projects without the assistance of private, usually profit-orientated firms. Even national and international non-governmental organisations (NGOs) that support communities in relief and recovery efforts are usually without the capacity and expertise to deliver the type of complex and capital intensive construction projects that is frequently required; instead they elect to procure major work from the private sector.

This chapter begins by investigating the nature and extent of the relationship between the private sector and post-disaster reconstruction efforts. The results of an empirical study are presented that considers what strategies and mechanisms might be developed to encourage an appropriate level of participation by the private sector.

11.2 Challenges in post-disaster reconstruction

Disaster recovery presents an opportunity to make situations better than they were before; built environment professionals are needed to achieve this aim. Mudge (2008) argues that immediate solutions are not sustainable. In support

Post-Disaster Reconstruction of the Built Environment: Rebuilding for Resilience, First Edition.
Edited by Dilanthi Amaratunga and Richard Haigh.
© 2011 Blackwell Publishing Ltd. Published 2011 by Blackwell Publishing Ltd.

of this, Amin and Goldstein (2008) note that humanitarian organisations require a global vision with an agreed shared response. A resource gap currently exists between emergency relief and longer-term recovery. The effectiveness of long-term reconstruction is currently constrained by the lack of planning, co-ordinated management and targeted funding. Furthermore, 'stable and secure post-disaster recovery is threatened by institutional constraints, gaps in communication, a lack of access to professional skills and knowledge to support local effort' (Lloyd-Jones, 2006). The results of these capacity constraints are evidenced in disasters around the world, from poor building standards exposed by the tsunami across South Asia in 2004 (Lloyd-Jones, 2006) to the high number of child casualties in China's earthquake in 2008 that have been attributed to substandard school buildings (BBC News, 2008). Development and strengthening of institutions, mechanisms and capacities at all levels is required (United Nations International Strategy for Disaster Reduction, 2005). Indeed, response agencies usually have only relatively small numbers of appropriately experienced personnel who can operate in an emergency at an international level (Telford and Cosgrove, 2006).

11.3   What is the role of the private sector?

According to Twigg (2001) there has been little private sector involvement in disaster risk reduction. However, the value of building partnerships with the private sector is being realised (Binder and Witte, 2007). Twigg adds that the private sector is often willing to provide technical expertise and volunteer labour. The reason for this input might, as Yamamoto and Ashizawa (1999) suggest, be because, 'corporations ... are under pressure to contribute more positively to the overall interests of communities ... and also to improve corporate governance'. Twigg (2001) concludes that a vehicle is needed to overcome the segregated nature between agencies. In summary, the role of business is becoming more prominent, but, as Binder and Witte (2007) suggest, 'it remains a niche phenomenon [and] more research is required'.

Gunewardena *et al.* (2008) concur and add that national and local governments need to embrace the private sector in order to help to tackle some of the challenges associated with reconstruction. The business community has a growing role in generating employment and wealth through trade, investment and finance. Doing this means exploring business models to work towards more sustainable urbanisation.

The business sector can be regarded as a valuable partner because it can bring to the partnership a sense of accountability and result-orientated attitudes (Yamamoto and Ashizawa, 1999). Businesses are also considered to be better at adapting to market conditions than government and international development agencies (Mehta, 2008). Although businesses focus on economic growth and profit, while humanitarian organisations work to promote peace and security, reduce poverty and ensure human rights, businesses and humanitarian organisations need each other. As the United Nations Secretary General Ban Ki Moon pointed out, 'the work of the United Nations can be viewed as seeking to create an enabling environment within which business can thrive' (UN-HABITAT, 2007). Similarly, the International Federation of Red Cross and Red Crescent Societies (IFRC) (2001) acknowledge that relief merely buys the time to make the right decisions. But relief is just the beginning of a bigger commitment. Prevention is not only more humane than cure; it is also more cost effective (Lim, 2003).

It would appear that humanitarian operations and the private sector should not be viewed as mutually exclusive. Indeed, partnerships with business can also bring needed technical expertise. Partner selection should be based on a match between identified gaps, the skills and capacities on offer and the ability of the agency to manage the partnership (Binder and Witte, 2007). Private sector partnerships are critical to practical sustainability in cities (UN-HABITAT, 2007).

The built environment professions have invaluable expertise and a key role to play (Lloyd-Jones, 2006), doing so by working with the established community of humanitarian organisations. Humanitarian organisations have decades of experience on the front lines of disasters and in long-term development initiatives (Thomas and Fritz, 2008). It is important to ensure that there are sufficient trained personnel in appropriate institutions (Amin and Goldstein, 2008).

11.4   Business and humanitarian collaborations

Businesses from all sectors have long provided services not available from other recovery stakeholders. Corporate support to humanitarian efforts has been steadily increasing (Binder and Witte, 2007). There are considered to be four generic categories of business/humanitarian collaborations: philanthropic, strategic, commercial, and political. Philanthropic collaborations advance social welfare by facilitating the delivery of humanitarian organisations' services. Strategic collaborations realise exclusive benefits for the firm, while advancing social welfare through the activities of the humanitarian organisation. Commercial collaborations increase revenues for both the company and the humanitarian organisation. Finally, political collaborations aim to reproduce or change institutional arrangements (Powell and Steinberg, 2006).

Traditionally, companies have occupied a secondary place in humanitarian efforts, providing services to dominant humanitarian organisations (Binder and Witte, 2007). This is due to the concern that business is an unreliable development partner and is not conducive to long range planning (Warhurst, 2005). New meta-initiatives are, however, beginning to emerge. These involve companies and other actors joining forces to enhance coordination in humanitarian work and share lessons learned (Binder and Witte, 2007). Many companies recognise that becoming involved in the management and implementation of humanitarian work can be difficult and risky. Companies are also frequently interested in building relationships with well-known and reputable humanitarian organisations in order to profit in terms of their image. An example approach is the Business Partnership for Sustainable Urbanization (BPSU) which is a strategic alliance of business companies, working towards sustainable urbanisation (UN-HABITAT, 2007).

It is important to note that single business engagement in humanitarian work is also beginning to develop. This is where initiatives are launched and implemented by a single corporation in response to a specific crisis. Binder and Witte (2007) note that many companies now recognise that, to make a difference, long-term engagement is critical.

Some other initiatives and examples of how business sectors are engaging with humanitarian operations are the UNHCR's Council of Business Leaders. This consists of top executives from five major corporations working together with the refugee agency to improve opportunities for refugees: Manpower, Microsoft, Nestlé, Nike and Pricewaterhouse Coopers. Pharmaceutical and medical support has also been given by companies such as Pfizer and Merck and various logistics and transport companies have also provided assets and expertise to help deliver support (Warhurst, 2005). In Tanzania, Ericsson has worked with the United Nations Development Programme's (UNDP) Growing Sustainable Business (GSB) initiative to expand its telecommunications infrastructure to rural areas. The Business Alliance for Food Fortification (BAFF), co-chaired by Unilever, Danone and Coca-Cola, has encouraged business leadership in eliminating vitamin and mineral deficiencies (Mehta, 2008).

According to the Disaster Resource Network (DRN) (2009), three sectors appear particularly prominent – logistics, information technology and telecommunications. This is likely due to their considered relevance of contributions. However, as Lloyd-Jones (2006) notes, construction skills can add value to disaster management and recovery by:

> 'assessing disaster related damage, land surveying, GIS and rapid mapping of disaster impacts and risks, monitoring funding, valuation, cost planning and spending priorities; development finance, procurement and project management, sourcing construction materials and equipment, building quality audits pre- and post-disaster, aiding logistical planning, aiding local government land administration, cadastral mapping, knowledge of land and property legislation, knowledge of regulatory frameworks and ways they could be improved, training and knowledge transfer, disaster risk management, links with other built environment professions; inter-disciplinary and team working, contacts with local businesses and industry, knowledge of appropriate forms of disaster-resistant construction and engineering'.

11.5   Corporate social responsibility

Businesses are considered to be most successful when they meet the expectations of their wider stakeholders within society (Warhurst, 2005); this is the concept of corporate social responsibility (CSR). In the construction sector there are six generic CSR issues, namely those relating to the environment, health and safety, human resources, supply chain management, customers and communities, and governance and business ethics (Jones *et al.*, 2006). It is the concept of business ethics which is deemed to motivate the private construction sector to contribute to recovery efforts. Twigg (2001) purports that CSR initiatives are more likely to develop where there is already a tradition of corporate philanthropy within a company. However, Mills (2008) clarifies this by adding that all construction companies have the power and technical knowledge to contribute.

There are many drivers for business embracing this broader view of CSR, including: organisational values, stakeholders, accountability, risk and opportunities and organisational continuity. There are currently growing stakeholder concerns about business ethics, particularly in developing countries. Most corporations want to get involved in well-publicised emergencies; however, they cannot be depended on to help the most needy in less 'popular' disasters (Thomas and Fritz, 2008). Furthermore, risk and opportunity are changing concepts and non-financial performance (social, ethical and environmental) now contributes to the intangible value of a company and is considered bottom line, relevant and material. In addition to these motivations, businesses can also use supporting initiatives as a way of gathering business intelligence and motivating staff (Binder and Witte, 2007). For example, *Building Magazine* has compared the experiences of young professionals working on similar projects in the UK and abroad, and found that working overseas brings job satisfaction and excitement (Leftly, 2008).

An alternative argument for the private sector supporting post-disaster reconstruction is that there are profits to be made. Overwhelmingly, the profits to be made from disaster capitalism are in the contracts that are available for rehabilitation and reconstruction. Most of the contracts are issued by governments, often with normal tendering processes short-circuited, giving immense opportunities for windfall profits. Neo-liberal disaster recovery strategies rest on the efficacy of the market resolving presumed economic inefficiencies and governance problems. It proposes macroeconomic growth as the route to poverty reduction (Gunewardena *et al.*, 2008). However, contributing goods to traditional humanitarian actors may simply hook the receiver into using a specific proprietary technology in which they will later have to invest further, thus benefiting the business but not the greater good. For example, this was a sticking point in negotiations between Ericsson and the UN (Binder and Witte, 2007).

The concern for accountability is a major constraint on private construction sector support (Lloyd-Jones, 2006). The most significant challenge relating to business engagement is business' short-term emphasis on quarterly profit targets and cost cutting. It is challenging to make it attractive to invest in preventing disasters that might not happen, even if retrospectively the costs of prevention can be demonstrated to have significantly outweighed the damage costs accrued (Warhurst, 2005).

The nature of disaster recovery and prevention necessitates the adoption of a more strategic and collaborative approach to the issues, plus the need for strong governance frameworks that put in place rules of engagement, for the development, implementation and termination of the partnership agreements.

11.6   Encouraging private sector participation

Large corporations are not going to solve the economic problems in developing countries by themselves (Mehta, 2008); working with multi-disciplinary teams and with local intermediaries is essential (Lloyd-Jones, 2006). The international humanitarian community needs to re-orientate itself from supplying aid, to supporting community recovery. As Telford and Cosgrove (2006) conclude, all recovery stakeholders should strive to improve the coherence between themselves and other actors in the international disaster response system. Since aid alone will never be enough, humanitarian organisations have a moral duty to engage with donors, host governments and the public, in an honest dialogue about the limits of humanitarian action (IFRC, 2001). Country leadership is also fundamental to effective disaster response (Amin and Goldstein, 2008). Central planning and coordination needs to be backed by mechanisms that allow funds, information and resources to flow down to the local level. Linking relief, rehabilitation and development is being seen as an increasingly important element of international development policy (Lloyd-Jones, 2006). An option could be for humanitarian organisations to identify relevant company core competences in order to best understand how different businesses can contribute most effectively (Warhurst, 2005).

Of increasing prevalence are international humanitarian organisations forging partnerships with local NGOs, companies, business coalitions, governments, relief NGOs and civil society organisations. International humanitarian organisations are also well aware that the private sector is a vital part that must be engaged, if the world's cities are to achieve sustainability (UNDP, 2009). There is, however, a concern that unnecessary duplication of relationships will undermine the capacity of external partners to engage constructively. As Campher (2005) suggests, this could be overcome by developing industry, or country-based, joint disaster response teams. There are two reasons why companies may prefer to act through partnerships rather than alone. Firstly, many companies recognise that becoming involved in the management and implementation of humanitarian work can be risky. Secondly, companies are frequently interested in building relationships with well-known and reputable inter-governmental humanitarian organisations in order to profit in terms of their image (Binder and Witte, 2007). Private sector organisations must make a decision whether they want to foster a deep partnership with a single humanitarian organisation, or to pool their resources with other companies in order to extend their impact to more humanitarian organisations (Thomas and Fritz, 2008).

Corporations and humanitarian organisations naturally have different perspectives on diverse issues (Yamamoto and Ashizawa, 1999). Large companies are most likely to be more geared towards large-scale operations, while smaller companies contribute through the rebuilding of the local economy. The company's core business and its resources are at the heart of the matter (Campher, 2005). Today, the UN, for example, looks for partnerships with private sector organisations that display corporate responsibility in the community, make a positive contribution to the urban environment, have a record of socially-responsive behaviour, and have responsive labour and environmental practices. The UNDP forges partnerships across diverse spheres of influence, from national, municipal and local governing bodies to non-governmental and civil society organisations (CSOs), including grassroots coalitions, faith-based groups, academia, as well as the private sector and international donors (UNDP, 2009). The purpose of these partnerships is to achieve a synergised, coordinated and effective recovery effort.

Recently, ties between companies and humanitarian organisations have expanded (Powell and Steinberg, 2006). A post-disaster reconstruction example is Bovis Lend Lease's partnership with the UN-HABITAT's reconstruction efforts in Sri Lanka following the tsunami. The UN-HABITAT's objectives included coordination of projects, project planning, construction management, logistics management and quality control of the reconstruction. The programme brought together NGOs such as the IFRC, Sri Lanka Red Cross Society, together with the UN-HABITAT agency, a private sector partner, the government of Sri Lanka and a community development council (Mudge, 2008). Bovis Lend Lease provided a key support role in assisting the UN in gaining additional funding support from the IFRC and other NGOs (MacFadyen, 2005).

Some larger scale partnership initiatives are also emerging within this sector. These include: The Partnership for Disaster Response by the 'Business Roundtable', 'The Growing Sustainable Business (GSB)' Initiative and the 'Disaster Resource Network', a creation of the World Economic Forum. The Business Roundtable is an association of approximately 160 CEOs of leading United States corporations; its goal is to coordinate private sector assistance. The Business Roundtable matches needs of international agencies to capabilities of the private sector. One example is the partnership between Dutch logistics giant TNT and the World Food Programme (Thomas and Fritz, 2008). The overall objective of the 'GSB' initiative is to contribute to poverty reduction and sustainable development, by promoting and facilitating business and investments by the private sector. The Growing Sustainable Business initiative leverages the UNDP's unique capacity to create a neutral 'space' at country level, where information can be shared, and appropriate local partners brought together to attack a specific problem. Key organisations involved at the global level include: UNDP, UN Global Compact Office, UNGC core agencies (ILO, UNEP, UNIDO, UNEP and OHCHR), private companies (Shell, ABB, Ericsson, Unilever, EDF and Total) (UNDP, 2004). Finally, the philosophy of the

'Disaster Resource Network' is specifically to make it easier for businesses in the engineering and the construction industry to donate expertise, services, product or financial support to disaster relief and recovery operations. The Industry Partnership Programme is a forum which hopes to increase the participation of global engineering and construction companies in post-disaster reconstruction (Disaster Resource Network, 2009). In summary, there is much scope for increased private construction sector support; the question remains as to which strategies prove best for all post-disaster stakeholders.

11.7   Integrating private construction sector support for post-disaster reconstruction

The remainder of this chapter presents the results of an empirical study aimed at developing a model arrangement and set of recommendations in order to better integrate the private construction sector into post-disaster reconstruction. The specific objectives of this empirical study were to: investigate the nature of the resource gap for post-disaster reconstruction; examine what support the private sector is *currently* providing to post-disaster reconstruction; explore the rationale behind the private construction sector supporting post-disaster reconstruction; and, to examine appropriate strategies to encourage an appropriate level of participation by the private sector.

11.7.1   *Methodology*

Semi-structured interviews were carried out with members of stakeholder organisations involved with post-disaster reconstruction. These were utilised as there were, 'a number of specific topics around which to build the interview' and the interview 'collected factual information as well as opinions . . . [and was] . . . designed to elicit answers pertinent to the research [propositions]' (Naoum, 2007).

A non-random, selected sampling design was used as it enabled the possibility of choosing respondents with desired specific characteristics (Naoum, 2007). This form of sampling is otherwise known as theoretical sampling, as the interviewees 'were chosen to help the researcher formulate theory' (Robson, 2002). The primary consideration with this form of sampling was the author's judgement, as to who can best provide information, in order to achieve the objectives of the study (Kumar, 2005).

The author's sample interviewees were selected key/senior level figures from the various stakeholder organisations, who could give relevant and strategic answers to questions on the research topic. It must be noted that, in qualitative research, the size of the 'sampling has little significance as the main aim . . . is to explore or describe diversity in a situation' (Kumar, 2005).

Eighteen interviews were conducted. The respondents represented private construction sector multinational corporations, members of a post-disaster reconstruction programme, and representatives from construction industry professional organisations.

Exploratory data analysis was used to interpret the results; the aim was to generate a theory to explain what was central to the interview data (Robson, 2002), as it captured emergent phenomena (Holstein and Gubrium, 1995). The authors adopted an approach of recording and describing the meaning behind the interviewees' responses (Weinberg, 2002), via interpretation and narrative analysis (Kvale, 1996). This process was conducted by 'systematically grouping and summarising the descriptions, and providing a coherent organising framework that encapsulated . . . [what] . . . respondents portrayed' (Kvale, 1996).

11.7.2   *Results*

The majority of interviewees identified that natural disaster losses are expected to increase over the next 50 years. This demonstrated that key people have an awareness of future disaster potential as outlined by Kreimer *et al.* (2003). Arguments discussed in the earlier part of this chapter, regarding the need for the utilisation of construction resources to fill the resource gap, were supported by all respondents. It was commented that construction expertise is required for the restoration of essential infrastructure. It was deemed that although the country in which a disaster occurs may have the capacity to commit resources, in many cases the scale of reconstruction will be beyond the capacity and capability of the affected country. Temporary support, logistics, project management, engineering and technical expertise could all be provided by the private sector. In addition to this, key delivery people are needed, alongside people who can harness the local population and communities. These points confirm Lloyd-Jones' (2006) propositions discussed earlier. However, an interesting argument was that major construction organisations have not conducted sufficient internal knowledge management processes to be aware of the capabilities of their organisation. Therefore, needs and capabilities need to be stated from both sides.

With regards to specialist skills, there is a need for project management and engineering expertise in delivering projects and re-establishing infrastructure, power and sanitation. It was also held that construction professionals would be in touch with labour and resource markets. These points further sustain Telford and Cosgrove (2006) in that support agencies only have a select number of suitably trained staff. Furthermore, the construction industry professional body suggested the possibility of knowledge sharing and training.

The identified advantages of private construction sector involvement included: obvious benefits to the local people due to higher efficiency in the construction works, in terms of personal development and experience of private sector employees due to their interaction, and also technological transfer to the local communities. These responses support Yamamoto's and Ashizawa's (1999) argument that business is a valuable partner. However, it was highlighted that there would be a commercial disadvantage if support was given on too large a scale. Concerns were also raised about health risks to employees, the private sector not being so adaptable in terms of organisational processes and the business' day-to-day responsibilities which it must fulfil. Other points made included the difficulty in responding in geographical locations where the business is not operating, having access to local knowledge, and the low cost focus struggling to entice big business. It was noted that engagements would have to be carefully managed.

With regards to collaborations, there is a realisation from both sectors that there is scope for support and that there is also a need to achieve a win-win scenario. Many respondents were not certain if potential support would separate into philanthropic, strategic, commercial and political reasons. The common conception was for a philanthropic aim (providing a few people on a supportive basis) which would provide short-term gains but with the correct motivations. It was held, however, that in general, construction firms cannot afford to be philanthropic. There was a realisation that it depends on the organisations' longer term objectives and a longer term focus would need to be strategic for disaster management stakeholders. In summary, commencing with philanthropic support and developing this into a longer term focus might be achievable. Explicit transparent arrangements are needed. These arguments support Twigg's (2001) viewpoint that business' responsibilities are to the bottom line of profitability, but they are now considering their CSR focus.

The majority of interviewees from both sectors referred to examples of private construction sector support to post-disaster reconstruction. This highlights how private sector support has previously been used and further substantiates Binder and Witte's (2007) view that the utilisation of private sector expertise is progressing. The specific examples discussed by respondents included: fundraising and charitable donations to help victims of the Australian bush fires; support being given following the Kobe earthquake in Japan; individual private sector employees belonging to the disaster management organisation REDR; secondment of employees to support reconstruction projects in Sri Lanka and the Maldives, through a partnership with the UN-HABITAT; and, a 'School for Life' support programme, in which the private construction organisation supports tsunami orphans.

With regards to the assistance discussed above, different methods of engagement were utilised. These included: unilateral donations, fundraising, support to international NGOs (INGOs), secondment of employees and sabbaticals. In summary, the support provided was voluntary and a form of philanthropy.

It was deemed that interactions could be further developed by engaging in a relationship with any number of organisations; the structuring of a network would enable the private construction sector to act sooner. In addition to this, secondments were held to provide overseas experiences, training and personal development opportunities; sabbaticals could prove to be a career development option. It was noted, however, that there needs to be a training and feedback process to inform all parties of the benefits being provided. Partnerships with

INGOs and NGOs could provide a suitable option. The interaction must, however, remain manageable in order to prevent overcomplexity; both sectors also need to broadcast their resource needs and capabilities.

In summary, very similar views were held by both sectors; links, partnerships and providing the support of people could be a suitable option. Stakeholders are thus able to join forces to enhance coordination and delivery.

The key motivations for the private sector offering support were philanthropic contributions, longer-term sustainability and a focus on benefiting communities. These conclusions support Twigg's (2001) reasoning that CSR initiatives will develop where there is a tradition for corporate philanthropy. The rationale for post-disaster reconstruction stakeholders wishing to engage with the private construction sector was for access to technology, expertise and knowledge sharing; all of which can be utilised to benefit the local communities. A notable point was the construction sector's focus on delivering projects on time, to the correct quality standards and within budget. It was held that on a small scale, philanthropy and business' commitment to community and staff development might provide a sustainable option in the longer term. This will, however, need to be driven by people within organisations who have significant influence. It was noted that, in the longer term, businesses will need to receive benefit through their support.

When asking respondents to summarise their motivations, CSR and organisational values were considered to be important by the majority of respondents. Both sectors also discussed the possibility of longer term opportunities. These findings support Warhurst's (2005) proposition that businesses are most effective when they meet expectations of wider industry stakeholders. In addition to this, staff motivation, sabbaticals, and longer-term profitability options were also noted to be important for private sector employees.

With regards to the media profile surrounding natural disasters, conclusive results indicated that the media profile would not affect support provided by the private construction sector due to copious good practices being conducted in industry and not being recognised. This challenges the view of Thomas and Fritz (2008), that the private sector is only motivated towards 'popular' disasters. The industry professional body did question, however, whether the support being offered could provide a marketing advantage for the private construction sector.

It must be noted that a number of reservations were raised regarding private construction sector support being supplied. In summary, concerns were raised regarding the longer term viability for the private construction sector, and for post-disaster reconstruction stakeholders: the private sector's effect on empowerment and local communities. The private construction sector held that if a large scale, longer-term contribution was required, it would have to add value and be viable. These points support Gunwardena et al.'s (2008) note that over the longer term, profits will be required. Another reservation was the risks to individual employees and the intermittency in support which could be provided. It was also raised that the private construction sector may want to offer support and may not be invited to do so. Finally, questions were also raised regarding 'outsiders' imposing their own rules and systems not being effectively planned and implemented.

When considering the most effective method of providing support, both sectors referred to the possibility of financial support, donating peoples' personal expertise and forming links with other organisations. It was deemed that providing support would be more effective compared with actually conducting construction work. Providing people, knowledge and experience may also be considered to be more effective as, when providing funds, it is difficult to establish how they will be effectively used. Furthermore, the contribution of people could make beneficial use of the organisational slack which is currently present in the industry. It is noted, however, that donations and fundraising can get people emotionally involved. Bilateral partnerships between construction multinational corporations and INGOs received significant attention from respondents, especially if facilitated workshops could explain each sector's differences. This would increase efficiency, but it was held that there would need to be a charter for the longer term engagement. These concepts support Lloyd-Jones' (2006) theory that multi-disciplinary teams are essential, and also further builds upon the IFRC's (2001) suggestion for INGOs to engage with other partners.

The most effective form of support was considered to be a national focus. In addition to this, the concept of construction organisations utilising a regional approach was also discussed; providing support, if operating within the country. It was noted, however, that while there needs to be an overriding strategy which integrates all stakeholders, ownership needs to be held by the host nation. Furthermore, a key point was raised regarding the importance of building relationships between professionals as opposed to organisational processes. Relationships must be emphasised between professionals in the various organisations, as people are a very important resource in the construction industry.

Realisations were also achieved regarding the collaboration needed between all post-disaster stakeholders and the engagement required with local governments and local communities. It was held that relationships would depend on the scale of the disaster and the longevity of support being provided. Interview results were, however, very conclusive; providing support and relationships with INGOs can be very effective. This supports Campher's (2005) view that partnerships are becoming more common.

The advantages of private construction sector support to post-disaster reconstruction are realised in terms of efficiency, expertise and project delivery capabilities. The difficulties in initiating support included whether the private construction sector has the ability to engage with the post-disaster culture and has enough presence in the disaster area. Comment was also made that there is a limit to how far philanthropic responses can be developed, if there would be no commercial gain. Another difficulty related to getting the right people within the different sector organisations to talk to one another, as engaging with one disaster management organisation (INGO) may divert the benefit of the private construction sector's support.

Leadership of the construction support would depend upon the organisation's motivations; whether they are to provide philanthropic support or to receive some profit, and whether or not expertise is being given on a large, long-term scale. There was, however, a realisation that government leadership is important. The majority of respondents held that disaster team led initiatives would be the most effective, as post-disaster stakeholders have the experience, infrastructure, and could coordinate the private construction sector skills.

There was great interest in various private sector and disaster recovery sector engagement models. There was, however, a significantly low level of awareness of these concepts from respondents in both sectors. For this reason, the responses given cannot be deemed to be conclusive, as they have been volunteered based upon a limited level of knowledge. It must be noted, however, that there was more interest from the private construction sector regarding the UNDP's Growing Sustainable Business Guidelines (GSB), as this could be regarded as a win:win scenario, thus indicating the private sector's underlying need for profitability in order to provide longer term support. In addition, the DRN would effectively integrate resources in a similar way to organisations such as REDR. Therefore, this questions whether the volume of networks needs to be reduced and consolidated in order to achieve more effective coordination.

All responses regarding future strategy development options were unique. Concepts included the need for better planning, faster reaction times, and the need to identify where the most vulnerable communities are located and to set up plans to assist in the event of disaster. An internal business framework for response within a private construction sector organisation might be required to achieve this. A pro-active stance is required, and questions were raised as to whether the private construction sector should be actively talking to post-disaster reconstruction stakeholders. The need to focus on training, knowledge transfer and up-skilling was recognised, as was the importance of collaborative approaches. This supports Campher's (2005) notes regarding coordinated responses and Binder's and Witte's (2007) note that businesses prefer to operate through partnerships than alone. More direction is needed from somebody, and the question remains: who is going to initiate an increased level of support?

11.8 Summary

This chapter has shown how the private sector can contribute effectively to post-disaster reconstruction. The results from the empirical study support using private construction sector resources to fill the professional resource gap, with a natural fit for project management and engineering expertise. The provision of temporary support, logistics management, technical expertise and key people is also essential. The results also established, however, that capabilities need to be stated from both sides. The advantages of using private construction sector resources included benefits to local people, increased efficiency in construction works and improved staff development and experience. The disadvantages included the limitation of the geographical locations in which the business is operating, the potential health risks to employees, businesses not being adaptable in terms of organisational processes, and the commercial disadvantage if support was provided on too large a scale. The research concluded that an initial philanthropic aim may prove to be most beneficial, but it would depend upon the organisation's objectives as over the longer term it would have to add value to the business. Therefore, transparent relationships are required and these would have to be carefully managed.

Discussions related to specific examples from both sectors have enabled an examination of current practices. Binder's and Witte's (2007) outlook is supported and the investigation has highlighted that the private construction sector has already interacted with post-disaster reconstruction efforts. In each scenario, different methods of engagement were used, including: unilateral donations, fundraising, support to INGOs, secondment of employees, consultancy and sabbaticals. It was held that a network is required in order for the private sector to respond more rapidly. The concept of secondments from the private sector would enable individuals to gain life skills, overseas experiences and training. It is also essential that a feedback process is incorporated.

The rationale regarding private sector involvement includes: philanthropy, sustainability and a focus on communities. The rationale for post-disaster reconstruction stakeholders to engage with the private sector was a focus on technology, expertise and knowledge sharing. On a smaller scale, philanthropy and businesses' commitment to communities and staff development might provide a sustainable focus for future interaction. CSR and organisational values were also noted to be an important rationale for engagement. It was held that media profile will not affect support given by the private sector to post-disaster reconstruction, which contradicts Thomas and Fritz's (2008) views. There were, however, reservations regarding the longer-term viability for the private construction sector in providing support, and also for post-disaster reconstruction stakeholders in the private sector's effect upon and level of empowerment to local communities. It was held that interactions need to be effectively planned for the longer term.

It has been concluded that the most effective form of support would be a national focus, especially if construction firms were operating in the area. There does, however, need to be an overriding integration strategy. There were realisations that collaboration between all post-disaster stakeholders and relationships with INGOs could be effective, and that people are the most important resource which would need to be integrated. Questions were raised as to the need for a charter for a longer term engagement.

The construction sector is deemed to be able to provide efficiency, expertise and project delivery skills. However, reservations included the limit to how far a philanthropic response can be maintained without value being added to the private sector organisation, and the difficulties in getting the right people to talk to each other was highlighted. In summary, both sectors agreed that private construction sector support would have to be led by disaster management stakeholders, but there is a need for a collaborative approach. With regards to private sector engagement mechanisms, there was interest from both sectors but a limited level of awareness, indicating that interaction frameworks need to be refined and published.

It can be concluded that views on future development options are varied. There should, however, be a focus on vulnerable communities, with organisations having an internal response framework within their organisation. In the longer-term, support needs to be better planned and it was proposed that the private construction sector should be actively talking to post-disaster reconstruction stakeholders. The private sector support should, through a collaborative approach, focus on training, knowledge transfer and up-skilling other disaster management stakeholders.

References

Amin, S. and Goldstein, M. (2008) *Data Against Natural Disasters: Establishing Effective Systems for Relief, Recovery, and Reconstruction.* Washington, DC: The World Bank.

BBC News (2008) China anger over 'shoddy schools'. From BBC News: http://news.bbc.co.uk/1/hi/world/asia-pacific/7400524.stm [accessed 02/04/2009].

Binder, A. and Witte, J.M. (2007) *Business Engagement in Humanitarian Relief: Key Trends and Policy Implications.* London: Overseas Development Institute.

Campher, H. (2005) *Disaster Management and Planning: An IBLF Framework for Business Response.* London: IBLF The Prince of Wales International Business Leaders Forum.

Disaster Resource Network. (2009) An Initiative of the World Economic Forum. From Disaster Resource Network: www.drnglobal.org/home [accessed 02/06/2009].

Gunewardena, N., Schuller, M. and Waal, A. (2008) *Capitalizing on Catastrophe: Neoliberal Strategies in Disaster Reconstruction.* Plymouth: AltaMira Press.

Holstein, J. A. and Gubrium, J.F. (1995) *The Active Interview.* London: Sage Publications.

International Federation of Red Cross and Red Crescent Societies (IFRC) (2001) World Disasters Report: Focus on Recovery. Switzerland: International Federation of Red Cross and Red Crescent Societies.

Jones, P., Comfort, D. and Hillier, D. (2006) Reporting and reflecting on corporate social responsibility in the hospitality industry. *International Journal of Contemporary Hospitality Management,* **18**, 329–340.

Kreimer, A., Arnold, M. and Carlin, A. (2003) *Building Safer Cities: The Future of Disaster Risks.* Washington, DC: World Health Organization.

Kumar, R. (2005) *Research Methodology: A Step-By-Step Guide For Beginners.* London: Sage Publications.

Kvale, S. (1996) *An Introduction to Qualitative Research Interviewing.* London: Sage Publications.

Leftly, M. (2008, May 22) Should I stay or should I go? *Building Magazine,* 1–5.

Lim, A.A. (2003) The Role of the Business Sector in Disaster Preparedness and Response. The International Conference on Total Disaster Risk Management, 2–4 December, 2003, Kobe, Japan.

Lloyd-Jones, T. (2006) *Mind the Gap! Post-Disaster Reconstruction and the Transition from Humanitarian Relief.* London: Royal Institution of Chartered Surveyors.

MacFadyen, T. (2005) *Tsunami Proposal.* Australia: Bovis Lend Lease.

Mehta, V. (2008) *The Role of Global Institutions in Tackling Poverty.* Oxford: VM Peace.

Mills, L. (2008) International Development Overview and Construction Organisations [Online]. From lucycmills@yahoo.co.uk [accessed 09/05/2008].

Mudge, E. (2008) *Sustainable Development Through Partnerships, Community Empowerment and Innovation.* India: Bovis Lend Lease.

Naoum, S.G. (2007) *Dissertation Research and Writing for Construction Students.* Oxford: Butterworth Heinemann.

Powell, W.W. and Steinberg, R. (2006) *The Non-Profit Sector: A Research Handbook.* New Haven, CT: Yale University Press.

Robson, C. (2002) *Real World Research.* Oxford: Blackwell Publishing Ltd.

Telford, J. and Cosgrove, J. (2006) *Joint Evaluation of the International Response to the Indian Ocean Tsunami: Synthesis Report.* London: Tsunami Evaluation Coalition.

Thomas, A. and Fritz, L. (2008) Harvard Business Review Disaster Relief Inc. From Harvard Business Review: http://harvardbusiness.org/ [accessed 02/02/2009].

Twigg, J. (2001) *Corporate Social Responsibility and Disaster Reduction: A Global Overview.* London: Benfield Grieg Hazard Research Centre, University College London.

UN-HABITAT (2007) *Business for Sustainable Urbanisation: Challenges and Opportunities.* Kenya: UN-HABITAT.

UNDP (United Nations Development Programme) (2004) The Global Compact: Growing Sustainable Business for Poverty Reduction. From UNDP: www.undp.org/business/gsb [accessed 02/04/2009].

UNDP (United Nations Development Programme) (2009) Partners: Private Sector. From UNDP: http://www.unhabitat.org/content.asp?cid=4227&catid=527&typeid=24&subMenuId=0 [accessed 02/02/2009].

United Nations International Strategy for Disaster Reduction (2005) Hyogo Framework for Action 2005–2015: Building the Resilience of Nations and Communities to Disaster. World Conference on Disaster Reduction, 18–22 January, 2005, Hyogo, Japan, UNISDR.

Warhurst, A. (2005) *Disaster Prevention: A Role for Business?* Warwick: MapleCroft.

Weinberg, D. (Ed.) (2002) *Qualitative Research Methods.* Oxford: Blackwell Publishing Ltd.

Yamamoto, T. and Ashizawa, K.G. (Eds) (1999) *Corporate–NGO Partnership in Asia Pacific.* Tokyo: Japan Centre for International Exchange.

12 Knowledge Management Practices and Systems Integration

Chaminda Pathirage

12.1 Introduction

As a consequence of knowledge being increasingly recognised as a valuable resource, there are greater efforts to deliberately manage knowledge in a systematic manner. This chapter focuses upon the importance of knowledge management in a disaster context. The basic concepts underpinning the management of knowledge are explored. The chapter moves on to specifically examine the management of knowledge across the pre- and post-disaster lifecycle. Examples of the knowledge generated and lessons learned in different disaster contexts are then presented using case studies. The chapter concludes by introducing a project, entitled ISLAND, which aims to increase the effectiveness of disaster management by facilitating the capture and sharing of appropriate lessons and good practices in land, property and construction.

Definitions of knowledge range from the practical to the conceptual to the philosophical, and from narrow to broad in scope. Knowledge is built from data, which is first processed into information. Researchers such as Simon (1957), Galbraith (1974) and Tushman and Nadler (1978) perceived organisations as essentially information processing entities, which accounted for the 'information age' of the 1980s and early 1990s. However, information processing across organisational boundaries presents significant barriers to effectiveness, mainly due to the absence of a cognitive dimension. Thereby, the various definitions of knowledge suggest that it is much more than information. As Grey (1996) contends, knowledge is the full utilisation of information and data, coupled with the potential of people skills, competencies, ideas, intuition, commitment and motivation. According to Nonaka and Takeuchi (1995), knowledge could be defined as a dynamic human process of justifying personal belief towards the truth.

Post-Disaster Reconstruction of the Built Environment: Rebuilding for Resilience, First Edition.
Edited by Dilanthi Amaratunga and Richard Haigh.
© 2011 Blackwell Publishing Ltd. Published 2011 by Blackwell Publishing Ltd.

Work by Polanyi (1958) and Nonaka and Takeuchi (1995) divided knowledge into tacit and explicit. Although knowledge could be classified into personal, shared and public, practical and theoretical, hard and soft, internal and external, foreground and background, the classification of tacit and explicit knowledge remains the most common. Tacit knowledge represents knowledge based on the experience of individuals, expressed in human actions in the form of evaluation, attitudes, points of view, commitments and motivation (Nonaka *et al.*, 2000). Since tacit knowledge is linked to the individual, it is very difficult, or even impossible, to articulate. Explicit knowledge, in contrast, is codifiable knowledge inherent in non-human storehouses including organisational manuals, documents and databases. Although scholars, including Nonaka and Takeuchi (1995), Spender (1996), and Baumard (1999), commonly offered a clear bounded distinction between tacit and explicit knowledge, primarily on the basis of codification and formalisation, or ease of transfer, it is difficult to find two entirely separate dichotomies of tacit and explicit knowledge; instead knowledge can fall within the spectrum of tacit knowledge to explicit knowledge. According to McDermott (1999), knowing is a human act and any discussion of knowledge is meaningless in the absence of a knower: 'the heart of knowledge is a community in discourse and sharing ideas, as such to leverage knowledge we need to focus on the community that owns it and the people who use it, not the knowledge itself' (McDermott, 1999, p.110). Thereby, information becomes knowledge when it enters the system and when it is validated (collectively or individually) as a relevant and useful piece of knowledge to implement in the system (Blumentritt and Johnston, 1999).

Discussions on knowledge emphasise two perspectives, as identified by Empson (2001): 'knowledge as an asset' and 'knowing as a process.' On the 'knowledge as an asset' perspective, knowledge is often viewed as an objectively definable commodity, which can be managed and controlled by certain mechanisms. For 'knowing as a process' perspective, knowledge is a social construct, developed, transmitted and maintained in social situations. As suggested by Snowden (2002), this evolution of knowledge-based views could be summarised into three generations. The first generation is the information processing view of the organisation ('information age'). The second generation begins with identification of knowledge as a key resource. The third generation goes beyond managing knowledge as an 'asset' or a 'stock' to managing knowledge as a 'flow' or a 'process'. When knowledge is seen as an 'asset', codification strategies, which especially disseminate explicit knowledge through person-to-document approaches, are considered. When knowledge is seen as a 'flow', personalised strategies, which especially disseminate tacit knowledge through person-to-person approaches, are considered.

Knowledge management (KM) definitions and tools further emphasise this dominant perspective of knowledge. Accordingly, KM discussion has focused into two principal camps. These fall mainly into the IT perspective (explicit knowledge) where authors focus on IT tools to deliver KM solutions, i.e. codification strategies, or the Human Resource (tacit knowledge) perspective that relies on the people aspect to provide KM solutions, i.e. personalised strategies. However, KM could be defined in general as 'the process of creating, acquiring, capturing, sharing and using knowledge, wherever it resides, to enhance learning and performance' (Scarborough et al., 1999). Any KM approach that is purely based on IT is bound to be less successful because people issues, which are not readily solved by IT systems, would need to be addressed. Thus, the process view critically considers the human element of knowledge sharing by adding the element of people experiences to the equation of knowledge (Kogut and Zander, 1992). This is very much applicable within the disaster management context, where the knowledge and experiences of disaster practitioners remain mainly in the individual domain.

With the realisation of the strategic importance of the people factor within the last decade, there has been an increasing interest in the tacit dimension of knowledge, which is perhaps hardest to manage as it cannot be formally communicated and is often embedded within human beings. As Herrgard (2000) suggests, tacit knowledge is obtained by internal individual processes like experience, reflection, internalisation or individual talents. Individuals are the primary repositories of tacit knowledge as due to its transparent characteristics, it is difficult to communicate and therefore cannot be managed and taught in the same manner as explicit knowledge. While highlighting the importance of tacit knowledge, Tiwana (2000) defines it as know-how that is stored in people's heads which is personal, acquired mainly through education, training and experience. In a similar sense, Saint-Onge (1996) describes tacit knowledge as an individual's intuition, beliefs, assumptions and values, formed as a result of experience. It is from these beliefs and assumptions, which make up an individual mindset, that decisions are made and patterns of behaviour developed. Hence, the strategic importance of tacit knowledge cannot be underestimated, especially within the disaster management context.

12.2   Disaster management cycle

Disasters, both natural and man-made, have been occurring with increasing frequency and effect in recent decades in many countries around the world. As such, disaster management efforts aim to reduce or avoid the potential losses from hazards, assure prompt and appropriate assistance to victims of disaster, and achieve rapid and effective recovery (Warfield, 2004). As shown in Figure 12.1, the process of disaster management is commonly visualised as a two-phase cycle, with a post-disaster recovery informing pre-disaster risk reduction and vice versa.

Warfield (2004) adds that the disaster management cycle illustrates the ongoing process by which governments, businesses and civil society plan for and reduce the impact of disasters, react during and immediately following a disaster, and take steps to recover after a disaster has occurred. The significance of this concept is in its ability to promote the holistic approach to

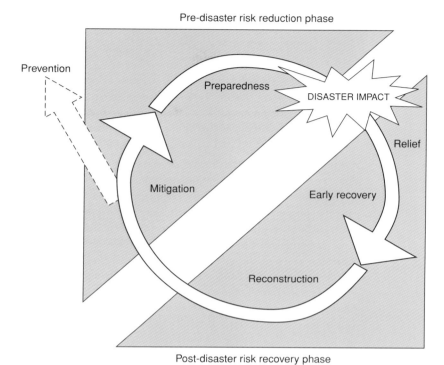

Figure 12.1   Disaster management cycle (source: this is a generic model used commonly in disaster management)

disaster management as well as to demonstrate the relationship of disasters and development (de Guzman, 2001). Once a disaster has taken place, the first concern is effective 'recovery' – helping all those affected to recover from the immediate effects of the disaster. 'Reconstruction' involves helping to restore the basic infrastructure and services which the people need so that they can return to the pattern of life which they had before the disaster (Davis, 2005). The complex and multi-faceted processes of post-disaster recovery and reconstruction extend well beyond the immediate period of restoring basic services and life-support infrastructure.

The RICS (2006) defines disaster mitigation as any structural and non-structural measures undertaken to limit the adverse impacts of natural hazards, environmental degradation and technological hazards. 'Mitigation' measures may eliminate or reduce the probability of disaster occurrence, or reduce the effects of unavoidable disasters. As such, these measures include: building codes; vulnerability analyses updates; zoning and land use management; building use regulations and safety codes; preventive health care; and, public education (Warfield, 2004). In the ideal case, mitigation eliminates the risk of future disasters by effective sharing of lessons learned through 'preparedness' planning. Thereby, knowledge, lessons and good practices learned during the

Figure 12.2   Components of a disaster (source: this is a generic model used commonly in disaster management)

post-disaster reconstruction phase should be shared and transferred to the pre-disaster risk reduction phase to reduce the risk associated with disasters.

Natural events, however, only become potential hazards when they threaten people, property or infrastructure (Davis, 2005). As de Guzman (2001) notes, natural hazards themselves do not necessarily lead to disasters. Natural hazards like earthquakes, however intense, inevitable or unpredictable, translate to disasters only to the extent that the population is unprepared to respond, unable to cope, and consequently, severely affected. An earthquake will cause little damage if it takes place in an empty desert. It may also cause little damage if it takes place where people can afford to be well protected. Hence, a natural event only causes serious damage when it affects an area where the people are at risk and poorly protected. Disasters occur when these two factors are brought together (as shown in Figure 12.2):

(1)   People living in unsafe conditions
(2)   A natural hazard such as a flood, tsunami, hurricane or earthquake

Thereby, the threat from natural hazards can only be minimised through the elimination of unsafe conditions, as much as possible, in terms of people, property and infrastructure. Within this context, sharing and transfer of knowledge in terms of lessons and good practices learned during the post-disaster reconstruction phase can help to eliminate unsafe conditions. It can thereby help to reduce the risk associated with future disasters, and help a society to take steps to recover efficiently after a disaster has occurred.

12.3   Knowledge management in disaster management context

In many countries there is a conscious effort for disaster management at a national, provincial and sub-provincial level. Despite this, knowledge appears fragmented, although there are undoubtedly many successful practices and lessons to be learned (Mohanty *et al.*, 2006). Hence, there is a perceived gap in information coordination and sharing, particularly relating to disaster mitigation. A lack of prior knowledge and proper points of reference have made most of the recovery plans guessing games, eventually failing without adding appropriate values to the recovery attempts (RICS, 2006). The lack of effective information and knowledge sharing, and dissemination on disaster mitigation measures, has thereby been identified as one of the major reasons behind the unsatisfactory performance levels of current disaster management practices.

It can be perceived that valuable knowledge on disaster mitigation is present at three different levels: institutional, group and individual, in the forms of both tacit and explicit knowledge. Thousands of organisations and institutions have been supporting efforts on disaster management all over the world. However, the linkage among all these agencies that are working on disaster management needs to be strengthened in order to derive regional good practice and coping mechanisms (RICS, 2006). In order to enhance the information sharing and management of the knowledge generated, it is essential to knit these organisations and institutions, and moreover groups and people working within these institutions (UNDP, 2005). There are many gaps that could be bridged by appropriate use of professional skills, but access to these by local organisations on the front line of the recovery effort is highly constrained by lack of recognition of their existence. Therefore, recognition needs to be given to the institutions and organisations operating not only at international and national level, but at the local level too. In addition, this local knowledge can reside among the groups operating within different communities; hence, the recognition can be extended to the existence of these formal and informal groups involved with the disaster management process. The knowledge and experiences of disaster practitioners remains mainly in the individual domain. Due to its large geography, the experiences, approaches and adopted modalities for disaster management are not codified and remain with individuals as tacit knowledge (Mohanty *et al.*, 2006).

In an organisational context, KM is about applying the collective knowledge of the entire workforce to achieve specific organisational goals and facilitating the process, by which knowledge is created, shared and utilised (Nonaka and Takeuchi, 1995). However, within the disaster management context, KM is all about getting the right knowledge, in the right place, at the right time (Mohanty *et al.*, 2006). As a strategic approach to achieve disaster management objectives, KM will play a valuable role in leveraging existing knowledge and converting new knowledge into action.

12.4 Sharing and transferring disaster management knowledge

The importance of improving the construction industries of hazard-prone countries is widely recognised, highlighting a need to equip them to manage the post-disaster scenario, especially during the post-disaster reconstruction process. There is growing recognition that the engineering community has a valuable role to play in finding and promoting rational, balanced solutions to what remains an unbounded threat and that the construction industry has a much broader role to anticipate, assess, prevent, prepare, respond and recover from disruptive challenges. Good practice improvements are likely to be required by the community in order to guarantee long-term sustainability of the reconstruction (Ofori, 2002) to ensure safe conditions for future disasters. Post-disaster management can have various approaches and different priorities in different countries. However, successful strategies for post-disaster management should be more-or-less compatible with disaster level, economic, social, cultural, institutional, technological, technical, cultural, environmental and legal or regulatory situations in the country under consideration (Kaklauskas *et al.*, 2009). As a result, international exchange of good practice and knowledge sharing among practitioners, authorities and institutions, particularly from the region, can significantly contribute to post-disaster reconstruction at all levels.

Disasters can strike at any time and it is the magnitude of the related impacts that will reflect the level of preparedness and awareness of the exposed country and community. KM can help communities in hazard-prone areas to gain a better grasp of the ways to cope with disaster risks. Accordingly, it is now widely agreed that achieving disaster-resilience is essentially a process of using knowledge at all levels (UNESCO, 2009). Generation, transfer and sharing of knowledge are key foundations for disaster risk management and mitigation. According to Egbu and Robinson (2005), processes such as knowledge generation, dissemination and sharing are considered to be important facets of a knowledge economy. Hence, there is a growing recognition that much more attention has to be paid to knowledge creation and sharing in the form of lessons learned and good practices in the disaster management field. Effective lessons and good practice sharing should reduce the risk of future disasters through well-informed mitigation and preparedness planning. Therefore, knowledge utilisation is a key factor in effectively executing post-disaster management. Ensuring the availability and accessibility of accurate and reliable disaster risk information when required, entails an efficient system for knowledge sharing. In this regard, an efficient disaster risk management knowledge system is vital.

Knowledge bases of good practice are knowledge capturing and sharing tools, which provide information on the best post-disaster management practice in different forms such as regulations, e-books, slide presentations, structural schemes, text, video and audio material (Kaklauskas *et al.*, 2009). The tacit knowledge base of good practice consists of informal and unrecorded procedures, practices, and skills. KM systems are of value to the extent that they can codify 'good practices' in post-disaster management, store them, and disseminate them as needed. However, tacit knowledge is highly personal, context-specific, and therefore hard to formalise and communicate. Therefore, capturing tacit knowledge is extremely important within the post-disaster management context because, once disaster consequences are eliminated, professionals tend to forget them and start something new.

However, due to the inherent difficulty in tacit knowledge diffusion, the construction industry faces a knowledge diffusion problem. According to the empirical findings of Gieskes and Broeke (2000, p.194), 'lessons learnt may become individual tacit knowledge when they are not captured on an organisational level: there are no signs of a systematic diffusion within the organisation or across organisational boundaries.' Hence, KM tools needs to be introduced to overcome this knowledge diffusion problem and to create a proper KM environment to share and disseminate knowledge. 'KM tools' are both IT and non-IT tools required to support sub-processes of KM such as generating, capturing, disseminating and sharing knowledge, although the term is too often used narrowly to mean IT tools. There are a large number of tools available to choose from in managing knowledge. Often, to distinguish between KM tools, the terms 'KM techniques' and 'KM technologies' are used to represent 'non-IT tools' and 'IT tools' respectively. Details of these two types of tools are given in Table 12.1.

KM techniques do not depend on IT, although it provides support in some cases. According to Bresnen *et al.* (2003), there are difficulties, challenges and limitations in attempting to capture and codify project-based learning by using technological mechanisms. Therefore, significant effort should also be directed towards exploiting KM techniques (non-IT) such as communities of practices, brainstorming sessions, action learning, post-project reviews and similar to facilitate person-to-person interactions, through which tacit knowledge could be generated and used. The UK Higher Education Disaster Relief Project Report (2007) highlighted the lack of mechanisms at a national level in the UK to link the expertise, skills and knowledge which reside within UK higher education, with those of the practitioners in the humanitarian agencies. Neither is there a comprehensive overview of the expertise which exists and who is willing to offer expertise. However, by using a knowledge base of experts it is possible to search for experts and facilitate communication with those experts using IT. By accessing an expert's knowledge base system, stakeholders can search for an expert with the relevant knowledge and background, and connect with an

Table 12.1   A comparison between KM techniques and technologies (adapted from Al-Ghassani *et al.*, 2005)

KM tools	
KM techniques	**KM technologies**
• Require strategies for learning/education	• Require IT infrastructure
• More involvement of people	• Require IT skills
• Affordable to most organisations	• Expensive to acquire/maintain
• Easy to implement and maintain	• Sophisticated
• More focus on tacit knowledge	• Implementation/maintenance
• Examples of tools:	• More focus on explicit knowledge
– Brainstorming	• Examples of tools:
– Communities of Practice	– Data and text mining
– Face to face meetings	– Groupware
– Recruitment	– Intranets/Extranets
– Training	– Knowledge bases
	– Taxonomies

expert in real time by using instant messaging, e-mail, telephone, or internet conferencing. As a result, stakeholders could receive direct tacit help from an expert who has experienced a similar problem in the past (Kaklauskas *et al.*, 2009).

The importance of training and education in order to capture required knowledge is well recognised within a disaster context (Yasemin, 1995). The scope of a training and education programme must include the identification of areas of vulnerability, mitigation measures (social, physical and organisational) that can be employed to reduce vulnerability and awareness of plans developed to manage post-disaster risk reduction activities (Jigyasu, 2002). Also it is important to capitalise on traditional local knowledge about hazards. Hence, education can involve the enhancement and use of indigenous knowledge for protecting people, habitat, livelihoods, and cultural heritage from natural hazards. Educational practices can be conducted through direct learning, IT, staff training, electronic and print media, and other innovative actions to facilitate the management and transfer of knowledge and information to citizens, professionals, organisations, community stakeholders and policymakers (Kobe Report, 2005). Education for disaster reduction and human security should not be a one-off affair, but rather a continuing process, offering individuals lessons in coping with hazards not just once but several times throughout their lives (UNESCO, 2009). Furthermore, education and raising awareness of disaster risks must respond to a society's changing needs and focus on empowering individuals throughout their lives, as knowledge is at its most effective when linked to community needs. Educating all sectors of society on disaster reduction actions that are based on application of sound scientific, engineering, and cultural principles to create sustainable systems should be the long-term goal for hazard-prone countries (UNESCO, 2009).

12.5   Case studies of good practices and lessons learned

This section presents four case studies of different types of disasters that have occurred in countries across the world. The cases consider the capture of lessons learned and good practices. In each case, a brief description of the disaster is provided. An overview of the challenges faced and actions taken are then discussed. At the end of each case the lessons that were learned are considered.

12.5.1   *Case Study 1 (Lawther, 2009): Tsunami, 2004, Maldives*

Theme: Community involvement in the post-disaster reconstruction process

The Maldives is a chain of 26 coral atolls southwest of Sri Lanka, extending across the equator in a north-south strip 754 kilometres (468 miles) long and 118 kilometres (73 miles) wide. Nearly 80% of the country is at 1 metre or less above sea level and the population is approximately 300 000. When a tsunami struck the Maldives on December 26, 2004, nearly one-third of the population was severely affected. The loss of life was relatively small (83 residents confirmed dead) in relation to victims in other countries in South-East Asia. However, the devastation to the environment, infrastructure, economy and human psyche was immeasurable. Nearly 39 islands were significantly damaged, and 14 islands were completely destroyed. Approximately 20 000 individuals were displaced from their homes (Pardasani, 2006).

The following case study is from the British Red Cross Society's (BRCS) Maldives post-tsunami recovery programme, which comprised the re-construction of 466 houses (216 houses across four islands in Phase 1 and 250 houses on one island in Phase 2) together with associated infrastructure.

Involvement of the community was not paramount in the initial stages of the reconstruction programme. Rather, it was an issue that evolved and grew as the project progressed. It did, however, become of some importance, particularly as BRCS received criticism from other non-governmental organisations (NGOs) for failing to include the community 'adequately' in the re-construction process. The level of involvement of the community in the BRCS programme was limited. There was no real involvement of the community in the beneficiary identification and selection process. This was done on the basis of government assessments of damaged houses. Disgruntlement with the beneficiary identification process was highlighted at the completion of the Phase 1 projects through housing satisfaction surveys.

In terms of the housing design process, the major design parameters of the houses (size, number of rooms, ceiling height) were decreed by the Government of Maldives and the detailed design of the houses was undertaken by a consultant, although this did to some extent take consideration of local norms for house design. However, identified beneficiaries were involved in the housing selection process. Through this process they were able to choose:

- A floor plan from a range of three
- A floor tile colour from a range of three
- An internal paint colour from a range of three
- An external paint colour from a range of three
- A roof sheeting colour from a range of three

The housing selection process was an interesting experience of community involvement. It became apparent that whilst beneficiaries were clearly concerned with the choice of floor plan (evidenced by the due consideration given to this aspect), there was far less concern for the 'colour choices'. Another variation to the model of direct employment of local people on the physical reconstruction occurred on Isdhoo Kalaidhoo. With construction to take place on a new land use-plan, this required the clearance of a large tract of virgin bush to enable construction to commence. Whilst this was the responsibility of the Government, a lack of capacity determined that BRCS pushed the process forward. This was done by engaging the community to undertake the bush clearing on a cash-for-work basis. This was a large project which required considerable planning and preparation with local authorities and the community prior to implementation. Eventually the project was completed as a collaboration between BRCS and the community. In addition to enabling the housing construction process proper to commence, this project developed an important platform of goodwill between BRCS and the community.

Lessons learned

While researchers have differing views on the level of community involvement required in redevelopment efforts, most agree that participation of affected individuals is key to the overall success of any such efforts (Coghlan, 1998; McCamish, 1998; Buckle and Marsh, 2002; McDowell, 2002). Buckle and Marsh (2002) assert that although the expert role in assessment is necessary and vital in planning and implementation efforts, local knowledge of needs, strengths and priorities cannot be dismissed or sidelined. Passive requirements for community involvement in the reconstruction process are particularly important and should not be underestimated, or given less prominence by comparison with the more visible active side of community involvement. If local communities and affected individuals are not involved in this process, there may be a disconnection between organisationally identified needs and those outlined by local communities and NGOs. Thus, it is imperative for local communities to be involved in the needs assessment process, and in the subsequent design and implementation of the redevelopment efforts. Such a community development model promotes the recognition, acquisition, maturation and connection of community assets, and produces self-reliant, self-sustaining and empowered communities (Kramer and Specht, 1983; Delgado, 2000; Homan, 2005).

12.5.2   *Case Study 2 (EPC, 2004): Earthquake, 2001, India*

Theme: Infrastructure planning for disaster mitigation

The Bhuj earthquake of 26 January 2001 was one of the most disastrous earthquakes experienced in India in recent history. The Bhuj Municipality was one of the hardest hit towns in Kutch. Almost half of its old walled city alone saw considerable damage to buildings and infrastructure. The death toll in Bhuj was over 7000. Most of the casualties were from the walled city area, where buildings made of stone and mud mortar came crashing down on very narrow streets. Along with the badly designed street pattern, the poorly framed regulations which had been loosely adhered to over the years accounted for the vast extent of destruction in Bhuj. The earthquake badly damaged social infrastructure (schools, hospitals, town halls, markets, libraries, colleges, a local gymkhana, an open-air theatre and religious buildings) and utilities (reservoirs, pipelines, telephone exchanges and power infrastructure). In the walled city, the serious damage to water supply and sewer networks was made worse by the movement of heavy machinery to demolish severely damaged buildings and remove debris. Many historic buildings were also destroyed. With the demolition and clearing of rubble, retracing the town's street form and architectural character proved very difficult.

With the unprecedented devastation of towns witnessed in Kutch and with no past experience with post-disaster urban reconstruction in India, putting in place a reconstruction strategy required careful thought. In the immediate aftermath of the earthquake, the discussion both on the ground and in government revolved around two alternatives – total relocation of the city ('New Bhuj') and *in situ* reconstruction. There were vocal proponents of both approaches among the public and among government officials. Having carefully considered all options, the government in April 2001 formulated a reconstruction package for the affected urban areas of Gujarat, which favoured partial reconstruction and partial relocation. To guide and regulate their reconstruction and future growth, the government package stipulated that town planning would be carried out and that the development control regulations in these towns would be revised.

The Government of Gujarat decided to use existing provisions in the Gujarat Town Planning and Urban Development Act, 1976, to undertake the preparation of the Development Plan for the towns of Bhuj, Bhachau, Anjar and Rapar. It decided to use two statutory planning instruments – the Development Plan and the Town Planning Scheme – to guide the planning and reconstruction of the four towns. It appointed the Gujarat Urban Development Company Limited as the Special Purpose Vehicle for the implementation of the urban reconstruction programme. For both town planning and infrastructure design and reconstruction, consultants were appointed. Through competitive bidding, EPC was selected as the Town Planning Consultants for Bhuj, considering its core competence in urban, regional and environmental policy and planning, as well as development research and management.

Lessons learned

The planning process carried out for the redevelopment of the walled city of Bhuj is perhaps one of the most complex, but very rewarding planning exercises attempted in India. The majority of deaths and injuries from earthquakes are caused by the damage or collapse of buildings and other structures. These losses can be reduced through documenting and understanding how structures respond to earthquakes. Gaining such knowledge requires a long-term commitment because large devastating earthquakes occur at irregular and often long intervals. Recording instruments must be in place and waiting, ready to capture the response to the next temblor whenever it occurs. The new information acquired by these instruments can then be used to better design earthquake-resistant structures. In this way, earth scientists and engineers help reduce loss of life and property in future earthquakes (Celebi *et al.*, 1996).

12.5.3   Case Study 3 (Delaney and Shrader, 2000): Hurricane Mitch, 1998, Central America

Theme: Gender involvement in the mitigation and post-disaster reconstruction process

In 1998 the most destructive hurricane to strike the western hemisphere in the last 200 years savaged the countries of Central America. Mitch thrashed the region with 180 miles per hour winds and dumped between 300 and 1800 millimetres of deluging rains. It caused tremendous loss to human life, property, livelihoods, and physical infrastructure throughout Central America. Its effects were greatest in two of the poorest countries in the region, Honduras and Nicaragua. Although the hurricane itself was a naturally-occurring phenomenon, the intense impacts of the storm were exacerbated by human actions, including incomplete development practices. A greater understanding of the social variables surrounding disaster vulnerability is a necessary precursor to operations that address the underlying causes of disaster. The purpose of this case study is to use a gender lens to examine the importance of social variables before and after Hurricane Mitch in Honduras and Nicaragua.

Despite the fact that Central America is one of the most disaster-prone regions in the world, relatively little prevention, preparedness or mitigation was in place prior to Hurricane Mitch. In the places in which disaster plans and other mitigating measures were in place, gender was only minimally considered. Most emergency committees were formed in a non-consultative manner and missed substantial opportunities to use local people's social capital, including coping skills. Some disaster committees in Honduras designed emergency plans in which women were to evacuate and take care of dependents while men were assigned the role of protecting assets, including land and animals. As a result, female-headed households were forced to choose between their children and their assets. Having learned the importance of community organising during Hurricane Joan, women's groups in Mulukutú, Nicaragua, developed their own preparedness plans. Mitigation projects targeted to the household level consciously included men, women and children, and achieved faster results than others in the region. Pilot programmes in community-based mitigation, incorporating women's explicit participation and social as well as geophysical vulnerabilities, succeeded in La Masica, Honduras.

Men and women usually have different priorities and are differentially engaged in the reconstruction process. National governments have presented reconstruction plans that place a heavy emphasis on public infrastructure, while NGOs and other actors in civil society have tended to prioritise housing, agricultural production, and political decentralisation. Many local governments and NGOs believe that major infrastructure has been overemphasised in national government reconstruction plans. Projects in sectors prioritised by marginalised groups, such as housing for the poor and income-generation for female-headed households, face the greatest challenges to implementation.

Where psychosocial counselling was included in other reconstruction activities, people were able to return to productive economic activities more quickly, and gender roles were carefully considered.

Many implementing agencies have not consciously engaged women because they assumed that their needs would be addressed in projects targeted to 'family wellbeing'. Ad-hoc assessments and a lack of gender analysis tools precluded the careful consideration of gender in reconstruction planning. Consultation with local populations, and with women in particular, has been limited. Women have been most involved in decision-making in instances where their participation was explicitly sought out. While men and women have been equally involved in hands-on project implementation when permitted to do so, many projects have been top-down and non-participatory due to the 'tyranny of the urgent'.

Lessons learned

Despite the substantial literature on gender and disasters, most actors in Central America did not consider the issue during their response to Hurricane Mitch. There are several reasons for this and they represent the major challenges to be overcome in order to mainstream gender considerations in disaster management. The tremendous time and resource pressures of Hurricane Mitch resulted in the 'tyranny of the urgent,' which overrides developmental concerns and sustainable approaches, including gender sensitivity. The lack of institutional familiarity with disaster management in general and the dearth of experience with post-disaster assessment methodologies led to a narrow view of disaster impacts as exclusively physical and precluded effective consideration of gender concerns. The lack of coordination between disaster response and long-term development was a significant limiting factor. The absence of institutional capacity in gender analysis and the apparent resistance to the inclusion of gender as an analytical construct further hampered the inclusion of gender. Gender analysis will enable agencies to contribute to the growing understanding of the links between disasters and development by identifying important considerations for social inclusiveness in the context of natural disaster mitigation, rehabilitation, reconstruction, and social transformation.

12.5.4   *Case Study 4 (LGA, 2007): Floods 2007, UK*

Theme: Infrastructure planning for disaster mitigation

The flooding that hit the UK in June and July 2007 was amongst the worst in the country's recorded history. Areas of Northern Ireland, north-east England, the Midlands and Wales were particularly badly affected. Thirteen people were killed and around 350 000 people in Gloucestershire were left for more than 2 weeks without clean water. The RAF mounted its biggest ever peacetime rescue operation to airlift stranded people to safety. By the end of the summer, the Association of British Insurers had put the cost of the damage at £3 billion.

A rising percentage of buildings are already at serious risk of flooding (500 000). With planned new development, up to one-third of development in the south-east will be on flood plains and thousands of additional homes, schools and businesses may be at risk. If building homes and businesses in high risk areas is to continue, the design of buildings, landscaping and flood defences should be re-visited and the creation of new flood plains should be considered. In cases where development is essential in flood risk areas, adaptive measures which make buildings more resilient to flood events should be encouraged. Even where development is proposed outside the floodplain, increased run-off rates from hard surfaces will add to flood risks elsewhere. Therefore, sustainable drainage solutions need to be incorporated as a requirement rather than an optional extra.

To effectively manage water supply and flood risk, the links to strategic planning, guidelines on the design, construction and location of buildings and infrastructure, and responsibility for drainage systems, should be looked at. Crucial infrastructure was at serious risk of flooding during summer 2007. The Gloucestershire floods demonstrated how vulnerable the UK is to loss of sub-stations and water infrastructure. It is crucial to establish more integrated water planning, so that the infrastructure, drainage systems and wider planning policies work together to promote sustainable drainage and water management and reduce the risk of flooding. To improve current flood defences, strategic planning should be linked to an investment plan and to a duty for all stakeholders to cooperate in development.

Lessons learned

Given recent extreme weather and the general acceptance that climate change will have a significant effect on future weather patterns, adapting buildings at risk of flooding is a priority. This includes looking at the way buildings are constructed, designed and used, and what materials, furnishings and fittings are used. Accordingly, a review of planning controls for construction on the flood plain or near watercourses should be undertaken, and guidelines prepared for the types of buildings that should be built in these places. Also, government needs to look at a duty on occupiers and the insurance industry to repair damaged buildings or infrastructure with the most resilient materials and designs, rather than simply reinstating fittings and materials that won't better resist any future flood damage. Central and local government need to work with the insurance industry to encourage home flood protection, possibly through insurance discounts and to raise awareness of simple actions that reduce the risk of local flooding. In adapting buildings to cope with an increased risk of flooding, existing and new buildings need to be considered in terms of how to fund adaptation and the incentive of initial costs versus long-term savings for occupiers, insurers, local and central government.

These four simple case studies highlight different lessons and good practices that could be captured and shared in view of eliminating unsafe conditions to reduce the risk associated with future disasters, and increase the efficiency and effectiveness of post-disaster reconstruction work after a disaster has occurred. As these cases illustrate, knowledge creation and sharing in the form of lessons learned and good practices in a post-disaster management context, have the potential to contribute towards improved resilience.

12.6 Capacity enhancing and knowledge strategies: the ISLAND project

In view of addressing the perceived need to share knowledge relating to disaster management strategies, the School of the Built Environment at the University of Salford undertook a 12-month research project '*ISLAND*' (Inspiring Sri-Lankan reNewal and Development), funded by the RICS Education Trust. The research aimed at increasing the effectiveness of disaster management by facilitating the capturing and sharing of appropriate knowledge and good practices in land, property and construction. Due to the broad scope of disaster management-related activities, this initial research focused on creating a knowledgebase on the post-tsunami response, with specific reference to case material in Sri Lanka. The broad aim of the research was addressed by:

- Creating an infrastructure for developing, sharing and disseminating knowledge about disaster management, particularly mitigation measures, for land, property and construction.
- Developing a knowledgebase on disaster management strategies arising from post-tsunami recovery efforts.
- Developing case materials on post-tsunami response in Sri Lanka.

Although the initial research focused on tsunami mitigation strategies and Sri Lanka in particular, the infrastructure developed during the project is scalable to permit growth in the knowledgebase to address other aspects of disaster management.

12.6.1   ISLAND website

The ISLAND web portal and knowledgebase was developed to capture, process, and disseminate the lessons learned from the Indian Ocean tsunami in the form of policy advice and good practices to guide future post-disaster interventions. Hence, the web portal provides an organised common platform to capture, organise and share the knowledge on disaster management strategies and to create a versatile interface among users from Government, professional bodies, research groups, funding bodies and local communities (visit http:/veber.buhu.salford.ac.uk/island/index.php).

Figure 12.3   ISLAND website home page

The knowledgebase (shown in Figure 12.3) was created to address several themes of disaster management based on published case materials collected on the Asian tsunami disaster of 2004, particularly from the Sri Lankan context. Case materials are organised into type of disaster, phase, country, source, the research methodology followed, scope and method of access. The cases are stored in a MySQL database using a PHP-Database interface. With the usage of SQL query, simple and advanced searches are provided to retrieve and view data. Separate descriptions were developed for each and every case material to provide an overview on the article, which is then linked to their respective original source. Article descriptions included the title, author, year, an overview, keywords and the link to the original article. Also a keyword search function is provided to search the relevant keywords in the provided description of the case materials.

In addition, the website provides an introduction to the ISLAND project and project output together with the publications of the project. The web portal acts as the public interface to share and disseminate the lessons learned, and good practices on disaster management.

12.6.2   Analysis on good practices and lessons learned

This section presents an analysis on the good practices and lessons learned from the tsunami disaster based on case material collated, with a particular emphasis on the Sri Lankan context. Good practices and lessons learned are summarised into several themes that emerged from the case material: social, technical, operational, legal and environmental.

Social issues

The importance of community participation within the reconstruction process, public awareness and education, and job creation programmes like Cash for Work (CFW) are emphasised in most of the case material collated. Within the last decade, the growing recognition of the necessity for community participation for sustainable disaster reduction was translated into actions to realise community-based disaster management. Major benefits of community-based risk assessment, mitigation planning and implementation processes included (Houghton, 2005) building confidence, pride in being able to make a difference, and enhanced capabilities to pursue disaster preparedness and mitigation, as well as bigger development responsibilities at the local level. Additionally, individual and community ownership, commitment and concerted actions in disaster mitigation, including resource mobilisation, produce a wide range of appropriate and innovative mitigation solutions, which can be cost-effective and sustainable.

As Doocy et al. (2006) note, job creation programmes have been used to provide aid to less well-off citizens, and can be considered as antecedents to CFW, which are an increasingly common element of humanitarian assistance in food-insecure settings, disaster-affected areas and post-conflict environments. The tsunami of 26 December 2004 caused massive devastation and hundreds of thousands of people were no longer able to participate in their routine employment activities. Considering the benefits of harnessing idle labour in the immediate post-disaster period, cash for work programmes can be recognised as a logical response that provide a structured mechanism to engage people in low-skilled constructive activities while injecting cash into the economy and promoting decision-making at the community and individual level.

Experiences in implementing large-scale CFW programmes (Houghton, 2005; Doocy et al., 2006) in the Asian post-tsunami phase have led to the following set of recommendations for CFW programmes:

- Communities need to be informed of benefits and limitations of CFW.
- It is helpful to identify potential community coordinators as well as to be aware of what other agencies are doing.
- Adequate attention should be paid early on to procurement, warehousing and the delivery of supplies and equipment in order to expedite CFW activities.
- Cash for planning is seen as a good way of working; aids participation in the planning process and promotes informed choice.
- CFW implementers either limit the need for technical expertise by providing simple project design or ensure the availability of skilled labour needed to complete CFW activities.
- Train local staff to lead these programmes and for community leadership.
- Monitoring through unannounced visits to work groups is an effective way of guaranteeing compliance and pinpointing problems. In cases of repeat

problems, the only solution may be to stop CFW activities so as to maintain a strong reputation across the implementation area.

- Weekly or less regular payments may be more workable from a management perspective once the immediate crisis is over and there is no longer a daily need for cash.
- Consider work groups with no more than 25 workers and a ratio of no more than four work groups to one supervisor (overall maximum ratio of 100 workers: 4 group leaders: 1 area supervisor) to ensure quality and efficient work.
- Slowing CFW activities as the programme nears completion rather than abruptly curtailing them is an option to consider as CFW programmes transition to more development-oriented activities.
- There is a need for synergy in communication and coordination between organisers and the community.

Public awareness and education is also a social issue and essential to protect people and property from disasters. A lack of awareness has been identified as a major reason behind the huge loss of lives and property after the 2004 Asian tsunami. As Briceno (2005) states, in Thailand more than 1800 people were saved because a tribal chief recognised that there was something wrong and decided to evacuate his people up to the hills. A 10-year-old girl from England saved 100 tourists on a beach in Phuket, Thailand after alerting her mother of the imminent tidal wave and prompting a speedy evacuation to safety. She recognised the signs after learning about tsunamis in her geography class. Knowing what to do and when to do it is the key. The media also have a social responsibility to promote prevention. Journalists need to be sensitised and maintain an ongoing focus on prevention aspects of disasters (Briceno, 2005). Disasters are happening on an almost daily basis around the world; the media's role in early warning systems should not to be overlooked.

Technical know-how

The tsunami affected two-thirds of the coastline of Sri Lanka, and it also resulted in the destruction of nearly 100 000 houses and infrastructure including roads and bridges (UNDP, 2005). Depending on the wave height, various types of structures were affected. Waves of up to 2 metres in height caused 1–2 metre high boundary walls to collapse. As wave heights increased, single-storey masonry structures were significantly damaged and were completely swept off their foundations at wave heights of around 4 metres. Buildings of two storeys and higher, especially those with concrete frames, had their infill masonry walls that were perpendicular to the waves knocked down by waves of up to 4 metres, but waves of even 5 metres did not cause the complete collapse of such buildings (Dias *et al.*, 2006). Partial collapse occurred, however, if foundations were undermined by waves of 3–5 metres in height.

According to Dias *et al.* (2006), there are two common threads that run through the structural failures. The first is that structures have to be tied down in addition to being held up. The latter is obviously the focus of everyday attention, since gravity loads will assert themselves almost immediately otherwise. However, when natural disasters such as cyclones and tsunamis occur, they have the effect of trying to lift up or push aside structures. Such actions can be resisted only by having a continuous chain of tying down from roof to foundation, and also by having sufficient gravity load to resist the overall upward or lateral forces. The second thread is that soil scouring has to be accounted for, or anticipated (Dias *et al.*, 2006). This can be done by: improving the soil properties, especially soil that has been backfilled; deepening foundations, whether in buildings or bridges; and also by providing sufficient structural redundancy to prevent catastrophic collapse even if some foundations fail. The strategic use of natural features such as sand dunes and provision of vegetation barriers are also ways of mitigating potential tsunami damage (University of Washington, 2007). In Sri Lanka, newly published national guidelines for reconstruction emphasise the importance of tying down structures against upward and lateral loads as well as the need to anticipate and reduce soil scour around foundations, especially of backfilled earth.

However, not only buildings were destroyed due to tsunami tidal waves; civil engineering structures like roads and bridges were also damaged. An investigation into infrastructural damage in Sri Lanka due to the tsunami (Kusakabe *et al.*, 2005) revealed the following:

- Damage to roads induced by the tsunami included erosion of embankments, erosion of abutment backfills, and collapse of bridges following the loss of stability of the abutments.
- Erosion of embankments tended to have occurred at locations where the land was relatively low, presumably because the back flow of the tsunami concentrated on those parts of the land.
- No bridge girders were washed away by the direct impact force of the tsunami. However, it is too optimistic to conclude that bridges are always safe against the impact force of a tsunami.
- Existence of detours and quick restoration weakened the socio-economical impact of the damage.

Operational issues

Coordination is often a scarce resource in disasters, yet remains the key operational principle for effective response. It is important in order to avoid duplication of effort so that resources are directed to those most severely affected by the disaster. Good coordination can also facilitate lesson learning. The importance of effective coordination of disaster management work at international, regional, national, organisational, group and individual level is overwhelmingly highlighted within case material. Reducing risk depends on communication and knowledge exchange between the scientific community and politicians. The Asian tsunami disaster showed that in the absence of an open dialogue, valuable knowledge and research from technical sectors is redundant. As Senanayake (2005) argues, there was a striking absence of expertise and professionals from the region in the post-tsunami operation in Sri Lanka, despite the stated aim to develop regional disaster response capacity in the Asia Pacific Region by a number of agencies. Hence, it is necessary to strengthen the link between scientific institutions and national and local authorities that need to react to avoid human, economic and social losses from disasters.

International, regional and national organisations should work better together and be better coordinated. Coordination of the entire UN system, governments and non-governmental organisations is an essential element of disaster prevention, mitigation, preparedness and response. Efforts need to be made to promote complementarity and avoid duplication (Briceno, 2005). A number of reports (Briceno, 2005; Senanayake, 2005; UNDP, 2005) emphasise the primary role of national authorities in coordinating and directing national and international assistance. Existing interagency coordination arrangements should be further strengthened, particularly concerning the sharing of information and knowledge in the early phases of disaster response. Mechanisms should be devised to ensure the participation of smaller organisations and institutions with less international experience in the coordination process. Further, governments need to demonstrate their political will and commitment to disaster-risk reduction through concrete measures, e.g. reserve national budget line for disaster reduction, and strategic donor funds to support and build capacity for disaster risk management.

Legal concerns

Coastal zones and small islands are often densely populated areas that increase people's risk and vulnerability. Nearly 3 billion people, or almost half the world population, live in coastal zones which in many cases are prone to hazards including tropical cyclones, floods, storms and tsunamis (Briceno, 2005). Often coastal populations are dependent on the sea for their livelihoods (e.g. fishing villages) and do not have the choice to live elsewhere. Small island countries such as the Nicobar and Andaman islands are barely a few metres above sea level, which means that evacuation to higher land is almost impossible. Governments and local authorities need to take human habitats into consideration in long-term development planning, thereby ensuring that risks are minimised.

Beyond preparing for evacuation and emergency response, communities can reduce their tsunami risk by modifying their land use planning and development approval practices. Although planning for tsunamis will not be a top priority for most coastal communities, relatively small efforts to plan for this hazard can significantly increase community safety. The US National Tsunami Hazard Mitigation Program's publication 'Designing for Tsunamis' stresses the importance of understanding site planning. Through zoning, creation of open space and not allowing new development in potential tsunami areas, safer land use will be better able to protect people and buildings. Specific site planning strategies to reduce tsunami risk can include (University of Washington, 2007):

- *Avoiding inundation areas*: Site buildings or infrastructure away from hazard area or locate on a high point.
- *Slowing water*: Forests, ditches, slopes, or berms can slow down waves and filter out debris. The success of this method depends on correctly estimating the force of the tsunami.
- *Steering*: Water can be steered to strategically placed angled walls, ditches and paved roads. Theoretically, porous dykes can reduce the impact of violent waves.
- *Blocking*: Walls, hardened terraces, berms and parking structures can be built to block waves.

Several reports (Briceno, 2005; Government of Indonesia and UN, 2005; Government of Sri Lanka and UN, 2005) emphasise the necessity for a national and institutional level legislative framework governing disaster management efforts. From the institutional point of view, the law should bring about a reform of the entire national institutional arrangement for disaster management, provide for the allocation of resources for preparedness and emergency response at all levels of governance, and create a permanent liaison mechanism with the international humanitarian community. Decentralisation of decision-making authority should feature prominently in the new set up. Administratively, such law should promote the development of detailed contingency plans at local level. These plans should include (Government of Indonesia and UN, 2005):

- Risk analysis, zoning and mapping.
- Comprehensive air, sea and road transportation arrangements (including stand-by agreements with the national air carrier and ship companies).
- The pre-positioning of relief supplies and, notably, of fuel.
- Backup emergency communications arrangements, notably assigning an institutional role to amateur radio communications.

Further, drafting of a National Disaster Management Bill is recommended, which should:

- Deal with the creation of policies/provisions/regulations at sectoral level to enable special conditions applicable for emergency response.
- Formulate operating policies for the mobilisation of military assets in disaster management and emergency response.
- Regulate the role of NGOs in the national setup for disaster response.
- Specify provisions for the request and reception of international assistance.

Environmental concerns

The tsunami reduced some coastal communities to piles of bricks, tin and wood mixed with car and boat parts, construction materials, ocean mud and dead bodies. While the December 2004 tsunami killed more than 250 000 people in 12 countries and left millions of homes and businesses in ruin, officials were most worried about the killer wave's environmental aftermath. In many cases, the tsunami worsened pre-existing environmental management problems on inhabited islands. The Joint United Nations Environment Programme (UNEP) and United Nations Office for the Coordination of Humanitarian Affairs's (OCHA) Environment Unit (Joint Unit), integrated in the Emergency Services Branch of the OCHA, is the principal United Nations mechanism mandated to assist countries facing environmental emergencies.

As Casey *et al.* (2005) argue, a common good practice approach to debris removal should be developed to minimise negative environmental impacts. Related guidance material should be translated into local languages and effectively disseminated. Re-mapping affected areas before redevelopment begins can ensure the identification of hazardous areas created by tsunami-induced changes, such as mass graves and locations vulnerable to flooding. Remapping is therefore an important tool to help to ensure that tsunami victims do not face new dangers when they resettle, and can also reassure affected populations of the safety of the locations where they rebuild. In this regard, local expertise and capacities in recycling, composting and environmental management can play a key part in clearing efforts. However, as Pasche and Kelly (2005) argue, immediately following the tsunami, the inclusion of environmental issues into disaster management efforts at the national level, was limited. Another revelation that emerged from several studies is that many operational agencies of the United Nations system have very little awareness of the potential environmental threats in the aftermath of disasters.

In Sri Lanka, the key environmental findings from the Rapid Environmental Assessment (REA) carried out on the 2004 tsunami disaster included the following (Casey *et al.*, 2005):

- While there is damage to the natural and built environment in affected coastal areas, there are no major life-threatening environmental emergencies as a result of the tsunami.
- Specific coordination needs to enhance environmental risk mitigation efforts.
- Remapping needs to ensure effective reconstruction efforts.
- Areas of acute environmental concern requiring immediate attention include management of tsunami waste and debris, and sanitation and sewage issues in settlements.

Calvi-Parisetti and Pasche (2005) suggests the process to be followed when assessing environmental impacts after a tsunami disaster. Assessments carried out in the first 48–72 hours after a major disaster should aim at identifying major secondary risks through a relatively simple checklist that should become a standard feature of the overall emergency assessments. If such risks are identified, expertise should be quickly mobilised for further assessments and a timely response. The initial environmental assessment should also look at those issues that are not immediately life-threatening but may become so at a later stage if not dealt with immediately. Once the most acute phase of the response is over, the environmental consequences of the disaster on the livelihood of the affected population should be assessed and programmes should be designed to address them. Finally, the environmental consequences of the disaster on ecosystems and habitats should be assessed so that they may be addressed through programmes in the reconstruction and rehabilitation phase.

This section presented an overview of the types of good practices and lessons learned that emerged from the 2004 Asian tsunami disaster. These examples demonstrate that vital knowledge to reduce the risk associated with future tsunami events is already available. However, if society is to effectively address disaster risk during post-disaster reconstruction work, it must share these good practices and integrate them into practice. Future research must aim at increasing the effectiveness of disaster-management activities through sharing and disseminating appropriate knowledge and good practices.

12.7 Summary

The reconstruction phase after a disaster, as a major driver for change and with the support of large investment, represents a significant opportunity for improved resilience. The techniques learned and the expertise developed could be applicable elsewhere in the country, within the region or across the world. It is important that effective mitigation actions are promoted as far as possible beyond the affected area so that other areas at risk from similar hazards, can address those challenges with the benefit of knowledge generated by others. The experiences of the disaster and the reconstruction and the mitigation measures it engenders, should be shared and exported with relevant adaptations to people and institutions that need it most. Hence, effective knowledge sharing should reduce the risk of future disasters through well-informed mitigation and preparedness planning. Ensuring the availability and accessibility of accurate and reliable disaster risk information when required, entails an efficient system for knowledge sharing. Tacit knowledge, although highly personal and context-specific, facilitates a capability for adapting to and shaping real-world environments, and is therefore an important determinant of effective performance in the dynamic, uncertain and unpredictable disaster management situations. International exchange of good practice and knowledge sharing among practitioners, authorities and institutions, can significantly contribute to the increased efficiency and effectiveness of post-disaster reconstruction at all levels after a disaster has occurred. However, disaster management knowledge remains fragmented and a lack of effective sharing and dissemination of knowledge is one of the major reasons behind the unsatisfactory performance levels present in current disaster management practices. In this regard, ISLAND represents an early attempt to increase the effectiveness of disaster management by facilitating the capture and sharing of appropriate lessons and good practices in land, property and construction.

References

Al-Ghassani, A.M., Anumba, C.J., Carrillo, P.M. and Robinson, H.S. (2005) Tools and techniques for knowledge management. In: Anumba, C.J., Egbu, C.O. and Carrillo, P. (Eds), *Knowledge Management in Construction.* Oxford: Blackwell Publishing Ltd.

Baumard, P. (1999) *Tacit Knowledge in Organisations.* London: Sage.

Blumentritt, R. and Johnston, R. (1999) Towards a strategy for knowledge management. *Technology Analysis and Strategic Management,* **11**, 287–300.

Bresnen, M., Edelman, L., Newell, S., Scarbrough, H. and Swan, J. (2003) Social practices and the management of knowledge in project environments. *International Journal of Project Management,* **21**, 157–166.

Briceno, S. (2005) *10 Lessons Learned from South Asia Tsunami of 26th December 2004.* Geneva: ISDR, UN.

Buckle, P. and Marsh, G. (2002) *Local Assessment of Disaster Vulnerability and Resilience: Reframing Risk.* Brisbane: International Sociological Association (ISA).

Calvi-Parisetti, P and Pasche, A. (2005) *Learning and Using Lessons: Environmental Impacts during the Indian Ocean Tsunami Disaster.* www.uneptie.org/pc/apell/events/pdffiles/6agee/lessons_learnde_present.pdf [accessed 13/03/2007].

Casey, E., Kelly, C., Negrelle, R. and Pasche, A. (2005) *Indian Ocean Tsunami Disaster of December 2004: Rapid Environmental Assessment in Sri Lanka.* Switzerland: UNEP/OCHA.

Celebi, M., Page, R. A. and Seekins, L. (1996) *Building Safer Structures.* http://quake.usgs.gov/prepare/factsheets/SaferStructures/ [accessed 21/10/2009].

Coghlan, A. (1998) Post-Disaster Redevelopment. Paper presented at the International Sociological Association (ISA), Montreal, Quebec.

Davis, I. (2005) *What Makes a Disaster.* http://tilz.tearfund.org/Publications/Footsteps+11-20/Footsteps+18 [accessed 18/12/2006].

De Guzman, M. (2001) *Towards Total Disaster Risk Management Approach.* Japan: Asian Disaster Reduction Center (ADRC).

Delaney, P. and Shrader, E. (2000) *Gender and Post-Disaster Reconstruction: The Case of Hurricane Mitch in Honduras and Nicaragua.* Decision Review Draft, Gender Team, World Bank.

Delgado, M. (2000) *Community Social Work Practice in an Urban Context: The Potential of a Capacity Enhancement Perspective.* New York: Oxford University Press.

Dias, P., Dissanayake, R. and Chandratilake, R. (2006) Lessons learned from tsunami damage in Sri Lanka. *Journal of Civil Engineering,* **159**, 74–81.

Doocy, S., Gabriel, M., Collins, S., Robinson, C. and Stevenson, P. (2006) Implementing cash for work programmes in post-tsunami Aceh: experiences and lessons learned. *Disasters,* **30**, 277–296.

Egbu, C.O. and Robinson, H. (2005) Construction as knowledge based industry. In: Anumba, C.J., Egbu, C.O. and Carrillo, P. (Eds), *Knowledge Management in Construction*. Oxford: Blackwell Publishing Ltd.

Empson, L. (2001) Introduction: human relations. *Professional Service Firms* (Special Issue on Knowledge Management), **54**, 811–817.

EPC-Environmental Planning Collaborative (2004) *Participatory Planning Guide for Post-Disaster Reconstruction, Indo-US Financial Institutions Reform and Expansion (FIRE-D) Project*. www.tcgillc.com/tcgidocs/TCGI%20Disaster%20Guide.pdf [accessed 21/10/2009].

Galbraith, J.R. (1974) Organisation design: an information processing view. *Interfaces*, **4**, 28–36.

Gieskes, J.F.B. and Broeke, A.M. (2000) Infrastructure under construction: continuous improvement and learning in projects. *Integrated Manufacturing Systems*, **11**, 188–198.

Government of Indonesia and United Nations (2005) *Post-Tsunami Lessons Learned and Best Practice Workshop: Report and Working Groups Output*. www.humanitarianinfo.org/sumatra/reference/workshop/docs/GoI-UN-LessonLearnedBestPracticesWorkshop_16-17May2005.pdf [accessed 13/03/2007].

Government of Sri Lanka and United Nations (2005) *Post-Tsunami Lessons Learned and Best Practices Workshop in Sri Lanka*. www.tsunami-evaluation.org/NR/rdonlyres/219D730D-DBD2-4604-98D1-07210642B127/0/OCHA_lessons_learned_workshop_20050609.pdf [accessed 13/03/2007].

Grey, D. (1996) What is knowledge. *The Knowledge Management Forum*. www.km-forum.org/what_is.htm [accessed 12/03/2004].

Herrgard, T. H. (2000) Difficulties in the diffusion of tacit knowledge in organizations. *Journal of Intellectual Capital*, **1**, 357–365.

Homan, M.S. (2005) *Promoting Community Change: Making it Happen in the Real World*. Belmont, CA: Wordsworth/Thompson Learning.

Houghton, R. (2005) *Key Findings and Lessons: The First 4-6 Months of the Tsunami Response*. Working document available online www.humanitarianinfo.org/srilanka/catalogue/Files/Info%20Centre/TEC/TEC001_Key%20Findings.pdf [accessed 13/03/2007].

Jigyasu, R. (2002) From Marathwada to Gujarat – Emerging Challenges in Post-earthquake Rehabilitation for Sustainable Eco-development in South Asia. Proceedings of the First International Conference on Post-disaster Reconstruction: Improving Post-Disaster Reconstruction in Developing Countries, 23–25 May 2002, Université de Montréal, Canada.

Kaklauskas, A. Amaratunga, D. and Haigh, R. (2009) Knowledge model for post disaster management. *International Journal of Strategic Property Management*, June **2009**, 12–26.

Kobe Report (2005) *Capacity Building and Disaster Reduction, World Conference on Disaster Reduction*. www.unisdr.org/wcdr/thematic-sessions/thematic-reports/report-session-3-10.pdf [accessed 21/10/2009].

Kogut, B. and Zander, U. (1992) Knowledge of the firm, combinative capabilities and the replication of technology. *Organization Science*, **3**, 383–397.

Kramer, R. and Specht, H. (1983) *Readings in Community Organization Practice*. Englewood Cliffs, NJ: Prentice-Hall.

Kusakabe, T., Matsuo, O. and Kataoka, S. (2005) Introduction of a methodology to mitigate tsunami disaster by the pre-evaluation of tsunami damage considering damage investigation of 2004 tsunami disaster in the Indian Ocean. *Technical Memorandum of Public Works Research Institute*, **4009**, 207–218.

Lawther, P.M. (2009) Community involvement in post disaster re-construction – case study of the British Red Cross Maldives recovery program. *International Journal of Strategic Property Management*, June **2009**, 37–51.

Local Government Association (LGA) (2007) *Flooding Lessons Learned Review*. www.lga.gov.uk/lga/core/page.do?pageId=431599 [accessed 20/10/2009].

McCamish, E. (1998) The Role of Community Recovery Workers in Development Following A Disaster. Paper Presented at the International Sociological Association (ISA) Montreal, Quebec.

McDermott, R. (1999) Why information technology inspired, but cannot deliver knowledge management. *California Management Review*, **41**, 103–117.

McDowell, C. (2002) Involuntary resettlement, impoverishment risks and sustainable livelihoods. *Australian Journal of Disaster and Trauma Studies*, **6**. www.massey.ac.nz/~trauma/issues/2002-2/mcdowell.htm [accessed 19/06/2008].

Mohanty, S., Panda, B., Karelia, H and Issar, R. (2006) *Knowledge Management in Disaster Risk Reduction: The Indian Approach*. India: Ministry of Home Affairs.

Nonaka, I. and Takeuchi, H. (1995) *The Knowledge Creating Company: How Japanese Companies Create The Dynamics Of Innovation*. New York: Oxford University Press.

Nonaka, I., Konno, N. and Toyama, R. (2000) Emergence of Ba. In: Nonaka, I. and Nishiguchi, T. (Eds) *Knowledge Emergence: Social, Technical and Evolutionary Dimensions of Knowledge Creation*. Oxford: Oxford University Press.

Ofori, G. (2002) Developing the Construction Industry to Prevent and Respond to Disasters. Proceedings of the First International Conference on Post-Disaster Reconstruction: Improving Post-Disaster Reconstruction in Developing Countries, 23–25 May 2002, Université de Montréal, Canada.

Pasche, A. and Kelly, C. (2005) *Indian Ocean Tsunami Disaster of December 2004: Rapid Environmental Assessment in Sri Lanka*. www.benfieldhrc.org/disaster_studies/rea/environmental_assessment_rapid_ocha_unep_sri_lanka_indian_ocean_tsunami_disaster_december2004.pdf [accessed 13/03/2007].

Pardasani, M. (2006) Tsunami reconstruction and redevelopment in the Maldives: a case study of community participation and social action. *Journal of Disaster Prevention and Management*, **15**, 79–91.

Polanyi, M. (1958) *Personal Knowledge Towards a Post-Critical Philosophy*. London: Routledge and Kegan Paul.

RICS (2006) *Mind the Gap! Post-Disaster Reconstruction and the Transition from Humanitarian Relief*. London: Royal Institution of Chartered Surveyors.

Saint-Onge, H. (1996) Tacit knowledge: the key to the strategic alignment of intellectual capital. *Strategy and Leadership Journal*, **24**, March/ April, 10–16.

Scarborough, H., Swan, J. and Preston, J. (1999) *Issues in People Management: Knowledge Management: A Literature Review*. Wiltshire: Institute of Personnel and Development, The Cromwell Press.

Senanayake, D.R. (2005) *Humanitarian Assistance and the International Aid Architecture after the Tsunami: Lessons from Sri Lanka and India*. Singapore: Institute of South East Asia Studies.

Simon, H.A. (1957) *Administrative Behaviour: A Study of Decision Making Processes in Administrative Organisation*, 3rd edn. New York: Collier Macmillan Publishers.

Snowden, D. (2002) Complex act of knowing: paradox and descriptive self-awareness. *Journal of Knowledge Management*, **6**, 100–111.

Spender, J.C. (1996) Making knowledge the basis of a dynamic theory of the firm. *Strategic Management Journal*, **17**, 45–62.

Tiwana, A. (2000) *The Knowledge Management Toolkit*. New Jersey: Prentice Hall.

Tushman, M.L. and Nadler, D.A. (1978) Information processing as integrative concept in organisational design. *Academy of Management Review*, **3**, 613–24.

UNDP (2005) *The Post-Tsunami Recovery in The Indian Ocean: Lessons Learned, Success, Challenges And Future Action*. Geneva: Bureau for Crisis Prevention and Recovery, UN.

UNESCO (2009) *Knowledge Management and Education for Disaster Reduction*. http://www.grida.no/publications/et/ep3/page/2618.aspx [accessed 15/10/2009].

University of Gloucestershire (2007) *UK Higher Education Disaster Relief Project: Report and proposals*.

University of Washington (2007) *Tsunami Mitigation and Prevention*. http://courses.washington.edu/larescue/precedents/prevention.htm [accessed 13/03/2007].

Warfield, C. (2004) *The Disaster Management Cycle*. http://www.gdrc.org/uem/disasters/1-dm_cycle.html [accessed 22/12/2006].

13 Restoration of Major Infrastructure and Rehabilitation of Communities

Kaushal Keraminiyage

13.1 Introduction

Irrespective of the cause of a disaster, impacts tend to be multifaceted. Loss of lives, destruction of properties and disturbance to the normal way of life are among some of the main areas that create high disaster impacts. Disasters often damage personal property such as dwellings, shops and farms, as well as common infrastructure such as roads, bridges, and electrical and telecommunication networks. In fact, infrastructure damage is typically a key component of total direct losses that result from a disaster (Freeman and Warner, 2001). Availability of proper infrastructure is vital to the wellbeing of all communities. Thus, restoration of infrastructure is a prerequisite in post-disaster recovery, especially to ensure meaningful rehabilitation of affected communities. Accordingly, the objective of this chapter is to explore challenges of major infrastructure restoration after a disaster and its impact towards rehabilitation of communities.

The chapter begins with a discussion about the nature and scale of damage caused to infrastructure by reference to past disasters in a variety of countries. It continues with a detailed discussion about various infrastructure restoration efforts post-disaster, and their cost and social impacts. The discussion then expands into post-disaster infrastructure reconstruction challenges, ranging from policy level economic and capacity problems, to operational level social problems.

Later in this chapter, the concept of 'disasters as an opportunity' is discussed, with examples from previous post-disaster infrastructure reconstruction projects to highlight how post-disaster infrastructure reconstruction is an opportunity to improve quality. Further, the impact of post-disaster infrastructure reconstruction towards rehabilitation of communities is explored.

Post-Disaster Reconstruction of the Built Environment: Rebuilding for Resilience, First Edition.
Edited by Dilanthi Amaratunga and Richard Haigh.
© 2011 Blackwell Publishing Ltd. Published 2011 by Blackwell Publishing Ltd.

13.2   Impact of disasters on infrastructure

In search for a definition of infrastructure, Grimsey and Lewis (2002) state that infrastructure is easier to recognise than to define. However, looking from the investment point of view, they recognise that the following can be categorised under infrastructure:

- Energy (power generation and supply)
- Transport (toll roads, light rail systems, bridges and tunnels)
- Water (sewerage, waste water treatment and water supply)
- Telecommunications (telephones)
- Social infrastructure (hospitals, prisons, courts, museums, schools and Government accommodation)

In a similar vein, Edwards (2003) notes that the US President's Commission on Critical Infrastructure Protection (PCCIP) considers the following as fundamental to the definition of infrastructure:

- Transportations
- Oil and gas, production and storage
- Water supply
- Emergency services
- Government services
- Banking and financing
- Electrical power
- Information and communications

From these examples it is evident that 'infrastructure' provides essential goods and services to a community so that it can continue with its normal way of life. PCCIP (1997) highlights the following:

'Infrastructure means a network of independent, mostly private owned, man-made systems and processes that function collaboratively and synergistically to produce and distribute a continuous flow of essential goods and services.'

This clearly signifies the ever important distribution and continuous flow of essential goods and services as a key feature of infrastructure. The attributes 'independent' and 'mostly private owned' are more contentious. Hudson *et al.* (1997, page 30) note that, '. . . infrastructure consisting of physical systems can be owned and managed by either or both public agencies and private enterprises'. Further, they provide a more generic definition to infrastructure. In addition to the aforementioned transportation, utilities, energy and telecommunications, they also suggest that infrastructure refers to: '. . . physical systems

Table 13.1   Effects of disasters on infrastructure (adapted from Freeman and Warner, 2001)

Type of disaster	Infrastructure impact
Hurricane, typhoon and cyclone	• Damage to buildings, distribution and high-tension lines (through gusty winds) • Damage to bridges, buildings and roads (through flooding)
Drought	• Shrinkage damage to building foundations and underground infrastructure • Disruption to water supply
Flood	• Softening of building foundations • Buries buildings and other structures such as roads • Affects the function of hydro-power dams, water management systems
Tsunamis	• Destruction or damage to buildings, bridges, irrigation systems, roads, power distribution • Water pollution

or facilities that provide essential public services such as waste disposal, park lands, sports, recreational buildings and housing facilities.'

For the purpose of this chapter, infrastructure is considered as the systems, processes and resources that function to produce and distribute a continuous flow of essential goods and services to a community. Essentially, this includes, but is not limited to, transportation, energy, water, telecommunication and social infrastructure, including the examples given above.

The type of damage caused to infrastructure by disasters is discussed by Freeman and Warner (2001), who succinctly summarise findings from the literature (Table 13.1).

Irrespective of the type of damage, the scale of the impact towards infrastructure has been found to be proportionally high. For example, Freeman (2000) notes that direct damage to infrastructure in Asia alone approximates to nearly 50% of the total lending activity of the World Bank over the past decade, which amounts to US$25 billion annually.

The following discussion explores the nature and scale of disaster impact on infrastructure in two large scale disasters: the 2004 Asian tsunami on Sri Lanka and the 2005 Hurricane Katrina on the United States.

13.2.1   Impact of the 2004 tsunami on infrastructure in Sri Lanka

On 26 December 2004, an earthquake on the West Coast of Northern Sumatra set off a series of other earthquakes lasting for several hours, which resulted in a tsunami in the Indian Ocean. This led to a widespread devastation, particularly in Sri Lanka, India, the Maldives, Indonesia and Thailand, with damage also in

Malaysia, Bangladesh, Somalia, the Seychelles and Kenya. Sri Lanka, the 'pearl of the Indian Ocean', generally a country to be considered at low risk to natural disasters, faced one of the worst natural disasters recorded in its history. The December 2004 tsunami struck a relatively thin but long coastal area stretching over 1000 kilometres – two-thirds of the country's coastline. At least 40 000 people are known to have died (BBC, 2005). It destroyed more than 100 000 houses and the number of homeless people was put at between 800 000 and 1 000 000, from a population of 19 million (UNEP, 2005). The overall cost of damage to Sri Lanka was estimated at US$1 billion, with a large proportion of losses concentrated in housing, tourism, fisheries and transportation (ADPC, 2005). Coastal infrastructure, namely roads, railways, power, telecommunications, water supply and fishing ports were significantly affected. Some of these infrastructure facilities had been continuously used without any significant capacity upgrade for nearly 100 years. For example, the railway network in Sri Lanka had not been extended since it was introduced under British rule. Moreover, based on a study conducted by the Industrial Development Board of Sri Lanka, micro, small, medium, and large scale industries affected by the tsunami need a minimum Rs1.2 billon to re-establish their businesses. This study surveyed 4389 manufacturing ventures across the island and found that the reinstatement of infrastructure, including the road, railway and telecommunications networks, was a pre-requisite to the restoration of these and other core manufacturing businesses, which were so fundamental to the Sri Lankan economy and livelihoods. In fact, the nation's very recovery from the trauma of the December 2004 tsunami was very much dependent on the successful and speedy recovery of its facilities and infrastructure base. As such, even though the immediate priorities of the post-tsunami recovery activities were centred on the provision of basic requirements such as food, shelter and medicines to affected communities, the necessity to recreate infrastructure facilities destroyed during the disaster was considered as a priority when establishing the longer term recovery measures (ADPC, 2005).

13.2.2   Impact of Hurricane Katrina on infrastructure

Hurricane Katrina originally formed near the Bahamas on 23 August 2005. It moved across southern Florida causing some deaths and flooding before developing to a strong hurricane in the Gulf of Mexico. On 29 August it hit South East Louisiana and caused severe destruction along the coast from central Florida to Texas. The worst hit areas were New Orleans and Louisiana affected mainly by the flood created by the hurricane. At least 1600 people lost their lives during the hurricane and resultant floods which made it the deadliest hurricane in the USA since 1928. In addition to the loss of life, the total economic loss caused by Hurricane Katrina were estimated at over US$150 billion (Burton and Hicks, 2005), which makes it a hurricane with one of the largest economic impacts in the history of the United States. As Burton and Hicks (2005) further show, out of the estimated US$150 billion, highway damages amounted to US$3 billion, sewer system damages were estimated at US$1.2 billion, and electrical utility damages at US$231 million. Further, they note that this estimate excludes some key infrastructure damages, including those to the water system. Similarly, the National Alliance to Restore Opportunity to the Gulf Coast and Displaced Persons (TNAROGCDP) (2006) reveal that the post-Katrina repair cost to the New Orleans flood protection system was US$800 million, US$62 million more than it actually cost to initially put in place. As a way of summarising these losses, it is interesting to note that the US Army Corps of Engineers (USACE) (2009) report that as of December 2009, 215 construction contracts have been awarded with an estimated value of US$5.6 billion. This signifies the actual economic loss and impact to the infrastructure caused by the Hurricane Katrina.

13.2.3   Impacts in developed and less developed economies

In addition to the high impact on infrastructure visible in the above two cases, a significant difference can be seen between the two cases, in terms of death tolls and total economic losses. In the case of Sri Lanka, the death toll was in excess of 40 000, whereas in the USA, it was about 30 times less. The Royal Institution of Chartered Surveyors highlight that 24 out of 49 low-income developing countries face high levels of disaster risk (RICS, 2006) and that when facing a disaster hazard, they experience higher levels of mortality when compared with developed countries. Taking another example, the earthquake that hit central California in 2003, with a magnitude of 6.5 on the Richter scale, killed two and injured 40 people, whereas the earthquake that hit Iran four days later with a magnitude of 6.6, killed at least 26 000 people (NEIC, 2003).

On the other hand, the total economic losses reported in the Sri Lanka example were about US$1 billion, whereas in the USA, they were in excess of US$150 billion. This comparison is also indicative of a wider trend: developed economies tend to suffer higher economic losses in absolute terms, although developing countries' economic losses are often higher as a proportion of their GDP.

It is also important to note the post-disaster scenario. Not only do developing countries experience higher levels of mortality during a disaster; they generally require longer periods for post-disaster recovery. One of the reasons behind this is that developing countries take more time to restore their major infrastructure due to reasons such as financial and intellectual resource availability.

13.3   Impact of the failure of infrastructure in disaster-affected communities and infrastructure interdependencies

It is apparent from the aforementioned examples that disasters frequently have major impact on a region's infrastructure. Further, the impact on infrastructure actually creates a vicious cycle, amplifying the impact of the disaster on the affected community.

For example, the financial institutions located within the affected area of Hurricane Katrina informed the Federal Financial Institutions Examination Council Member Agencies and the Conference of State Bank Supervisors that they faced a range of challenges in responding to Katrina (FFIEC, 2009):

- Communications outages made it difficult to locate missing personnel.
- Access to and reliable transportation into restricted areas was not always available.
- Lack of electrical power or fuel for generators rendered computer systems inoperable.
- Multiple facilities were destroyed outright or sustained significant damage.
- Some branches and ATMs were underwater for weeks.
- Mail service was interrupted for months in some areas.

In addition to the direct inconvenience resulting from infrastructure failure, some infrastructure has interdependencies: the failure in one results in the failure of other interdependent infrastructure. Peerenboom *et al.* (2002) explain four types of infrastructure interdependencies. These are:

(1) Physical interdependency – material output of one infrastructure is used by another infrastructure.
(2) Cyber interdependency – infrastructure depends on information transmitted through information and communication infrastructure.
(3) Geographic interdependency – two or more infrastructures are co-located in the same areas that can be affected by a local event.
(4) Logical interdependency – state of one infrastructure on the state of another infrastructure in a way that is not physical, cyber or geographic (e.g. financial or government funding).

One or more of these interdependencies can exist between two infrastructures. For example, the public transport system in a particular area may have a physical interdependency with the local road network, while cyber interdependency may also co-exist whereby the local transport system is planned and managed using information transmitted through information and communication networks. Not only may multiple interdependencies exist between various infrastructures, but also interdependencies may also create a 'multiplier' effect. Consider an example where the power supply infrastructure has failed in a particular area. This may lead to the failure of communication infrastructure. Failure of communication infrastructure will lead to the failure of cyber interdependent infrastructure such as some local transport systems (e.g. underground system). Lack of local transport system may lead to the failure of other interdependent infrastructure creating a multiplier effect.

Leavitt and Keifer (2006) note that the increased use of advanced technologies and computer-based automation systems has both positive and negative impacts towards infrastructure: the increased reliability and efficiency that advance technologies can contribute in creating and managing infrastructure may actually make infrastructure more vulnerable in the event of a disaster, as the use of advance technologies increases the cyber interdependency. As noted above, infrastructure interdependencies make the impact of disasters on infrastructure more complex and unpredictable. Understanding these inter-dependencies will lead to better disaster preparedness and better post-disaster recovery. Since infrastructures in a particular locale may have evolved over a long period of time, studying and predicting the interdependencies of existing infrastructure may be a complex issue. However, from the opportunist's point of view, post-disaster infrastructure reconstruction and restoration provides a good starting point to plan and management infrastructure interdependencies effectively. The next section elaborates more on post-disaster infrastructure reconstruction and restoration.

13.4   Post-disaster infrastructure reconstruction and restoring major infrastructure

Restoration of major infrastructure is key in achieving post-disaster recovery. The impact of a disaster exists while the capability of the affected community cannot cope with the demand of an event or a series of events. As appropriate infrastructure is vital in building the capability of such communities and emergency systems, it is essential that the restoration of such infrastructure is one of the early activities of post-disaster recovery.

Reconstruction is one of four identifiable post-disaster stages: emergency, restoration, reconstruction and resettlement. Within the emergency stage, of-ten the main focus is on immediate relief. Activities such as search and rescue take place and provision of emergency shelter and feeding to the affected community is of prime importance. Often this is the stage where most of the funds, aid and the attention of the international community are attracted. The restoration stage focuses on the establishment of repairable essentials just after a disaster. Often activities such as removal of debris and establishment of tem-porary access routes are among the first activities within this stage. This stage can last weeks or months depending on the nature and scale of the devasta-tion. Towards the end of this stage, the pre-disaster capabilities of the affected community are partially restored; however, the community will not have suf-ficient resilience to face any further disturbances. The community resilience to disasters is often regained during the reconstruction stage, where capabilities of the community can be reinstated in a permanent manner. During this stage, a major focus is on rebuilding a community's infrastructure, which is required to ensure the permanent resettlement of the affected communities.

13.4.1   Post-Hurricane Katrina reconstruction challenges

Kates *et al.* (2006) provide an insightful evaluation of the reconstruction of New Orleans in the aftermaths of Hurricane Katrina. They reveal that the emergency phase for Hurricane Katrina lasted for about 15 weeks, whereas the restoration phase spanned from around week 3 until about week 45. Within this long restoration period, the reconstruction work co-existed, starting around as early as week 4 and continuing beyond the 1 year period. In 2006, based on historical data, Kates *et al.* predicted that the construction phase will last nearly 10 years, while the re-settlement phase may match that, resulting in a total post-disaster recovery period of 20 years.

Although infrastructure restoration of New Orleans began in the second week, the return rate of the evacuated local population was low. For that reason, the population-dependent infrastructure was not restored or used at the expected pace. Utilities and other related infrastructure such as electricity, gas, public transportation, schools, hospitals, and food stores were functioning at less than half of pre-Katrina capacity (Liu *et al.*, 2006). On a more positive note, Kates *et al.* (2006) found that the actual restoration period for the pre-existing levee system was 40 weeks, substantially shorter than the expected 60 weeks. The restoration of the pre-existing levee system, in this instance, indicated a significant restoration of the capabilities of the affected community, thus signalling the completion of the restoration phase. Despite this, the National Alliance to Restore Opportunity to the Gulf Coast and Displaced Persons (TNAROGCDP, 2006) noted that the pre-Katrina levee system was inadequate to face the level of challenges posted by hurricanes towards the City. It was further stated that while a category five storm protection system has been estimated to cost US$30 billion, no commitments were made to meet this need. Indeed, as the inadequacy of the pre-Katrina levee system has already been established, restoration of the existing levee system, without an attempt to upgrade, may amount to little more than a waste of resources, as it leaves communities in the region just as vulnerable to future disasters.

This example highlights the importance of effective reconstruction strategies to ensure that effective decisions are made that are in the best long-term interests of disaster-affected communities. Often post-disaster infrastructure reconstruction strategies are considered at a national level. However, attention must always be given to the special needs and circumstances of local communities. The United States' health care system is built upon a health insurance-based approach. Historically, New Orleans has been noted as an area with a high level (28%) of uninsured population (Rowland, 2009). Due to this reason, a part of the New Orleans community was traditionally dependent on the Charity and University Hospitals for their healthcare needs. As Rowland (2009) noticed, the impact of Hurricane Katrina made the situation even more demanding, as the Charity Hospital was damaged irreparably and the University Hospital was forced to close. This local demand for public healthcare meant that restoration of health-care infrastructure became a priority in post-Katrina New Orleans. Accordingly, the Department of Health and Human Services released US$100 million to progress in restoring healthcare facilities in post-Katrina New Orleans. This exemplifies the requirement and importance of considering the local priorities in infrastructure restoration.

It is also important to consider that post-disaster infrastructure reconstruction activities are often influenced by the level of urgency created by a disaster. Under the pressure of urgency to restore normalcy in a post-disaster situation, reconstruction projects may not always be implemented in the most economical manner and quality control may not always be given the highest priority. For example, it is often determined that not all post-disaster reconstruction projects can go through lengthy competitive tendering processes or comprehensive design stages; the situation sometimes demands that construction be started as early as possible and speedy completion is of paramount importance. In a disaster situation, bringing back the road network to operational level as quickly as possible is vital. This is largely due to the fact that most of the other infrastructures have a strong physical interdependency on the road network. In relation to this requirement for the speedy restoration of the road network after a disaster, Philips (2005) notes the lessons that can be learnt from previous disasters such as the 1994 Northbridge earthquake. In specific terms he suggests that competitive bidding and enforcement of labour standards such as the Davis-Bacon prevailing wage law can help ensure that work is done expeditiously, safely, cost effectively, and with maximum benefit to the local population. As noted earlier, while the general tendency is to think that strategies such as competitive bidding and enforcement of labour standards would slow down reconstruction, Philips's (2005) argument reveals the positive impact of the same. The argument is that the desired speedy construction can be achieved through the implementation of strategies such as competitive bidding and enforcement of labour standards under post-disaster situations, as those often result in securing capable and skilled labour forces. In addition, these strategies will bring the additional benefits of competitive pricing and high quality end products.

13.4.2 *Post-tsunami reconstruction in Sri Lanka*

When compared with the case of post-Hurricane Katrina in New Orleans, the case of infrastructure reconstruction in developing countries takes a different route. Infrastructure development in post-tsunami Sri Lanka provides some valuable insights. Developing countries often receive financial and other humanitarian support as immediate relief aid from international communities, non-governmental organisations and donor agencies. However, infrastructure reconstruction and restoration is often identified as primarily a national, sub-national and local government-led matter. As such, the donors and other organisations that work towards humanitarian relief, pay less attention to the restoration and reconstruction aspects of infrastructure within the disaster management lifecycle. Thus, not surprisingly, developing countries that witness disasters often fail to launch successful infrastructure restoration and reconstruction programmes due to a lack of financial and intellectual resources.

One of the problems that the governments of developing countries often face in relation to post-disaster infrastructure reconstruction is their response capacity. Generally, capacity at a local government level to plan and implement recovery strategies is usually very limited and often incapacitated as a result of the disaster itself. In light of this, strengthening national capacities for the restoration and reconstruction process, including disaster preparedness and long-term disaster risk management, has been identified as a main strategic objective in a United Nations report on post-tsunami recovery and reconstruction (UNDP, 2006).

Strengthening national capacities of developing countries towards post-disaster infrastructure reconstruction predominantly demands financial incentives. Since the financial incentives given by donors during a disaster generally go towards short-term relief efforts rather than long-term recovery programmes, often the governments of affected countries (specifically developing countries) are financially incapacitated to launch successful recovery programmes. As such, it has been identified by various agencies (e.g. RICS, 2006; UN, 2006) that the main focus of the donor organisations should be to achieve the appropriate balance of fund allocations between the immediate/short-term relief and the medium/long-term recovery. Similarly, while the UN system capacities for disaster response and humanitarian assistance are widely recognised as well developed, there is currently a vacuum in terms of capacities and accepted system wide mechanisms for post-disaster recovery, particularly those with a risk reduction focus (UNEP, 2005).

A further contributory factor that has hindered the implementation of successful post-disaster recovery plans is the lack of appropriate intellectual capacity. This refers to a lack of expertise and lack of training related to post-disaster recovery. With specific reference to post-tsunami recovery in Sri Lanka, a report by the Ministry of Disaster Management and Human Rights, Sri Lanka (MDMHR, 2006) stresses that an important aspect of any disaster management strategy is to anticipate the requirements for disaster-related awareness, education and training. A major programme identified under this theme includes increasing capacity among key institutions through training of officials and training aids or tools.

Although it is frequently overlooked in the relief phase of a disaster, construction's role as an essential part of infrastructure reconstruction activities following disasters, both natural and human-caused, is well documented. In particular, infrastructure reconstruction has been the subject of a continuous discussion with particular emphasis on developing countries that are less able to deal with the causes and impacts of disasters (see Karim, 2004; Lizarralde and Boucher, 2004; Nikhileswarananda, 2004; Young, 2004). Thus, the importance of improving the construction industries of developing nations is widely recognised, highlighting a need to equip them to manage the infrastructure reconstruction effectively (Ofori, 2002).

Construction is typically engaged in a range of critical activities: temporary shelter immediately after the disaster; restoration of public services such as hospitals, schools, water supply, power, communications, and environmental infrastructure, and state administration; and securing income earning opportunities for vulnerable people in the affected areas (World Bank, 2001). Further, there is growing recognition that the engineering community has a valuable role to play in finding and promoting rational, balanced solutions to what remains an unbounded threat (Sevin and Little, 1998) and that the construction industry has a much broader role to anticipate, assess, prevent, prepare, respond and recover from disruptive challenges. Peña-Mora (2004) suggests construction professionals have a key role to play because they are involved in the construction of the infrastructure, and therefore should also be involved when an event destroys that infrastructure. Construction engineers possess valuable information about their projects, and that information can be critical in disaster preparedness, as well as response and recovery. The information they possess may be the difference between life and death. Sevin and Little (1998) further suggest that computerised building plans, structural analysis programmes, and damage assessment models may all facilitate rapid rescue and recovery of victims in the aftermath of a disaster, and that these all require the active involvement of the construction professions.

Moreover, the impact of built environment-related disciplines to disaster recovery management are not limited to the so-called hard sciences. Managerial and strategic issues, such as the management of public and commercial infrastructure facilities, require a fine mixture of knowledge with regard to the technological and human aspects of the built environment. Even though the knowledge related to these aspects of the built environment is significantly developed within the West, developing countries such as Sri Lanka usually do not possess this knowledge at the desired level.

13.5 Post-disaster infrastructure reconstruction for improved quality of life

The opportunist sees the bright side of disasters, with the chance of a fresh start and an invaluable opportunity to rectify past mistakes. This is especially true when it comes to issues such as relocation of vital infrastructure and town and country re-planning. Despite the importance of timely post-disaster reconstruction and restoration of infrastructures, one should not rush to initiate 'restoring' disaster-damaged infrastructure, without giving due consideration to the issues that existed with the pre-disaster infrastructure. In addition, the likely opportunities should also be considered, given by the disaster, to rectify such issues. In some instances, especially within developing country settings, construction of infrastructure (pre-disaster) may have been influenced by factors such as community (population) build up or location-specific demands such as tourism. This often results in the unplanned establishment of infrastructure. These unplanned settlements may lead to problems such as dense residential settlements, congested road networks and uneven distribution of infrastructure and social facilities. Based on this scenario, the opportunity provided by the destruction of a disaster can be used to rectify the above ex-plained anomalies in unplanned infrastructure developments. In turn this can improve the quality of life of the affected community.

Poverty is the main challenge to quality of life. Poverty exists, in vary-ing degrees, both in developing and developed countries. As Freeman (2000) explains, clear links exist between disasters, poverty and infrastructure. He explains the linkages in three components as:

'. . . access to infrastructure is often a measure of poverty, infrastructure is a key component of economic growth, and the loss of infrastructure may have significant indirect and secondary costs that directly impact the poor'.

The World Bank considers lack of access to basic infrastructure – sanitation, electricity and clean water – as measures of poverty. Further, access to basic infrastructure usually determines the revenue generation capability of local communities. For example, access to water is a prime infrastructure require-ment to maintain the livelihood of paddy farmers. Providing this essential infrastructure at acceptable levels helps to facilitate economic growth of the county and contribute to a reduction in poverty. Infrastructure development plays an essential role in reducing poverty. In a study by Freeman (2000), the World Bank reported that greater infrastructure development for agricultural development was associated with a one-third increase in average household incomes, a 24% increase in crop income, a 92% increase in wage income, and income for livestock and fisheries increased by 78% (World Bank, 1991). Further, the Bank estimates that a 1% increase in the stock of infrastructure translates to a 1% increase in GDP.

This shows both the challenges and opportunities that disasters create to influence quality of life through poverty reduction. The opportunity is the mere elimination of issues attached with the ad-hoc infrastructure and the chance to re-plan the infrastructure to create and elevate impact to remove poverty through better availability of essential infrastructure to the local com-munities. The challenge is to overcome the issues attached with post-disaster infrastructure reconstruction.

13.6 Summary

This chapter began by establishing a working definition of infrastructure, thereby highlighting its importance to the functioning of society. Examples were given of the nature and the scale of damage caused to infrastructure by two previous disasters. The impact of this damaged infrastructure was then evaluated in respect of the affected communities.

The importance of post-disaster infrastructure reconstruction was established as a part of the post-disaster recovery process. The cases of post-tsunami Sri Lanka and the reconstruction of New Orleans after Hurricane Katrina were cited to examine some of the challenges associated with effective reconstruction.

Finally, the opportunist's view about disasters was examined to establish how post-disaster infrastructure reconstruction can help to improve the quality of life. The opportunity for better planned infrastructure and a consequential reduction of poverty were suggested as a counterargument to urgent reconstruction without due consideration to the long-term impact of decision making.

References

ADPC (2005) *Report on Post Tsunami Rapid Assessment in Sri Lanka.* Bangkok: Asian Disaster Preparedness Center (ADPC).

BBC (2005) At a glance: Countries hit. Available online. http://news.bbc.co.uk/1/hi/world/asia-pacific/4126019.stm [accessed: 20/02/2006].

Burton, M.L. and Hicks, M.J. (2005) *Hurricane Katrina: Preliminary Estimates of Commercial and Public Sector Damages.* Center for Business and Economic Research, Marshall University, Huntington, VA. www.marshall.edu/cber/research/katrina/Katrina-Estimates.pdf

Edwards, P.N. (2003) Infrastructure and modernity: force, time, and social organization in the history of sociotechnical systems. In: Missa, T.J, Bery, P. and Feenberg, A. (eds), *Modernity and Technology.* Cambridge, MA: MIT Press. pp. 185–225.

FFIEC (2009) Lessons Learned From Hurricane Katrina: Preparing Your Institution for a Catastrophic Event. The Federal Financial Institutions Examination Council [online]. http://www.ffiec.gov/katrina_lessons.htm [accessed 10 April 2009].

Freeman, P. and Warner, K. (2001) Vulnerability of Infrastructure to Climate Variability: How Does This Affect Infrastructure Lending Policies. Report commissioned by the Disaster Management Facility of the World Bank and the ProVention Consortium, Washington, DC.

Freeman, P.K. (2000) Infrastructure, natural disasters, and poverty. In: Kreimer, A. and Arnold, M. (eds), *Managing Disaster Risk in Emerging Economies.* Washington DC: World Bank. pp. 55–61.

Grimsey, D. and Lewis, M.K. (2002) Evaluating the risks of public private partnerships for infrastructure projects. *International Journal of Project Management,* **20,** 107–118.

Hudson, W.R., Haas, R. and Uddin, W. (1997) *Infrastructure Management.* New York: McGraw-Hill.

Karim, N. (2004) Options for Floods and Drought Preparedness in Bangladesh. Proceedings of the Second International Conference on Post-disaster Reconstruction: Planning for Reconstruction, 22-23 April 2004, Coventry University, UK.

Kates, R.W., Colten, C.E., Laska, S. and Leatherman, S.P. (2006) Reconstruction of New Orleans after Hurricane Katrina: a research perspective. *Proceedings of the National Academy of Sciences,* **103,** 14653.

Leavitt, W.M. and Kiefer, J.J. (2006) Infrastructure interdependency and the creation of a normal disaster: the case of Hurricane Katrina and the City of New Orleans. *Public Works Management and Policy*, **10**, 306.

Liu, A., Fellowes, M. and Mabanta, M. (2006) Special Edition of the Katrina Index: A One-Year Review of Key Indicators of Recovery in Post-Storm New Orleans. Brookings Institution Metropolitan Policy Program. http://www.brook.edu/metro/pubs/2006_KatrinaIndex.pdf.

Lizarralde, G. and Boucher, M. (2004) Learning from Post-Disaster Reconstruction for Pre-Disaster Planning. Proceedings of the Second International Conference on Post-disaster Reconstruction: Planning for Reconstruction, 22–23 April 2004, Coventry University, UK.

MDMHR (2006) A road map for disaster management. In: *Towards a Safer Sri Lanka*. Colombo: Ministry of Disaster Management and Human Rights.

NEIC (2003) US Geological Survey, National Earthquake Information Center World Data Center for Seismology, Denver. http://www.earthquakeadvisor.com/articles/00069.html [accessed 27/12/2006].

Nikhileswarananda, S. (2004) Post-Disaster Reconstruction Work in Gujarat on behalf of Ramakrishna Mission. Proceedings of the Second International Conference on Post-disaster Reconstruction: Planning for Reconstruction, 22–23 April 2004, Coventry University, UK.

Ofori, G. (2002) Developing the Construction Industry to Prevent and Respond to Disasters. Proceedings of the First International Conference on Post-Disaster Reconstruction: Improving Post-Disaster Reconstruction in Developing Countries, 23–25 May 2002, Université de Montréal, Canada.

PCCIP (1997) Critical Foundations: Protecting America's Infrastructures. Report of the President's Commission on Critical Infrastructure Protection [online]. http://www.fas.org/sgp/library/pccip.pdf

Peerenboom, J., Fisher, R.E., Rinaldi, S.M. and Kelly, T.K. (2002) Studying the chain reaction. *Electric Perspectives*, **27**, 22.

Pena-Mora, W. (2005) Collaborative First Response to Disasters Involving Critical Physical Infrastructure. O'Neal Faculty Scholar Seminar from University of Illinois, September 19, 2005.

Philips, P. (2005) Lessons for Post-Katrina Reconstruction. Washington DC: Economic Policy Institute, Briefing paper. http://www.policyarchive.org/handle/10207/bitstreams/8035.pdf

RICS (2006) *Mind the Gap! Post-Disaster Reconstruction and the Transition From Humanitarian Relief.* London: Royal Institution of Chartered Surveyors.

Rowland, D. (2009) Health Care in New Orleans: Progress And Remaining Challenges. Post Katrina Recovery: Restoring Healthcare in New Orleans Region. [online]. Committee on Oversight and Government Reform. http://www.kff.org/uninsured/upload/8026.pdf

Sevin, E and Little, R. (1998) Mitigating terrorist hazards. *The Bridge*, **28**, 3–9.

TNAROGCDP (2006) *The Response since Katrina: Accomplishments and Challenges. Gulf Coast Revival Fact Sheet*, The National Alliance to Restore Opportunity to the Gulf Coast and Displaced Persons [online]. http://www.linkedfate.org/documents/Factsheet%20B_Response%20Since%20Katrina.pdf

UNDP (2006) *Post-Tsunami Recovery and Reconstruction Strategy.* Colombo: United Nations Development Programme.

UNEP (2005) Natural Rapid Environmental Assessment – Sri Lanka. UNEP, Sri Lanka, Country Report. Colombo, Sri Lanka: United Nations Environment Programme.

USACE (2009) *Greater New Orleans Hurricane and Storm Damage Risk Reduction System. Fact and Figures.* US Army Corps of Engineers.

World Bank (2001) World Bank and Asian Development Bank Complete Preliminary Gujarat Earthquake Damage Assessment and Recovery Plan. http://www.worldbank.org/gujarat; http://www.adb.org/

Young, I. (2004) Monserrat: Post Volcano Reconstruction and Rehabiliation – A Case Study. Proceedings of the Second International Conference on Post-disaster Reconstruction: Planning for Reconstruction, 22–23 April 2004, Coventry University, UK.

14 Sustainable Post-Disaster Waste Management: Construction and Demolition Debris

Gayani Karunasena

14.1 Introduction

A significant portion of the cost associated with recovery and reconstruction is spent on debris management (Pike, 2007). The increasing cost of waste disposal and a political willingness towards greater sustainability and a consequent desire to see a reduction in the amount of waste generation, have elevated waste management to a priority for effective and efficient post-disaster reconstruction (Morrissey and Phillips, 2007; Sokka *et al.*, 2007). Others such as Ardani *et al.* (2009) also note that disasters often present opportunities to salvage materials through reusing and recycling, which help to create advanced eco-industrial development. Despite such drivers, population growth, a waste disposal habit, the attitude of workers, a lack of resources and corruption, have also been highlighted as key barriers for the development of effective and sustainable waste management (Adewole, 2009).

Sustainable waste management encourages the generation of less waste, and the reuse, recycling and recovery of waste (Yahya and Boussabaine, 2006). Further, Blakely (2007) argues that, by focusing on long-term debris planning and setting measures for ecological and economic sustainability, it is possible to improve a region's resilience to future disasters. Furthermore, Deutz and Gibbs (2004) indicate that the expansion of recycling capabilities and eco-industrial planning results in more job creation and promotes partnerships.

Within this context, this chapter focuses on the challenge of developing more sustainable disaster waste management. Specifically, this gives more priority to construction and demolition (C&D) debris management post-disaster. First, the importance of C&D debris management post-disaster is discussed

Post-Disaster Reconstruction of the Built Environment: Rebuilding for Resilience, First Edition.
Edited by Dilanthi Amaratunga and Richard Haigh.
© 2011 Blackwell Publishing Ltd. Published 2011 by Blackwell Publishing Ltd.

and followed by C&D debris management strategies and models adopted in the past. Finally, the challenges towards sustainable post-disaster C&D debris management is discussed.

14.2 Construction and demolition debris management in post-disaster situations

In a disaster, the generation of waste is unavoidable. Generally, waste is defined as any losses produced by activities that generate direct or indirect costs but do not add any value to the product from the point of view of the client, or any substance or object which the holder intends or is required to discard (Formoso *et al.*, 1999). Peterson (2004) indicates that disaster waste has become critical as it differs from the normal situation which tends to generate waste in stable quantities and composition; whereas in a post-disaster scenario the type and quantity of waste radically changes. Specifically, disaster waste may contain or be contaminated with certain toxic or hazardous constituents.

Kobayashi (1995) classifies disaster waste as: rubble and other waste accumulated on roads; demolition and dismantling waste from buildings; bulky waste and raw materials; and, items in processes or other substances. In 2004, Baycan refined the classification more comprehensively as: recyclable materials (concrete, masonry, wood, metal, soil and excavated material), non-recyclable materials (household inventory, organic materials, and other inert materials) and hazardous waste (asbestos, chemicals). In a similar vein, in 2008 the Environment Protection Agency (EPA) in the USA identified a range of waste types generated in many post-disaster circumstances. These included soil and sediments, building rubble, vegetation, personal effects, hazardous materials, and mixed domestic and clinical wastes. They also noted that human and animal remains frequently pose a risk to human health (Environmental Protection Agency, 2008).

Although such categorisation of disaster waste exists, statistics on the extent of each type of disaster waste generated are not available; most data that is available focuses on the approximate overall quantities generated. For example, approximate total waste was reported as: Marmara earthquake in 1999 (13 million tons); Kobe earthquake in 1995 (20 million tons); Hurricane Katrina in 2005 (22 million tons); Kosovo earthquake in 1999 (10 million tons) (Kuramoto, 1996; Lauritzen, 1998; Baycan and Petersen, 2002; Shaw and Goda, 2004; Zeilinga and Sanders, 2004; Ardani *et al.*, 2009). A report on managing disaster debris by Luther (2008) highlights the importance of estimating the total volume of debris in order to manage disaster waste appropriately; it provides an early indication to identify appropriate staging grounds to separate waste, the necessary landfill space, and the required contract services and anticipated special handling applicable to hazardous debris. Factors such as the number of households affected, the amount of vegetative cover, the commercial density and a storm's intensity can be used to estimate disaster waste (Federal Emergency Management Agency of USA, 2007).

In a typical classification, C&D waste comprises asphalt, brick, tiles, concrete, mortar, reinforced concrete, rock, rubble, sand, soil, bamboo, ferrous metal, non-ferrous metal, glass, junk, fixtures, plastic, slurry, mud, trees, wood and other organics and garbage (Poon *et al.*, 2001). The US Environmental Protection Agency emphasises that C&D debris is a common type of waste generated in all types of disasters. Similarly, Kourmpanis *et al.* (2008) suggest that C&D should be a priority waste stream and appropriate action needs to be taken with respect to its effective management due to non-degradable components which lead to environmental degradation and health problems.

If C&D waste management is deemed to be important, it is necessary to understand how it fits into the broader plans for pre- and post-disaster management. Disaster management is a collective team effort which seeks to improve measures to prevent, respond and recover from the effects and consequences of a disaster (Amarasinghe *et al.*, 2006; GDRC, 2007). Adams and Wisner (2003) and Amarasinghe *et al.* (2006) emphasise that it is a cycle which leads to greater preparedness, better warnings, reduced vulnerability or prevention of disasters during the next iteration. Moreover, it includes the shaping of public policies and plans that either modify causes of disasters or mitigate their effects on people, property and infrastructure. Martin *et al.* (2006) note that disaster management requires comprehensive management across all sectors and more sustainable solutions to deal with disasters. In a similar vein, Moe and Pathranarakul (2006, p. 412) identify a range of critical factors for successful disaster management. These include: effective institutional arrangement; coordination and collaboration; supportive laws and regulations; effective management information systems; competencies of managers and team members; effective consultation with key stakeholders and target beneficiaries; effective communication mechanisms; clearly defined goals and commitments of key stakeholders; effective logistics management; and, sufficient mobilisation and disbursement of resources. It is evident that any attempt to initiate sustainable debris management will require a comprehensive set of measures across the disaster management lifecycle. Figure 14.1 illustrates the relationship of C&D debris with post-disaster management.

In this chapter, C&D debris is referred to as waste generated from totally damaged or partially damaged buildings and infrastructure as a result of the direct impact of a disaster, as well as demolition of partially damaged structures at rehabilitation and reconstruction stages, as illustrated in Figure 14.1. Some measures may aim at controlling disaster waste generation, for example through building regulations and codes at the mitigation phase of the cycle. Strategies for the management of debris when it is unavoidable will include collecting, transporting, reusing, recycling, land filling or disposing, and these strategies need to be in place at the rehabilitation and reconstruction phases. For example, a European Commission progress report on the post-tsunami rehabilitation and reconstruction programme in 2006 highlighted waste management as a key issue to be addressed among several dimensions of rehabilitating the environment after emergency relief had finished (European Commission, 2006). Similarly, Srinivas and Nakagawa (2007) also indicated

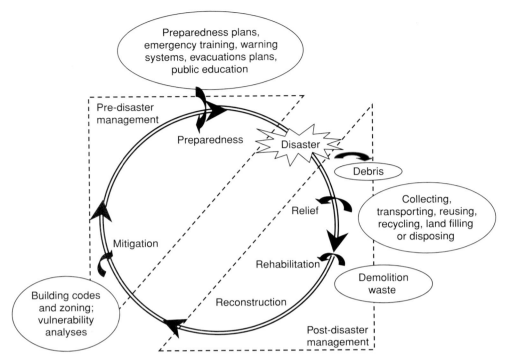

Figure 14.1   **Relationship of disaster waste with post-disaster management**

that disaster debris represented the most critical environmental problem facing countries affected by the Indian Ocean tsunami in 2004. Likewise, the General Accounting Office report on Hurricane Katrina: Continuing Debris Removal and Disposal Issues, indicated that failures in disaster debris management continued to impact the environmental health of citizens 3 years later (General Accounting Office of USA, 2008).

By way of contrast, others have highlighted the potential for benefits that can be achieved by effective waste management. Raufdeen (2009) highlights benefits such as conservation of virgin resources, economic use of landfill, environmental and economic sustainability, reduction of illegal and non-authorised dumping, reduced energy usage, cost recovery and financial incentives (Keeping and Shiers, 2004).

14.3   C&D debris management

The General Secretary of the International Federation of Red Cross stated that, 'Disasters are first and foremost a major threat to development and specifically a threat to the development of the poorest and most marginalized people in the world. Disasters seek out the poor and ensure that they remain poor' (Twigg, 2004, page 9). As Eceberger (2006) notes, it is therefore vital that disaster management strategies focus on development opportunities. The importance of focusing on long-term ecological and economic sustainable debris management strategies for resilience to future disasters is emphasised by Blakely (2007). Further, Lauritzen (1998), Baycan and Petersen (2002) and the Alameda Country Disaster Waste Management Plan (1998) highlight the need to design early stage strategies to manage debris in the most environmentally sound manner, through maximising source reduction and recycling options while minimising land disposal. Further, in order to design prior strategies, the following key requirements are highlighted for consideration (Baycan and Petersen, 2002):

- Procedure of handling waste stream including disposal sites and possible recycling materials.
- The quantity of waste generated including composition and source.
- The capacity of local areas, including number and types of trucks, condition of disposal site and opportunities to recycle.
- The scope of reconstruction works expected in order to identify opportunities for utilisation of recycled building waste.
- Understanding governmental and local authority structure in order to place responsibility for building waste management with the right office.

The importance of anchoring these strategies to national waste management and environmental policies is also noted (Baycan and Petersen, 2002; Shaw and Sinha, 2003). For example, following the 2004 tsunami, the United Nations Environment Programme (UNEP) launched the UN Post-Asian Tsunami Waste Management Plan in the Maldives and Indonesia, which supported removal of disaster debris. In Indonesia over one million cubic metres of tsunami waste was cleared, while in the Maldives 16 waste management centres were constructed for waste collection and disposal. Further, preparations were made for construction of another 22 waste management centres and a regional waste management facility (Progress Report on the European Commission Post Tsunami Rehabilitation and Reconstruction Programme, 2006).

14.3.1 C&D debris management strategies

Disaster debris management commences immediately following a disaster and continues during longer term reconstruction. The first phase of debris management is dedicated to immediate disaster relief and is focused on removing debris from access routes and residential and commercial areas. The second phase of debris management is the long-term removal of debris, which assists reconstruction. In the short term, removal of debris is necessary to facilitate recovery of a geographic area, whereas in the long term, it should not pose future threats to human health or environment (Blakely, 2007). The Republic of the Philippines defines debris management as a 'discipline associated with control of the generation of waste, storage, collection, transfer and transport, processing, reuse and recovery and disposal of waste in accordance with best principles of public health, economics, engineering, conservation of nature, aesthetics and environmental, while considering the general public attitude' (Republic of the Philippines, 2000).

Controlling of generating waste

Although total prevention may not be feasible, measures can be taken to reduce the generation of debris. The USA's Environment Protection Agency (2008) notes the importance of appropriate building codes and planning legislation to ensure that a community is prepared for any hazards that affect an area. Compiling hazard mitigation plans that discuss preventative measures may help to reduce the generation of disaster debris. This might include educating home owners on building homes to resist damage from disasters or the reuse of waste materials.

By way of example, 'safe islands' was set up by the Maldives Government to protect communities vulnerable to disasters by measures such as specially constructed areas of high ground, taller buildings, and buffer stocks of provisions (Srinivas and Nakagawa, 2007). Further, the Disaster Management Plan in California recommends that houses be anchored to their foundations to help prevent them from moving, which otherwise could result in lifting and placing the houses back on their foundation or their total demolition. Strengthening weak walls, foundations and chimneys can prevent horizontal movement of homes and reduce subsequent damage. Countries like Indonesia, which was severely affected in the tsunami in 2004, has been noted for having successful waste reuse and recycling plans (Srinivas and Nakagawa, 2007)

Collecting waste

The Federal Emergency Management Agency (FEMA) (2007) indicates that debris removal operations generally occur in two phases: initial debris clearance and debris removal. The transition period from initial clearance to debris removal depends on the magnitude of disaster impact. Typically, debris removal begins after the emergency access routes are cleared. Two methods of debris collection are suggested: curb side collection (mixed debris collection, source-segregated debris collection) and collection centres. For example, after the Kobe earthquake in Japan, decentralised depot collection and storage of C&D waste were seen.

Transporting waste to relevant sites

In a disaster situation, it may not be practical to employ a system of waste separation due to the amount of debris, time and labour it would require (Treloar *et al.*, 2003; Bekin *et al.*, 2007). According to Selvendran and Mulvey

(2005), waste separation may become impractical as clean-up and recovery became the first priority. However, a commonly suggested way to improve debris management is to pre-select temporary sites that can be used for storing, sorting and processing debris. These sites can be used to temporarily store debris before transferring to another recycling plant. Alongside the need for storing, sorting and processing waste, the site should be selected to ensure access by heavy equipment, protection of environmentally sensitive areas and logistical efficiency.

Processing of waste

Waste can be processed using two broad approaches: composting and recycling. Composting is most appropriate in the case of mixed debris or in situations where segregation is costly. Biodegradable materials can be easily composted by means of home composting, or where necessary, due to space problems, by centralised composting (Practical Action, 2008). Other non-degradable materials can be recycled: concrete following crushing, pre-sizing, sorting, screening and contaminant elimination; metal according to ferrous, non-ferrous and aluminium; timber according to solid and soft woods; and, rubber (Kindred Association, 1994; Eerland, 1995; Construction Material Recycling Association, 2008). Stationary and mobile recycling plants have been successfully used to manage waste, and specifically C&D debris, following earthquakes in Beirut, Kosovo and Marmara (Baycan and Petersen, 2002; Ardani *et al.*, 2009). Baycan and Petersen (2003) specify the importance of sorting debris into recyclables and non-recyclables prior to processing, which helps to ensure effective and efficient recycling.

Disposing of waste

The most common practice for disposing of disaster waste is using land fill. The United Nations Environment Program (2005) cautions that landfill is at the bottom of the waste management hierarchy from an eco-industrial perspective. Similarly, Peng *et al.* (1997), Faniran and Caban (1998) and Jayawardane (2007) advise that landfill should be the last option and only considered when all other options have been exhausted. Although land filling might be warranted immediately following a disaster, continued use of landfills throughout the reconstruction period exacerbates environmental hazards such as methane generation and ground water contamination (Peng *et al.*, 1997). Baycan and Petersen (2003) add that it often hinders sustainable development and causes significant economic and environmental damage. Despite these warnings, Ajayi *et al.*, (2008) found that most C&D waste goes to landfills, thereby increasing the burden on landfill loading and operation, particularly in third world countries. However, the problem is not confined to the third world; after the Marmara earthquake in Turkey 17 dump sites were used and after Hurricane Katrina in the USA 7 dump sites were used to dispose of disaster debris (Kuramoto, 1996;

Lauritzen, 1998; Baycan and Petersen, 2002; Zeilinga and Sanders, 2004; Ardani *et al.,* 2009). Pasche and Kelly (2005) add that collected waste is often disposed of into unplanned landfills that are located in environmentally sensitive sites.

The EPA propose incineration (uncontrolled open air incineration, controlled open air incineration, air curtain pit incineration) and chipping and grinding (rubber and metal materials) prior to disposal, in order to limit damage to the environment when using landfill (FEMA, 2007).

14.3.2   C&D debris management models

A number of disaster waste management models have been adopted in response to disasters. This section discusses their application in relation to C&D debris.

Lauritzen (1998) detail a C&D disaster waste management system adopted in Bosnia Herzegovina after armed conflict as a useful example for effective debris management. In Mostar, Bosnia Herzegovina in 1993, more than 1000 buildings needed to be demolished, which created 200 000 tons of C&D waste (Caloa and Parise, 2009). In this model disaster waste is sorted into four categories: unexploded munitions; reusable material; rubble material; and waste material. The sorted debris is subsequently sent to a range of processing facilities, such as recycling workshops, scrapyards and crushing plants, with the aim of using the debris for repair and reconstruction work (Lauritzen, 1995).

After the Northridge Earthquake, USA in 1994 18 recycling facilities were established. This reduced the total number of landfills required from three to one through salvaging and recycling materials (EPA, 1995). Further, recycled debris was used for reconstruction: crushed concrete and asphalt for roads; ground brick for baseball fields; dirt for landfill cover; and woody debris for landscaping, fuel cogeneration, and compost. Despite this, the non-existence of a pre-disaster debris management plan meant that debris was initially directed to temporary landfills (EPA, 1995).

The waste management model developed for the earthquake in Kobe in 1995 mainly focused on the separation of waste into specific components. Initially the waste was distinguished using separation plants with the capacity of 50 tonnes per hour. Screening, wind sifting, hand picking and belt separators were some of the technologies used. Finally, the separated materials were sent for reprocessing and reuse (Eerland, 1995).

Despite this plan, only a small proportion was actually recycled due to the high costs and time that would be involved in the separation of recycling material. Consequently, the majority of debris was either disposed of to landfill or used for land reclamation (Kuramoto, 1996; Zeilinga and Sanders, 2004).

In 1999, following the Marmara earthquake in Turkey, a rubble management plan was developed in collaboration with the United Nations Development Program with the intention of source reduction and sorting at the point of collection, followed by reuse, recycling, and disposal (Esina and Cosgun, 2005). However, high levels of reinforcement bars in demolition waste caused operational problems at recycling plants, which led to 17 landfill sites ultimately being used (Baycan and Petersen, 2002; Zeilinga and Sanders, 2004; Ardani *et al.*, 2009). Lauritzen (1998) found that a similar recycling facility in Beirut encountered the same operational difficulties when processing post-conflict waste. Such examples highlight both the importance and difficulty associated with sorting debris prior to processing to ensure effective and efficient recycling programmes (Baycan and Petersen, 2003).

In 2004, Baycan introduced a model for demolition waste management based on experience of the Marmara earthquake, Turkey. Accordingly, rubble was collected and transported to temporary dump sites during the emergency period and subsequently transported to recycling or disposing sites. The key principles of this model were: conservation of natural resources; reduction of quantities of waste for final disposal; and, minimisation of negative environmental impacts. An identified problem of this model was the double handling of waste, which led to high transport costs. This also highlights the importance of evaluating the full environmental impact of any debris management system. Accordingly, Bohne *et al.* (2008) advise that the environmental impact of waste management systems must take account of processing and disposal methods, transportation types and distances for all disposals, recycling and reuse options.

14.4   C&D debris management and sustainability

Disaster waste management, when responding to a disaster, has emerged as a significant weakness internationally (Pilapitiya *et al.*, 2006). This is despite solid waste and disaster debris being identified as the most critical environmental problem faced by countries following the 2004 Indian Ocean tsunami (Srinivas and Nakagawa, 2007). Baycan and Petersen (2002), and Mensah (2006) both recognise that waste management is a considerable challenge for national and local institutions during both the rehabilitation and reconstruction stages, and that this challenge is compounded in low income countries with a lack of available financial and human resource. Blakely (2007) also highlights the importance of focusing on long-term ecological and economic debris management strategies that are sustainable and therefore increase a community's resilience when facing future hazard events. Thus, it is important to maximise sustainable environmental values while also minimising disaster waste generation (Ajayi *et al.*, 2008). De Pauw and Lauritzen (1994), Lauritzen (1998) and Shaw and Goda (2004) highlight the economic benefits that can be achieved through use of recycled materials in large scale reconstruction; they cite specific examples in response to the disasters in Kobe, Beirut and Mostar. The important role that preparedness and mitigation measures have towards sustainable waste management has also been noted. Well maintained coral reefs, mangroves, sand dunes and other coastal ecosystems such as peat swamps, provided protection against the 2004 tsunami by acting as natural barriers (IUCN, 2005).

Although there is a lack of agreement as to what is meant by sustainability (Shediac-Rizkallah and Bone, 1998), the term has been interpreted as ensuring the adaptation and maintenance of communities and local organisations to cope with future challenges, while achieving set objectives (Schwartz *et al.*, 1993; Bracht *et al.*, 1994). The report of the World Commission on Environment and Development (1987) defined sustainable development as being, 'development that meets the needs of the present without compromising the ability of future generations to meet their own needs.'

Although sustainability is a broad and challenging concept, it is important to recognise that it does not mean self-sufficiency. Instead, communities need to exchange good practices. Second, sustainability does not mean that there are no environmental impacts, where population growth and economic decentralisation may impact in the absence of any adaptation. Finally, sustainability does not imply a change of human sprit where it only motivates humans towards actions which sustain the community (Braden and Van Ierland, 1999). Serageldin (1994) clarifies that the most important element of sustainability is to *get institutions right* in the sense of engaging all people to overcome the consequences in short- and long-term impacts. Hayles (2003) further suggests that sustainability could be described in terms of social, economic and environmental states that are required in order for overall sustainability to be achieved.

The importance of the individual homeowner has also been noted. There is considerable evidence of recycling of C&D debris by the individual homeowner. Re-use of material in reconstruction and also cash for work programmes organised by NGOs, are two such examples, which were reported as environmentally beneficial and helped to restore livelihoods (UNDP, 2006).

While it is now widely acknowledged that waste management will have an important role if society is to develop sustainably, it is important to recognise that there are many challenges associated with managing large volumes of waste; in particular, enabling property owners to return to an area following a disaster and assist with cleaning up, separating and managing hazardous and non-hazardous waste (Luther, 2008). In addition, deconstruction, establishment of permanent recycling infrastructure, and enhancement of eco-industrial networks through strategic planning have been identified as some of the key barriers in C&D debris management that need to be addressed (Baycan and Petersen, 2002; Zeilinga and Sanders, 2004; Ardani *et al.*, 2009).

Among these challenges, Ardani *et al.* (2009) argue that the lack of funds to acquire the required technology and equipment remains a major barrier, and is evident in most disasters. In a similar vein, inadequate resources to deconstruct, a lack of capital for appropriate equipment, and scarce space for recycling negatively impact on the ability to salvage materials for recycling and reuse in disasters (UNEP, 2005; Srinivas and Nakagawa, 2007). Other reported obstacles include a lack of knowledge, relatively new practice, limited recycling markets, limited market awareness and high costs (Arslana and Cosgunb, 2008; Raufdeen, 2009).

These types of challenges, when responding to Hurricane Katrina and the Marmara earthquake, led to land filling and illegal dumping. Correspondingly, following the Beirut and Kobe earthquakes, insufficient time was apportioned to sort and clean materials, and therefore much of the waste was unsuitable for recycling (Shaw and Goda, 2004). The Europe Aid Co-operation Office (2006) reported similar problems with C&D debris in Sri Lanka following the tsunami, and consequent over-use of landfill sites. Likewise, structural debris following Hurricane Andrew (1992) was deemed to be too contaminated to recycle; the subsequent disposal in South Florida used the equivalent of five year's landfill (Tansel *et al.*, 1994).

Other than a shortage of capital, most authors interpret a lack of capacity, within both local and national level institutions, as the other key barrier for sustainable C&D debris management. Kennedy *et al.* (2008) highlight the importance of integrating relief and development together by introducing capacity building and capacity development of local and national partners within post-disaster programmes. Its importance becomes even more pronounced in developing and less developed countries, and needs to be addressed within policy and practice (Webb and Rogers, 2003). In particular, local capacity building is seen as a way of increasing a community's resilience to natural hazard events. Within this context, a lack of vulnerability and risk assessment data, inadequate environmental baseline data or technology know how, and insufficient communication and coordination have been highlighted as capacity gaps for post-disaster waste management that require urgent attention (ICUN, 2005; UNEP, 2005; UNDP, 2006). By way of example, a lack of environmental codes, inadequate standards for proper enforcement and monitoring, and a lack of collaboration between local institutions and NGOs, were found to hinder C&D debris management across the Asian region following the 2004 tsunami (Shaw, 2003; UNEP 2005; UNDP, 2006; EC, 2006; Martin, 2007). More specifically, an in-depth review of national level polices for disaster management (Disaster Management Act, 2005) and waste management (National Environmental Act, 1981) in Sri Lanka, revealed that there was no provision for disaster waste management (Shaw, 2003; UNEP, 2005; EC, 2006; Martin, 2007), which may partially explain the open and illegal dumping that was visible. The consequences of such failings are significant and widespread. Jayawardena (2007) reports that the uncontrolled open dumping of contaminated waste had significant negative public health and environmental impacts. This was found to be due to contaminants leaking into soil and groundwater. Looking forward, the establishment of a separate ministry for disaster management provides the legal platform for disaster risk management activities and a multi stakeholder approach to disaster reduction in Sri Lanka that encompasses environmental management (Disaster Management Centre, 2009). Elsewhere, a lack of preplanned waste management strategies and poor awareness of recycling among public and private contractors led to similar environmental consequences following Hurricane Iniki (1992) in Hawaii (Reinhart and McCreanor, 1999). This was is in spite of documents such as Planning for Disaster Debris, which was published as far back as 1995 by the USA's Environmental Protection Agency as a framework for the development of a disaster response plan that provides information on what type of waste is expected from different types of disasters and recommended planning actions.

With a different emphasis, Srinivas and Nakagawa (2007) in a paper entitled 'Environmental Implications for Disaster Preparedness: Lessons Learned from the Indian Ocean Tsunami', describe environmental impacts in four countries: Indonesia, Thailand, Sri Lanka and Maldives. From these experiences, they conclude that Rapid Environmental Assessment (REA) and Environmental Impact Assessment (EIA) are vital at the early stages of disaster recovery. Others highlight evidence from responses to Hurricane Georges (1998) and Opal (1996), as well as the earthquake in Hyogoken-Nambu (1995) to suggest that pre-planned measures for the management of disaster debris are critical (Reinhart and McCreanor, 1999). Elsewhere, Ardani *et al.* (2009) propose a number of important measures that need to be adopted to achieve sustainable waste management:

- Involvement of community for selection of sites for segregation and sorting.
- Predefined prospective roles and functions of the stakeholder's involvement in debris management.
- Establishment of a hierarchy of debris management, such as controlling, reuse, recycling and land filling.
- Identification of local resources; plant, equipment, budget and expertise and material processing options.
- Enhancement of the institutional capacities.
- Establishment of commercial relationships for resource recovery activities.

It is apparent that, if it is to be successful, any attempt to introduce sustainable post-disaster waste management must address a wide range of policy areas that cross pre- and post-disaster planning.

14.5   Summary

Generation of waste during and following a disaster is a huge environmental and social problem. Moreover, it is a critical challenge due to the complexity of handling waste that includes a wide variety of organic, inorganic and even hazardous materials. The situation is further aggravated by the presence of non-degradable components that may lead to environmental degradation and health problems. Effective disaster waste management therefore has a vital role in helping disaster-affected communities to develop sustainability. If this is to be achieved, well thought-out policies and strategies are vital for success. Strategic planning for C&D debris management that is built on previous lessons, along with the collaboration of new technology and expertise, will provide such a basis.

Disasters are also an opportunity. The challenges associated with sustainable waste management are significant and wide-ranging, and many are linked to the high level of capital investment required. However, although opportunities to implement resource recovery and eco-industrial programmes may not be economically feasible prior to a disaster, the volume of potentially reusable debris and the availability of additional funding following a disaster, often change the cost–benefit equation. For example, in the aftermath of the 2004 tsunami, the Galle Municipal Council initiated a C&D waste recycling project named COWAM (Construction Waste Management) which was donated by a German company (COWAM, 2008; Raufdeen, 2009). There was neither the political will nor the resources to implement such a scheme beforehand.

References

Adams, J. and Wisner, B. (2003) Environmental Health in Emergencies and Disasters: A Practical Guide [online]. World Health Organization. Available from: http://www.who.int/water_sanitation_health/hygiene/emergencies/emergencies2002/en/ [accessed 19/05/2007].

Adewole, A.T. (2009) Waste management towards sustainable development in Nigeria: a case study of Lagos state. *International Non Government Organisation Journal*, **4**, 173–179.

Ajayi, O.M., Soyingbe, A.A. and Olandirn, O.J. (2008) The Practice of Waste Management in Construction Sites in Lagos State: Nigeria. RICS Construction and Building Research Conference, London 4–5 September 2008, RICS, pp. 1–9.

Alameda County Waste Management Authority (1998) Alameda County Disaster Waste Management Plan [Online]. Alameda County Waste Management Authority with assistance from EMCON. Available at: http://www.stopwaste.org/docs/d-plan.pdf [accessed 13/06/2008].

Amarasinghe, M.K.D.W., Pathmasiri, H.T.K. and Sisirakumara, M.H.A. (2006) The Public Sector Role in a Disaster Situation: An Empirical Analysis Based on Tsunami Disaster on 26 December 2004. ICBM – 2006, University of Sri Jayewardenepura, Sri Lanka, pp.372–379.

Ardani, K.B., Reith, C.C. and Donlan, C.J. (2009) Harnessing catastrophe to promote resource recovery and eco-industrial development. *Journal of Industrial Ecology*, **13**, 579–591.

Arslana, H. and Cosgunb, N. (2008) Reuse and recycle potentials of the temporary houses after occupancy: example of Duzce, Turkey. *Building and Environment*, **43**, 702–709.

Baycan, F. (2004) Emergency Planning for Disaster Waste: A Proposal Based on the Experience of the Marmara Earthquake in Turkey. International Conference and Student Competition on Post-Disaster Reconstruction, April 22–23, Coventry, UK.

Baycan, F. and Petersen, M. (2002) Disaster waste management-C&D waste. In: ISWA, (ed.), Annual Conference of the International Solid Waste Association, 8–12 July 2002. Istanbul, Turkey, pp. 117–125.

Baycan, F. and Petersen, M. (2003) *Disaster Waste Management: C&D Waste*. Ankara, Turkey: Ministry of Environment.

Bekin, C., Carrigan, M. and Szmigin, I. (2007) Caring for the community an exploratory comparison of waste reduction behaviour by British and Brazilian consumers. *International Journal of Sociology and Social Policy*, **27**, 221–233.

Blakely, E. (2007) Collaborating to Build Communities of Opportunity. Roosevelt Institute Symposium, 31 November, New Orleans, Louisiana.

Bohne, R.A., Bratteb, H. and Bergsdal, H. (2008) Dynamic eco-efficiency projection for construction and demolition waste recycling strategies at the city level. *Journal of Industrial Ecology*, **12**, 53–68.

Bracht, N., Finnegan, J.R., Rissel, C., Weisbrod, R., Gleason, J., Corbett, J. and Veblen-Mortenson, S. (1994) Community ownership and program continuation following a health demonstration project. *Health Education Research*, **9**, 243–255.

Braden, J.B. and Van Ierland, E.C. (1999) Balancing: the economic approach to sustainable water management. *Water Science Technology*, **39**, 17–23.

Caloa, F. and Parise, M. (2009) Waste management and problems of groundwater pollution in Karst environments in the context of a post-conflict scenario: the case of Mostar (Bosnia Herzegovina). *Habitat International*, **33**, 63–72.

Construction Material Recycling Association (2008) Concrete Recycling [Online]. Available at http://www.concreterecycling.org [accessed 01/10/2008].

Construction Waste Management (COWAM) (2008) Strategy for Sustainable Construction and Demolition Waste Management in Galle, Sri Lanka [Online]. Available at: http://www.cowamproject.org/cms/Content/download/Interim_Report_Vision_2018.pdf [accessed 11/10/2008].

De Pauw, C. and Lauritzen, E.K. (1994) Disaster Planning, Structural Assessment, Demolition and Recycling. RILEM Report 9. London: E&FN SPON.

Deutz, P. and Gibbs, D. (2004) Eco-industrial development and economic development: industrial ecology or place promotion. *Business Strategy and the Environment*, **13**, 347–362.

Disaster Management Act, No 13 of 2005 (2005) Published as a Supplement to Part II of the Gazette of the Democratic Socialist Republic of Sri Lanka of May 13, 2005. Colombo: Government Publications Bureau.

Disaster Management Centre (2009) *Newsletter* (2nd Quarter): Preparedness through awareness. Colombo: Disaster Management Centre (DMC).

European Commission (EC) (2006) Progress Report on Post Tsunami Rehabilitation and Reconstruction Program. [Online] European Commission (EC). Available at: http://ec.europa.eu/comm/world/tsunami/index.html [accessed 5/08/2008].

Eceberger, D.L. (2006) *How the Sustainable Solid Waste Management can be Achieved in Sri Lanka.* Sri Lanka: School of International Training in Colombo.

Eerland, D.W. (1995) Experience with the Construction and Demolition Waste Recycling in the Netherlands – its Application to Earthquake Waste Recycling in Kobe. In: IETC (ed.) *International Symposium on Earthquake Waste,* 12–13 June, Osaka. Shiga: UNEP, pp. 72–85.

Environmental Protection Agency (1995) Characterization of Building Related Construction and Demolition Debris in the United States. EPA 530-R-98-010, 1998.

Environmental Protection Agency (2008) Planning for Natural Disaster Debris [online]. Available at: http://www.epa.gov/CDmaterials/pubs/pndd.pdf [accessed 11/04/2008].

Esina, T. and Cosgun, N. (2005) A study conducted to reduce construction waste generation in Turkey. *Building and Environment,* **42**, 1667–1674.

Europe Aid Co-operation Office (2006) Construction Waste Management in Sri Lanka. [Online]. Available at http://www.cowam-project.org [accessed 22/09/2008].

Faniran, O.O. and Caban, G. (1998) Minimizing waste on construction project sites. *Engineering, Construction and Architectural Management,* **5**, 182-188.

Federal Emergency Management Agency (2007) Public Assistance: Debris Management Guide [online]. Available at: http://www.fema.gov/government/grant/pa/demagdes.htm [accessed 5/072008].

Formoso, C.T., Isatto, E.L. and Hirota, E.H. (1999) Method for Waste Control in the Building Industry. *Proceedings IGLC-7, 7th Conference of the International Group for Lean Construction,* 26–28 July. Berkeley, California.

General Accounting Office (2008) General Accounting Office Reports and Testimony. Hurricane Katrina: Continuing Debris Removal and Disposal Issues. GAO-08-985R. Washington, DC: GAO.

Global Development Research Centre (GDRC) (2007) Disaster Management. Available from http://www.gdrc.org/uem/disasters/1-introduction.html [accessed 3/03/2008].

Hayles, C. (2003) The Role of Value Management in the Construction of Sustainable Communities. A World of Value Conference. Hong Kong: Hong Kong Institute of Value Management.

International Union for Conservation of Nature (IUCN) (2005) Series on Best Practice Guidelines (Sri Lanka). After the Tsunami: Solid Waste Management. Available from http://www.iucn.org/ [accessed 11/04/2008].

Jayawardane, A.K.W. (2007) Recent Tsunami Disaster Stricken Sri Lanka and Recovery. International Seminar on Risk Management for Roads. Organised by PIARC and Ministry of Transport, 26–28 April, Vietnam.

Keeping, M. and Shiers, D.E. (2004) *Sustainable Property Development.* Oxford: Blackwell Science.

Kennedy, J., Ashmore, J., Babister, E. and Kelman, I. (2008) The meaning of build back better: evidence from post-tsunami Aceh and Sri Lanka. *Journal of Contingencies and Crisis Management,* **16**, 24–36.

Kindred Association (1994) *A Practical Recycling Handbook.* London: Thomas Telford Services.

Kobayashi, Y. (1995) Disasters and the Problems of Wastes. In: IETC (ed.), *International Symposium on Earthquake Waste,* 12–13 June, Osaka. Shiga: UNEP, pp. 6–13.

Kourmpanis, B., Papadopoulos, A., Moustakas, K., Stylianou, M., Haralambous, K.J. and Lolijodou, M. (2008) Preliminary study for the management of construction and demolition waste. *Waste Management and Research,* **26**, 267–275.

Kuramoto, N. (1996) The Actual State of Damage and Measures Undertaken in Hyogo Prefecture Earthquake. In: IETC (ed.) *International Symposium on Earthquake Waste,* 12–13 June, Osaka. Shiga: UNEP, pp. 5–23.

Lauritzen, E.K. (1995) Solving Disaster Waste Problem. In: IETC (ed.), *International Symposium on Earthquake Waste,* 12–13 June, Osaka. Shiga: UNEP, pp. 60–70.

Lauritzen, E.K. (1998) Emergency construction waste management. *Safety Science,* **30**, 45–53.

Luther, L. (2008) Managing Disaster Debris: Overview of Regulatory Requirements, Agency Roles, and Selected Challenges, Congressional Research Service [online]. Available at: http://wikileaks.org/wiki/CRS-RL34576 [Accessed 7/07/2009].

Martin, N. (2007) The Asian tsunami: an urgent case for improved government information systems and management. *Disaster Prevention and Management,* **16,** 188–200.

Martin, S.F., Fagen, P.W., Poole, A. and Karim, S. (2006), Philanthropic Grant-making for Disaster Management: Trend Analysis and Recommended Improvements [online]. Available at http://wikileaks.org/wiki/CRS-RL34576 [accessed 8/08/2009].

Mensah, A. (2006) People and their waste in an emergency context: the case of Monrovia, Liberia. *Habitat International,* **30,** 754–768.

Moe, T.L. and Pathranarakul, P. (2006) An integrated approach to natural disaster management. *Journal of Disaster Prevention and Management,* **15,** 396–413.

Morrissey, A.J. and Phillips, P.S. (2007) Biodegradable municipal waste (BMW) management strategy in Ireland: a comparison with some key issues in the BMW strategy being adopted in England. *Resources, Conservation and Recycling,* **49,** 353–371.

National Environmental Act, No 47 of 1980 (1980) Colombo: Government Publications Bureau.

Nordas, R. and Gleditsch, N. (2007) Climate conflict: common sense or nonsense? *Political Geography,* **26,** 627–638.

Pasche, A. and Kelly, C. (2005) *Concept Summary: Disposal of Tsunami Generated Waste.* Sri Lanka: UNDAC.

Peng, C.L., Scorpio, D.E. and Kibert, C.J. (1997) Strategies for successful construction and demolition waste recycling operations. *Construction Management and Economics,* **15,** 49–58.

Petersen, M. (2004) Restoring Waste Management Following Disasters. International Conference on Post Disaster Reconstruction, 22–23 April, Coventry, UK. IF Research Group.

Pike, J. (2007) *Spending Federal Disaster Aid: Comparing the Process and Priorities in Louisiana and Mississippi in the Wake of Hurricanes Katrina and Rita.* Baton Rouge, LA: Nelson A. Rockefeller Institute of Government and the Public Affairs Research Council of Louisiana.

Pilapitiya, S., Vidanaarachchi, C. and Yuen, S. (2006) Effects of the tsunami on waste management in Sri Lanka. *Waste Management,* **26,** 107–109.

Poon, C.S., Yu, T.W. and Ng, L.H. (2001) *A Guide for Managing and Minimizing Building and Demolition Waste.* Hong Kong: The Hong Kong Polytechnic University.

Practical Action (2008) Planning for Municipal Solid Waste Management[Online]. The Schumacher Centre for Technology and Development, UK. Available at http://www.practicalaction.org [accessed 23/09/2008].

Raufdeen, R. (2009) *Construction Waste Management: Current Status and Challenges in Sri Lanka.* Sri Lanka: COWAM Publication.

Reinhart, D.R. and McCreanor, P.T. (1999) Disaster Debris Management — Planning Tools. Final Report: US Environmental Protection Agency Region IV, USA.

Republic of the Philippines, RA 9003 (2000) Ecological Solid Waste Management Act of 2000. Available at http://eia.emb.gov.ph/nswmc [accessed 8/08/2009].

Schwartz, R., Smith, C., Speers, M.A., Dusenbury, L.J., Bright, F., Hedlund, S., Wheeler, F. and Schmid, T.L. (1993) Capacity building and resource needs of state health agencies to implement community-based cardiovascular disease programs. *Journal of Public Health Policy*, **14**, 480–493.

Selvendran, P.G. and Mulvey, C. (2005) Reducing solid waste and groundwater contamination after the tsunami, *Daily News*, Tuesday, 15 February, 2005.

Serageldin, I. (1994) *Water Supply Sanitation and Environmental Sustainability*. Washington, DC: World Bank.

Shaw, R. (2003) Role of non-government organizations in earthquake disaster management: on Asian perspective. *Regional Disaster Dialogue*, **24**, 117–129.

Shaw, R. and Sinha, R. (2003) Towards sustainable recovery: future challenges after Gujarat earthquake. *Risk Management*, **5**, 35–51.

Shaw, R. and Goda, K. (2004) From disaster to sustainable civil society: the Kobe experience. *Disaster*, **28**, 16–40.

Shediac-Rizkallah, M.C. and Bone, L.R. (1998) Planning for the sustainability of community-based health programs: conceptual frameworks and future directions for research, practice and policy. *Health Education Research*, **13**, 87–108.

Sokka, L., Antikainen, R. and Kauppi, P.E. (2007) Municipal solid waste production and composition in Finland — changes in the period 1960–2002 and prospects until 2020. *Resources, Conservation and Recycling*, **50**, 475–488.

Srinivas, H. and Nakagawa, Y. (2007) Environmental implications for disaster preparedness: lessons learnt from the Indian Ocean Tsunami. *Journal of Environmental Management*, **8**, 4–13.

Tansel, B., Whelan, M.P. and Barrera, S. (1994) Building performance and structural waste generation by Hurricane Andrew. *Housing Science*, **18**, 69–77.

Treloar, G.J., Gupta, H., Love, P.E.D. and Nguyen, B. (2003) An analysis of factors influencing waste minimization and use of recycled materials for the construction of residential buildings. *Management of Environmental Quality*, **14**, 134–145.

Twigg, J. (2004) *Disaster Risk Reduction: Mitigation and Preparedness in Development and Emergency Programming*, London: Humanitarian Practice Network at ODI.

United Nations Development Programme (UNDP) (2006) Post-Tsunami Recovery and Reconstruction Strategy. Colombo: United Nations Development Programme.

United Nations Environment Programme (UNEP) (2005) Sri Lanka Post Tsunami Environmental Assessment. United Nation Environment Programme (UNEP). Geneva: UNEP (DEP/0758/GE).

Webb, P. and Rogers, B. (2003) *Addressing the 'In' in Food Insecurity*. Washington, DC: Food and Nutrition Technical Assistance (FANTA) Project, Academy for Educational Development (AED).

World Commission on Environment and Development (1987) *Our Common Future (The Brundtland Report)*. Oxford: Oxford University Press.

Yahya, K. and Boussabaine, A.H. (2006) Eco-costing of construction waste. *Management of Environmental Quality: An International Journal*, **17**, 6–19.

Zeilinga de Boer, J. and Sanders, D.T. (2004) *Earthquakes in Human History: The Far-reaching Effects of Seismic Disruptions*. Princeton, NJ: Princeton University Press.

15 Linking Reconstruction to Sustainable Socio-Economic Development

Roshani Palliyaguru and Dilanthi Amaratunga

15.1 Introduction

The aim of this chapter is to identify the possible ways of linking reconstruction efforts to sustainable socio-economic development, i.e. the means of achieving or contributing to sustainable socio-economic development through post-disaster reconstruction efforts. 'Reconstruction' is one of the phases of the disaster management cycle, which is principally a post-disaster phase. It is defined in the *Oxford English Dictionary* as 'the action or process of reconstructing something', 'the rebuilding of something natural, artificial, or abstract' (Thompson, 1995). The term 'reconstruction' is well defined by the United Nations Development Programme and United Nations Disaster Relief Organisation (UNDP and UNDRC, 1992) as 'the action taken to re-establish a community after a period of rehabilitation subsequent to a disaster and these actions include the construction of permanent housing, a full restoration of all services, and complete resumption of pre-disaster state'. Thus, reconstruction does not stand only for reconstruction of physical stock. Reconstruction after a natural or a man-made disaster should address all physical, social, economic, political and environmental sectors. However, this chapter uses the term 'reconstruction' to denote 'built-environment specific reconstruction', which is 'physical reconstruction of housing and infrastructure' in consequence of natural disasters. It analyses how this built-environment specific reconstruction could address the developmental needs in such a way to contribute to sustainable socio-economic development as its primary focus. Accordingly, the overall aim of this chapter is shown in Figure 15.1.

In order to achieve this aim, the chapter will be set out under three headings:

Post-Disaster Reconstruction of the Built Environment: Rebuilding for Resilience, First Edition.
Edited by Dilanthi Amaratunga and Richard Haigh.
© 2011 Blackwell Publishing Ltd. Published 2011 by Blackwell Publishing Ltd.

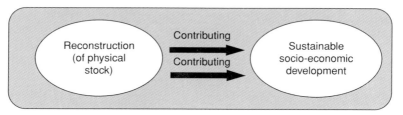

Figure 15.1 Aim of the chapter

(1) Post-disaster reconstruction as a window of opportunity for development.
(2) Millennium development goals as a framework of actions for sustainable socio-economic development and infrastructure reconstruction.
(3) Post-disaster infrastructure reconstruction as a sustainable socio-economic development strategy.

15.2 Post-disaster reconstruction as a window of opportunity for development

15.2.1 The way the link between disasters and development is seen

The close relationship between development and disasters is a much discussed topic in the current literature. This is reportedly a two-way relationship. As explained by many scholars and institutions such as the UNDP (2004) and McEntire (2004a), development can increase and/or reduce disaster risk, while disasters may halt development and/or provide new opportunities for progress. The following subsection elaborates on this: how disasters have an effect on the development process.

Impact of disasters on development

Disasters alone can set back development (Cuny, 1994 cited in McEntire, 2004a). Disasters sometimes put development gains at risk, for instance, meeting the millennium development goals is extremely challenging for many communities and countries due to losses from disasters triggered by natural hazards (UNDP, 2004). Disaster losses appear in various forms: as reduced human capital as a result of deaths, injuries and long-term trauma suffered by affected individuals; destruction of physical capital by destroying the housing stock and infrastructure and further loss of livelihoods, ecosystems etc. (Bendimerad, 2003). These losses result in major funding requirements to restore them, which is mostly fulfilled through using investments originally planned for development activities, Thus, disasters have the potential to delay or interrupt pre-planned development programmes by reducing available assets (Bendimerad, 2003; UNDP, 2004). Disasters are potential events which are able to harm the economic capacities of society by exacerbating poverty, disrupting small business and industry activities, and disabling lifelines vital for economic activity and service delivery (Bendimerad, 2003).

Not only do disasters set back development processes, they can make a considerable contribution to the economy of the affected place, together with surrounding economies; disasters can become a vehicle for major development programmes. As Albala-Bertrand (1993) reports, a major natural disaster can generate a construction-led economic boom in the longer term (Benson *et al.*, 2001), for example in Sri Lanka after the tsunami in 2004 and in Japan after the Kobe earthquake in 1995, etc. The Kobe earthquake in 1995 killed over 6000 people, and destroyed more than 100 000 homes, but the economic recovery not only of Japan but also of the Kobe economy was rapid (Becker, 2005). The tsunami in 2004 is distinctive among other natural disasters due to the extent of the economic impact it caused in a developing economy such as Sri Lanka and the construction-led economic boom that resulted. However, the lasting economic effects are small for most natural disasters that have occurred during the past centuries (Becker, 2005). Davis (2005) states that looking at disasters as development opportunities is becoming one of the core principles of disaster and emergency management (Asgary *et al.*, 2006). It is mainly through reconstruction whereas many scholars assert that post-disaster operations should include a developmental perspective (Berke *et al.*, 1993; McEntire, 2004a; Asgary *et al.*, 2006). Though conventionally reconstruction and development were considered as linear events, in reality, they are simultaneous with each 'stage' overlapping with others regardless of the same or different disasters they are responding to (Lewis, 1999). The 'disaster continuum' approach positively attempts to align post-disaster assistance with development, recognising the intervening stages of recovery, rehabilitation and reconstruction, as each stage should lead to the other in that sequence (Lewis 1999). Thus, disasters can enhance the development discourse, mainly through post-disaster reconstruction activities. The next sections provide a detailed synthesis of this issue with some case study examples where appropriate.

15.2.2 Shaping reconstruction for development: theoretical understanding and some practical examples

It is common for affected communities to demand a return to normalcy almost immediately, although experts often recognise that disaster is an opportunity to 'build back better' (Kates *et al.*, 2006; Thiruppugazh, 2007). For instance, New Orleans's disaster reconstruction programmes have suffered from discrepancies in decision making and rebuilding due to those involved in the planning process not sharing the same urgency as the affected communities to rebuild the familiar (Kates *et al.*, 2006). These sorts of circumstances give rise to a key decision about reconstruction. There is a choice between whether reconstruction is just for restoration to the status quo or for enhancing development. The first method involves replacement of damaged or destroyed assets, often called 'replacement-reconstruction' while the later method involves adjusting the reconstruction efforts towards the future, which is 'development-reconstruction' (Alexander, 2004; Mitchell; 2004; Thiruppugazh, 2007). The initiatives taken at the reconstruction phase which consider disasters as opportunities for development are often called disaster-inspired development initiatives. One of the major deciding factors is the availability of financial resources for the reconstruction work, which is a major concern in developing countries. For instance, following the major tsunami in 2004, most parts of the coastal belt road network were devastated in Sri Lanka. However, when it came to reconstruction of this devastated road network, the government and the relevant institutional bodies had to make the decision to prioritise the most significant roads for improvement during reconstruction. The others were restored to the status quo: they were replaced without fulfilling future needs. This was a 'replacement-reconstruction' activity. In addition, many decisions during reconstruction were based on budgetary limitations, especially in improving the additional facilities, etc. Conversely, whenever major disasters take place, the developed world and all other capacitated nations are always able to provide necessary financial aids for reconstruction activities. For example, in Indonesia, as a country highly prone to natural disaster, certain regions such as Nias and Nias Selatan have received high levels of foreign funding for reconstruction activities. The European Commission (EC) provided a total of €60 million for emergency assistance in Aceh and Nias followed by €210 million in post-tsunami rehabilitation and reconstruction. Responding to the earthquake in Java, the EC provided €9.5 million for emergency assistance and €35.4 million for reconstruction (Delegation of EC, 2009). The primary aim of these funds is to strengthen rehabilitation and reconstruction with the focus of activities being re-directed towards development. These resources provide an opportunity to address the developmental needs of communities through physical reconstruction in the region. Availability of financial resources is a major concern in development-reconstruction as others are the choices between rapid recovery versus safety, the betterment of some segments versus equity as reported by Kates *et al.* (2006), the differential ability of people to recover and the social and cultural changes due to the recovery process (Oliver-Smith, 1996 cited by Thiruppugazh, 2007). However, there are certain aspects which are particularly important in disaster-inspired development, which will be elucidated throughout this chapter with a specific focus on built environment-related reconstruction and infrastructure reconstruction in particular.

Within the widespread perception that disasters provide future windows of opportunity for development (Lewis, 1999; Jigyasu, 2002; UN/ESCAP, 2006; Thiruppugazh, 2007), there are certain major developmental issues identified by Thiruppugazh (2007). Addressing risk reduction, ensuring equity and the safety of the built environment, addressing social and economic changes and the betterment of communities during the reconstruction phase are major priorities and are ways of turning mere reconstruction into developmental opportunities. However, it is important to look for the best possible opportunities to achieve development needs throughout reconstruction.

'Disasters are commonly known as sudden events, which bring serious disruption to society with massive human, property, livelihood, industry and environmental losses, which exceed the ability of the affected society to cope using its own resources' (UN/ISDR, 2004a; Lloyd-Jones, 2006; Shaluf and Ahmadun, 2006; Quarantelli, 1998 cited in Eshghi and Larson, 2008; UN/ISDR, 2009). However, it is now a well accepted norm that disasters are the disruptive and/or deadly and destructive outcome of triggering agent(s) when they interact with, and are exacerbated by, various forms of vulnerability (McEntire, 2001). UN/ISDR (2004a), UN/ISDR (2004b), the Asian Disaster Reduction Centre (ADRC) (2005) and the Department for International Development (DFID) (2005a) also describe disaster risk results from a combination of hazards (they may originate from the natural environment, human activity or a combination of the two) and people's vulnerability to those hazards. According to McEntire (2001), vulnerability is the dependent component, which is determined by the degree of risk, susceptibility, resistance and resilience while the triggering agent stands as the independent component of a disaster. Thus, prevention or mitigation of disaster risk can be achieved by prevention or mitigation of hazard and/or prevention or mitigation of vulnerabilities. However, there are preventable and also unpreventable hazards (Cannon, 1993 cited in McEntire, 2005). For example, certain hazards such as floods are preventable whereas hazards such as earthquakes are unpreventable. Whether these hazards are preventable or not, their effects and losses can be prevented or mitigated. The best way of preventing or mitigating disaster losses is therefore by preventing (eliminating) or mitigating (reducing) vulnerabilities, which is commonly called 'vulnerability reduction'. McEntire (2004b) acknowledges that we can certainly limit, although not completely eliminate, our vulnerability to disasters. 'Vulnerability' does not only stand for vulnerability to hazard exposure but it represents a series of resultant states of social, economic, political, cultural, environmental, physical, technological underdevelopment processes, before, during and after disaster situations (McEntire, 2001; Jigyasu, 2004). On the other hand, disasters can highlight particular areas of vulnerability, for example in areas where there is serious loss of life and physical structures this indicates the general level of underdevelopment (Stephenson and DuFrane, 2005) because the losses from natural disasters are sometimes viewed as results of development that is unsustainable (Mileti et al., 1995 cited in McEntire, 2004a). Inappropriate development interventions result in an accumulation of disaster risks, which can be further excacerbated as various development decisions made by individuals, communities and nations can pave the way for unequal distribution of disaster risks (UNDP, 2004). This underdevelopment may be due to social factors (social vulnerabilities), economic factors (economic vulnerabilities), or another factor. Reconstruction can therefore be used as a development opportunity to help reduce these various vulnerabilities (Jayaraj, 2002).

Disasters provide opportunities for risk reduction (risk mitigation) (Christoplos et al., 2001; Alexander, 2006; Christoplos, 2006; Thiruppugazh, 2007); it is largely because the concept of disaster risk reduction can be more easily promoted after a disaster than before (Christoplos, 2006). This is due, for example, to: a new awareness of risk that leads to a broad consensus; the revealing of weaknesses in development policies; exposure of institutional weaknesses (such as corruption, lack of human resources and otherwise weak institutional structures that allowed high risk planning and discouraged appropriate monitoring before the disaster); much of the badly designed infrastructure being destroyed; weakened old vested interests; availability of money to make improvements; and enhanced political will.

Accordingly, disaster risk reduction is a key priority within reconstruction projects. Furthermore, it is emphasised that in the rehabilitation and reconstruction phases, considerations of disaster risk reduction should form the foundations for all activities (ADRC, 2005). Taking appropriate measures based on the concept of disaster risk management in each phase of the disaster risk management cycle can reduce the overall disaster risk (ADRC, 2005). This means that 'disaster risk reduction' deserves an important place in the pre-disaster phase of the disaster management cycle, and also an important position in post-disaster reconstruction. As acknowledged by Warfield (2008), not only do opportunities arise to enhance prevention and mitigation during the reconstruction period but also to increase preparedness and emergency management. When a reconstruction project is planned, constructed, operated and maintained by a set of multi-disciplinary stakeholders, an opportunity arises to plan preparedness actions necessary to prevent or mitigate losses for that particular built-environment structure. The community can also plan emergency management activities for possible future disasters. A former Chairman of the Road Development Authority in Sri Lanka commented on some of the opportunities provided by post-tsunami road reconstruction activities, which facilitated active participation of built-environment and property management professionals in the efficient preparedness for and emergency management of future potential disasters. As he states:

> 'it is not entirely possible to prevent major disasters such as tsunami, whereas minor disasters are somewhat preventable. Bearing this in mind, after the tsunami disaster in 2004, we have been able to develop a database which consists of construction professionals' contact details which will be very useful in future disaster events. This was basically done through the identification of skilled personnel, who were actively involved in the 2004 tsunami reconstruction activities in Sri Lanka. Now we are in a position to call upon these skilled professionals for their active support when and where it is necessary in another future emergency situation, which indeed became one of the hardest tasks during the most recent post-tsunami case' (PhD data collection interview, 2009).

It is important to understand the characteristics of the scope of reconstruction activities which could be undertaken to extend the opportunities for development (Thiruppugazh, 2007). In doing so, it is essential to construct the characteristics, occurrence, and interrelatedness of, and the opportunities for, developing mutually reinforcing preventive and mitigation measures which actually help to sustain positive impacts (Thiruppugazh, 2007). In actual practice, each post-disaster reconstruction project oriented towards development varies from project to project. However, there are instances where reconstruction has positively and also negatively affected development. Reconstruction following the Latur earthquake in 1993, the Orissa super cyclone in 1999 and the Gujarat earthquake in January 2001 shows that reconstruction may serve to reinforce and sometimes even increase the vulnerability of local communities (Jigyasu, 2004). More importantly, reconstruction projects may be more development-oriented if they can share characteristics and strategies of developmental projects. A reconstruction project should not aim for mere physical reconstruction of the structure. For instance, providing new employment opportunities, improving quality of life and maintaining equity of resource and service distribution should be prioritised during reconstruction.

Two popular theses of development offered by Karl Marx and Max Weber are worth mentioning here as they provide some important insights into disaster scholarship (McEntire, 2004a). Karl Marx believed that the tensions created by capitalism would lead to class conflict which, in turn, would result in a fundamental and complete change of social, political and economic relations. Max Weber declared that societies could take many forms depending on the organisation and legitimacy of authority and so form a society takes result from the ideas and values of its citizens (McEntire, 2004a). However, in the modern era, it is not only poverty, values and the beliefs of people that decide development, but both are fairly important to the process. In the social sciences, the term 'development' is used to cover a wide range of spheres: community, socio-cultural, economic, human, rural, social and sustainable development are just a few.

As far as traditional development theories are concerned, the radical thesis asserts that poverty is a major causal explanation of disaster and therefore restructuring of social, political and economic relations is vital so that calamities may be reduced. The conservative thesis asserts that culture is the determinant of disaster, and recommends alterations in beliefs or behaviour and increased rationalisation and bureaucratisation as a means to reduce the effects of hazards (McEntire, 2004a). However, both of these theses provide critical insights into disaster-inspired development initiatives. The importance of facilitating all social, political, economic and cultural changes through post-disaster reconstruction is emphasised.

15.3   Millennium development goals as a framework of action for sustainable socio-economic development and infrastructure reconstruction

15.3.1   Introduction to Millennium Development Goals

There are eight Millennium Development Goals (MDGs) to be achieved by 2015 which respond to the world's main development challenges (UN, 2007).

MDGs range from halving extreme poverty to halting the spread of HIV/AIDS and providing universal primary education.The MDGs are drawn from the actions and targets contained in the Millennium declaration that was adopted by 189 nations and signed by 147 heads of state and governments during the UN Millennium Summit in September 2000. These eight MDGs, which were agreed upon in 2000 and are listed below in turn, have been broken down into 18 quantifiable targets with 48 indicators for progress (UNDP, 2004):

- *Goal 1*: Eradicate extreme poverty and hunger
- *Goal 2*: Achieve universal primary education
- *Goal 3*: Promote gender equality and empower women
- *Goal 4*: Reduce child mortality
- *Goal 5*: Improve maternal health
- *Goal 6*: Combat HIV/AIDS, malaria, and other diseases
- *Goal 7*: Ensure environmental sustainability
- *Goal 8*: Develop a global partnership for development

15.3.2 *Influence of MDGs on sustainable socio-economic development*

Socio-economic development is the process of social and economic development in a society. From a policy perspective, *economic development* can be defined as 'efforts that seek to improve the economic well being and quality of life of a community' (Hayami and Godo, 2005). *Social development* is a 'process which results in the transformation of social structures in a manner which improves the capacity of the society to fulfill its aspirations'. A variety of indicators can be used to measure socio-economic development. The key indicators are: GDP per capita, life expectancy, literacy rates, poverty, levels of employment, quality of infrastructure and access to safe infrastructure (Sen, 1998; Meier and Rauch, 2000; Hayami and Godo, 2005). Less tangible factors are also considered, such as personal dignity, freedom of association, personal safety and freedom from fear of physical harm, and the extent of participation in civil society. As far as development issues are concerned, MDGs form an effective framework of actions for planning and implementation of so-called socio-economic development.

The term 'sustainable' stands for 'capable of being borne or endured; supportable, bearable', 'capable of being upheld or defended; maintainable', 'capable of being maintained at a certain rate or level' (*Oxford English Dictionary*). *Sustainable development* has been defined as 'the development that meets the needs of the present without compromising the ability of future generations to meet their own needs' (Anderson, 1995; UNDP, 2004). Therefore, sustainable development activities are conceptually considered as activities which result in environmental protection, economic growth and social equity. On the other hand, sustainability is related to the quality of life in a society. It focuses on whether the economic, social and environmental systems of a society are providing a productive and meaningful life for its people, present and future; sustainable development is essentially the use resources that aims to meet human needs while preserving the environment so that both present and future needs are met. Yaoxian and Okada (2002) present a snapshot of the components of a 'sustainable community': technology exists within the economy, both technology and the economy exist within society, and all three exist within the environment. It depicts the mutual interaction of every sector.

These definitions imply that so-called developmental activities should be sustainable enough to protect the environment, and maintain social equity and economic growth. Sustainable development contains within it two key concepts: the concept of 'need', in particular the essential needs of the world's poor, to which overriding priority should be given; and the idea of limitations imposed by the state of technology and social organisation on the environment's ability to meet present and future needs (UNDP, 2004). In this context, sustainable socio-economic development can be sustained in the long term. Therefore, the main constituents of sustainable socio-economic development are attaining socio-economic development of communities while ensuring equity, quality, environment preservation and growth. The MDGs provide a framework of action for the achievement of essential socio-economic needs in the 21st century.

15.3.3 *Case study (adapted from Willoughby, 2004)*

Attainment of almost all MDGs depends in significant part, and in some cases critically, on improvements in economic infrastructure services such as transportation, water supply and sanitation, energy services, etc. Research has been undertaken to clarify the nature and significance of the linkages between each main type of economic infrastructure and the various MDGs. It is worth presenting a few findings from empirical research from the last thirty related predominantly to rural areas, where the vast majority of very poor people are to be found, in Asia and sub-Saharan Africa. Evidence is based on detailed post-evaluations of individual cases, national and international econometric analyses, and statements of professional opinion based on practical experience in developing countries.

As far as local transport services (village to township or main road, low-volume local roads and associated networks of village tracks/paths) are concerned, significant improvements in this field can:

- Make a large contribution towards increasing income, reducing poverty and hunger especially by significantly reducing poor farmers' transaction costs and expanding their production possibilities (including non-farm).
- Moderately increase primary education coverage by improving school enrolment and attendance.

- In terms of the third MDG of gender equality in education, moderately increase girls' attendance.
- Result in increased use of primary healthcare facilities and improve access to better water (fourth MDG of reducing child mortality) but on a smaller scale.
- In terms of the fifth MDG of improve maternal health, positively affect antenatal care and increase the number of professionally attended deliveries, but again on a smaller scale.

15.4   Post-disaster infrastructure reconstruction as a sustainable socio-economic development strategy

15.4.1   Socio-economic needs/problems after a disaster

The needs or problems after a disaster primarily occur due to a wide range of consequences in a wide range of areas such as the human, economic, social, political, psychological and environmental concerns of communities. Disaster losses generally take two forms: human and economic, while social, political, psychological and environmental losses have links with economic losses. Economic losses of disasters can be mainly classified into three groups (UNDP, 2004):

- *Direct losses*: Comprise physical damage to productive capital and stocks, economic infrastructure and social infrastructure.
- *Indirect losses*: Formed of downstream disruption to the flow of goods and services (e.g. lower output from damaged or destroyed assets and infrastructure, and the loss of earnings as income-generating opportunities are disrupted), disruption of the provision of basic services, such as telecommunications or water supply, the costs of both medical expenses and lost productivity arising from the increased incidence of disease, injury and death (UNDP, 2004).
- *Secondary effects*: Include short- and long-term impacts of a disaster on the overall economy and socio-economic conditions (e.g. fiscal and monetary performance, levels of household and national indebtedness, the distribution of income and scale and incidence of poverty, the effects of relocating or restructuring elements of the economy or workforce). These secondary effects, however, have a significant impact on long-term human and economic development (UNDP, 2004).

Overall, the other socio-economic needs created by the above mentioned losses are jeopardised income, increased poverty, poor access to infrastructure, lack of quality of infrastructure, marginalisation of women, and insufficient knowledge about disasters, etc.

In this context, it is of utmost importance to map out an inclusive, holistic approach to address the needs of the disaster-affected communities. Thus, a plan for recovery and reconstruction can focus on these socio-economic needs/problems which may have occurred due to a particular disaster or may be long held. The focus should be on the immediate needs of communities (reasonable access to basic needs such as health care, nutrition, education and housing) and long-term development projects, including establishing or re-establishing the basic underpinnings of the economy (UNDP, 2006). Addressing both socio-economic deficits through post-disaster reconstruction presents both opportunities and challenges (UNDP, 2006). However, it is vital to convert these challenges into opportunities. The following section elaborates on the role of post-disaster infrastructure reconstruction in overcoming these socio-economic needs/problems and the ways that it could contribute to sustainable socio-economic development.

15.4.2 *Role of infrastructure reconstruction in sustainable socio-economic development*

The term *infrastructure* has different meanings in different fields. Infrastructure appears in many forms, e.g. economic, social, IT, etc. *Critical infrastructures* are 'physical and cyber-based systems that are essential to the key operations of the economy and government' (Bosher *et al.*, 2007). Types of critical infrastructure include water supply, transport networks, health services, energy systems such as electrical power, oil and gas production and storage, information and communications networks, etc. These systems are deemed vital as their incapacity or destruction would have debilitating effects on the defence, economic security and health of local or national administrations and populations (Bosher *et al.*, 2007).

The role of infrastructure reconstruction in sustainable socio-economic development is dependent on efficiency and whether it is effective in contributing towards fulfilling the socio-economic needs of disaster-affected communities. This is explored below. In linking infrastructure reconstruction with sustainable socio-economic development, it is imperative that the reconstruction programme works towards eradicating extreme poverty and hunger, achieving universal primary education, promoting gender equality and empowering women, reducing child mortality, improving maternal health, combating HIV/AIDS, malaria, and other diseases, ensuring environmental sustainability and developing a global partnership for development while ensuring equity, quality, environment preservation and growth.

Access and quality of infrastructure are key indicators of poverty in communities and improving them through reconstruction is vital. Proper access to infrastructure should be available at all times: before, during and after disasters. Therefore, reconstruction programmes should explore the best possible ways of increasing access and quality of infrastructure facilities. More importantly, due to limited funding, a vital step in reconstruction programmes is to prioritise which infrastructure facilities need immediate reconstruction. This is a very familiar situation in developing countries that have encountered major natural and long-term man-made disasters. This prioritisation should be based on persistent development needs, for example the economic or social development needs of affected communities, which in turn would address one of the cornerstones of sustainability: ensuring equity in service distribution. A high-profile official's (attached to a post-tsunami road network reconstruction project in Sri Lanka) comments on this matter are particularly informative:

'We carried out a study on economic feasibility and submitted a report to the donor agency before commencing the road network reconstruction work. We investigated the economic viability of the project through this feasibility study. Factors such as traffic volume on the roads, industries in the area, the existing developed industries, for example Hambantota Harbour and the airport, availability of factories, the present status of agricultural activities in the area, present status of transportation facilities for agricultural products distribution and the availability of health facilities in the area, were taken into account in the feasibility study. In this project there are about 30 roads being reconstructed under the scheme, which were selected from a range of about 200 roads. The feasibility study was done to select 30 roads from a total of 200. Factors such as economic growth and poverty reduction were considered in the feasibility study of this project. There was a marking scheme to prioritise the most needed roads. Let's say the cut off point is 75 marks. The roads that received more than 75 marks were selected for reconstruction. However, there were also some political influences, where some roads were reconstructed mainly based on these political influences. This project was commenced as a capacity building project, primarily in terms of enhancing the capacity building of staff in using IT facilities, bill preparation, using new construction methods, etc.' (source: PhD data collection interview, 2009).

In post-disaster contexts, the infrastructure planners need to cope with the already existing problems of infrastructure before being affected by disasters, which in turn facilitates integration of disaster-risk reduction with reconstruction (Anand, 2005). The process needs to address not only the infrastructure that may have been damaged but also the infrastructure that was never constructed or had been damaged due to lack of maintenance over the years. Adequate and pertinent post-disaster infrastructure reconstruction will particularly reduce the need for infrastructure reconstruction after a second disaster.

Addressing factors affecting various vulnerabilities is a good method of achieving sustainable development, which of course addresses issues relating to improving access and quality of infrastructure. This can be called 'vulnerability reduction', through which the concept of disaster risk reduction could be achieved. UN/ISDR (2004a) defines disaster risk reduction as 'the conceptual framework of elements considered with the possibilities to minimise vulnerabilities and disaster risks throughout society, to avoid (prevention) or to limit (mitigation and preparedness) the adverse impacts of hazards, within the broad context of sustainable development'. UN/ISDR (2009) defines it as 'systematic development and application of policies, strategies and practices for the same purpose as above'. Incorporating further clarification, UN/ISDR (2009) views disaster reduction as 'taking measures in advance to address vulnerabilities, reduce risk and anticipate hazards, which involve environmental protection, social equity and economic growth, the three cornerstones of sustainable development, to ensure that development efforts do not increase the vulnerability to hazards.' Thus, disaster risk reduction is emerging as an important requisite for sustainable development (UN/ISDR, 2003). Disaster risk reduction entails measures to curb disaster losses by addressing hazards and the vulnerability of people to them (DFID, 2005a). Good disaster risk reduction happens well before disasters strike, but also continues afterwards, building resilience to future hazards (DFID, 2005a). Reducing disaster risk is not just about additional investments — it is also about ensuring that development interventions are sound; for example, ensuring appropriate construction of physical infrastructure in highly vulnerable areas (DFID, 2006). Disaster risk reduction strategies can be categorised in various ways. Nateghi-A, (2000) classifies the range of techniques that the city of Teheran is considering after a strong earthquake in June 1990 as:

- Engineering and construction measures
- Physical planning measures
- Economic planning measures
- Policy guidance measures
- Public response measures

Concern Worldwide (2005) points out that the term 'disaster risk reduction' includes the three aspects of a disaster reduction strategy:

- Mitigation
- Preparedness
- Advocacy

DFID (2005b) has a similar classification of disaster risk reduction strategies:

- Policy and planning measures
- Physical preventative measures
- Physical coping and/or adaptive measures
- Community capacity building measures

All such categories of measures are of paramount importance for post-disaster infrastructure reconstruction because this integration brings developmental benefits. It is evident from all these classifications that the possible areas to link infrastructure reconstruction with sustainable socio-economic development are at the national level, institutional level, project level and community/individual level. There would be more political backing for a higher rate of economic and social change in the aftermath of disasters (Stephenson and DuFrane, 2005). National level policy changes may take place and infrastructure reconstruction would benefit. In this context, disaster risk reduction classifications address different levels in disaster-affected communities/countries: national, institutional, project and community/personnel level. They can be planned and implemented primarily within these four levels.

(1) The national level measures are mainly the disaster risk management policies, relevant guidelines and legal frameworks, and provision of relevant legal backing.

(2) Institutional level measures are the disaster risk management guidelines and planning measures which are in existence for the improvement of reconstruction project planning, implementation, operation, maintenance, disaster preparedness and emergency planning, e.g. the mainstreaming of disaster risk assessment into existing development instruments is critical (UNDP, 2004). This includes opportunities for building on existing risk impact assessment tools and examining opportunities for integration into activities such as housing and infrastructure development, industrial and agricultural development and the introduction of new technologies (UNDP, 2004). On the one hand, risk information can be used through instruments such as land-use planning and building regulations to increase the resistance, safety and sustainability of development interventions. Contingency planning is one measure which is similarly applicable to infrastructures such as road network systems, water supply and sanitation projects, etc. Measures such as contingency planning for road networks improve the access to such roads during and after disasters.

(3) Project level measures are mainly physical and engineering measures such as construction of flood defences, etc.

(4) Community/personnel level measures constitute capacity building of project stakeholders and the community as a whole for better reconstruction of infrastructure including resilience to future disasters, and preparedness and emergency management with regard to predicted and unpredicted future disasters.

In this context, infrastructure reconstruction programmes should aim to change the vulnerable conditions for the development of the country. Vulnerability is used in the field of risk, hazard, and disaster management as well as in the areas of global change, and environment and development studies (Weichselgartner, 2001). However, the meanings of vulnerability are still unclear as there is no common conceptualisation (Cutter, 1996 cited in

Weichselgartner, 2001). Weichselgartner (2001) sees vulnerability from perspectives as described below:

- *Individual vulnerability*: Personal or individual potential for, or sensitivity to, losses that have both spatial and non-spatial domains.
- *Social vulnerability*: Susceptibility of social groups or society at large to potential losses from disastrous events.
- *Technical vulnerability*: The vulnerability of a house, an electricity grid, or transport infrastructure in a technical sense.

If this classification is carefully observed, it is obvious that it is based on vulnerability of humans and of built-environment structures. Therefore, physical, social, cultural, political, economic and technological vulnerabilities can be seen as either the vulnerability of human or built-environment structures. On the other hand, vulnerability is viewed as a product of four components by McEntire (2001): risk, susceptibility, resistance and resilience. Thiruppugazh (2007), agreeing with this model of viewing vulnerability as a product of these four components, recommends that this model includes both the positive and negative aspects of the physical and social arenas. McEntire (2005) views the concept of vulnerability reduction as assessment of all types of liabilities and an assessment of the plethora of capabilities and proactive efforts to reduce liabilities and raise capabilities of people (social assets) and built-environment assets. Therefore, infrastructure reconstruction should be able to address both the built-environment structure and the community being benefited through the particular infrastructure facility. For instance, a risk-prone infrastructure may degrade individual and community resilience because it is unable to withstand the powerful forces of hazards (McEntire, 2005). It is role of disaster risk reduction to address both the built-environment structures and the surrounding community's various vulnerabilities.

Some report that reconstruction work should be community managed, controlled and owned, so that it is socially and culturally acceptable (Jayaraj, 2002). Also it is emphasised that the locally existing ecologically friendly, low cost materials should be made use of, which is a possible way of ensuring environmental sustainability. Also the reconstruction programmes should facilitate skill development, upgrading of traditional skills and encouragement of learning by doing (Jayaraj, 2002). This can be achieved through active participation of the affected community in planning and physical reconstruction of infrastructure. This will often enhance the capacities of local communities in construction activities and provide them with possible employment opportunities which facilitates eradication of extreme poverty and hunger, and affects growth.

Another possible way of linking infrastructure reconstruction with sustainable socio-economic development is to promote gender equality and empower women by involving them in the planning and implementation of disaster risk reduction activities within the infrastructure reconstruction processes. This is one way of maintaining equity, which is a key component of sustainable development. Infrastructure such as health and sanitation facilities can help to reduce child mortality, improve maternal health, combat HIV/AIDS, malaria, and other diseases, etc., while educational infrastructures such as schools can be of real importance in achieving universal primary education and improving access to education in the aftermath of disasters.

15.5 Summary

Natural disasters appear to be increasing in both their frequency and intensity. Many of these disasters occur in developing countries and their infrastructure has been repeatedly affected. The affected infrastructure needs to be reconstructed effectively and efficiently as it is often essential to sustain recovery after major disasters. However, one of the key challenges in the post-disaster infrastructure reconstruction sector is to link programmes with sustainable socio-economic development. Sustainable socio-economic development was defined as socio-economic development which can be sustained in the long term. The main constituents are attaining socio-economic development while ensuring equity, quality, environment preservation and growth – basically fulfilling the social and economic needs of disaster-affected communities. This chapter elucidates the possible areas of linking infrastructure reconstruction programmes with sustainable socio-economic development. Considering MDGs as a good framework of action for socio-economic development, it was identified that effective integration of disaster risk reduction at the national, institutional, project and community level is one of the possible means of reducing the vulnerability of a community and built-environment structure. In addition, prioritisation of infrastructure reconstruction projects, facilitation of community managed, controlled, owned and socially and culturally acceptable infrastructure reconstruction, and utilisation of locally existing ecologically friendly, low cost materials for reconstruction work were identified as other potential ways of linking infrastructure reconstruction with sustainable socio-economic development. Also the reconstruction programmes should facilitate skill development, upgrade traditional skills and encourage learning by doing, promote gender equality and empower women, which are possible through active participation of affected communities for planning and physical reconstruction of infrastructure.

References

Alexander, D. (2004) Planning for Post-Disaster Reconstruction. Proceedings of the Second International Conference on Post-disaster Reconstruction: Planning for Reconstruction, 22–23 April 2004, Coventry University, UK.

Alexander, D. (2006) Globalization of disaster: trends, problems and dilemmas. *Journal of International Affairs*, **59**, 2–22.

Anand, P.B. (2005) Getting Infrastructure Priorities Right in Post-Conflict Reconstruction. Proceedings of UNU/WIDER Jubilee Conference on the Future of Development Economics, WIDER, Helsinki, Research Paper No. 2005/42.

Anderson, M.B. (1995) Vulnerability to disaster and sustainable development: a general framework for assessing vulnerability. In: Munasinghe, M. and Clarke, C. (Eds), *Disaster Prevention for Sustainable Development: Economic and Policy Issues*. Washington, DC: International Decade for Natural Disaster Reduction (IDNDR). pp. 41–59.

Asgary, A., Badri, A., Rafieian, M. and Hajinejad, A. (2006) Lost and Used Post-disaster Development Opportunities in Bam Earthquake and the Role of Stakeholders. Proceedings of the International Conference and Student Competition on Post-Disaster Reconstruction: Meeting Stakeholder Interests, 17–19 May, 2006, Florence, Italy.

Asian Disaster Reduction Center (ADRC) (2005) Total disaster risk management – Good practices 2005. http://www.adrc.asia/publications/TDRM2005/TDRM_Good_Practices/GP2005_e.html [accessed 20/06/2007].

Becker, G.S. (2005) . . . And the economics of disaster management. *The Wall Street Journal*, New York, N.Y. p. A.12. http://home.uchicago.edu/˜gbecker/ [accessed 10/09/2007].

Bendimerad, F. (2003) Disaster Risk Reduction and Sustainable Development. World Bank seminar on the role of local governments in reducing the risk of disasters, Istanbul, Turkey, 28 April – 2 May (2003). Washington, DC: World Bank.

Benson, C., Michael F.V., Robertson, A.W. and Clay, E.J. (2001) Dominica: Natural Disasters and Economic Development in a Small Island State. Disaster Risk Management Working Paper Series no. 2. Washington DC: World Bank.

Berke, P.R., Kartez, J. and Wenger, D. (1993) Recovery after disaster: achieving sustainable development. *Disasters*, **17**, 93–108.

Bosher, L., Carrillo, P., Dainty, A., Glass, J. and Price, A. (2007) Realising a resilient and sustainable built environment: towards a strategic agenda for the United Kingdom. *Journal of Disasters*, **31**, 236–255.

Christoplos, I. (2006) The Elusive 'Window Of Opportunity' for Risk Reduction in Post-Disaster Recovery. Provention Consortium Forum 2006: Strengthening Global Collaboration in Disaster Risk Reduction, Session 3, Discussion Paper, 2–3 February, 2006, Bangkok.

Christoplos, I., Mitchell, J. and Liljelund, A. (2001) Re-framing risk: the changing context of disaster mitigation and preparedness. *Disasters*, **25**, 185–198.

Concern Worldwide – Emergency Unit (2005) *Approaches to Disaster Risk Reduction*. Ireland: Concern Worldwide.

Delegation of the European Commission (2009) Post-disaster reconstruction. http://eeas.europa.eu/delegations/indonesia/eu_indonesia/cooperation/sectors_of_cooperation/post_disaster_reconstruction/index_en.htm

Department for International Development (DFID) (2005a) *Disaster Risk Reduction: A Development Concern*. London: DFID.

Department for International Development (DFID) (2005b) Natural Disaster and Disaster Risk Reduction Measures: A Desk Review of Costs and Benefits. Draft final report. London: DFID.

Department for International Development (DFID) (2006) Reducing the Risk of Disasters – Helping to Achieve Sustainable Poverty Reduction in a Vulnerable World. A DFID Policy Paper. London: DFID.

Eshghi, K. and Larson, R.C. (2008) Disasters: lessons from the past 105 years. *Disaster Prevention and Management*, **17**, 62–82.

Hayami, Y. and Godo, Y. (2005) *Development Economics: From the Poverty to the Wealth of Nations*, 3rd edn. New York: Oxford University Press.

Inter-agency Secretariat for the International Strategy for Disaster Reduction (2002) *Women, Disaster Reduction and Sustainable Development*. Geneva: UN/ISDR.

Jayaraj, A. (2002) Post-Disaster Reconstruction Experiences in Andhra Pradesh, in India. Conference Proceeding of Improving Post-Disaster Reconstruction in Developing Countries, 23 May-25 May, 2002, Université de Montréal, Canada. Available from http://www.grif.umontreal.ca/pages/i-rec%20papers/annie.pdf

Jigyasu, R. (2002) From Marathwada to Gujarat – Emerging Challenges in Post-Earthquake Rehabilitation for Sustainable Eco-Development in South Asia. Proceedings of the First International Conference on Post-Disaster Reconstruction: Improving Post-Disaster Reconstruction in Developing Countries, 23–25 May, Université de Montréal, Canada.

Jigyasu R. (2004) Sustainable Post Disaster Reconstruction Through Integrated Risk Management – The Case of Rural Communities in South Asia. Proceedings of International Conference and Student Competition on Post-Disaster Reconstruction 'Planning for Reconstruction', 22–23 April, Coventry University, Coventry, UK.

Kates, R.W., Colten, C.E., Laska, S. and Leatherman, S. P. (2006) Reconstruction of New Orleans after Hurricane Katrina: A Research Perspective. *Proceedings of the National Academy of Sciences of the United States of America*, **103**, 14653–14660.

Lewis, J. (1999) *Development in Disaster-Prone Places: Studies of Vulnerability*. London: Intermediate Technology.

Lloyd-Jones, T. (2006) *Mind the Gap! Post-Disaster Reconstruction and the Transition from Humanitarian Relief*. London: RICS.

McEntire, D.A. (2001) Triggering agents, vulnerabilities and disaster reduction: towards a holistic paradigm. *Disaster Prevention and Management*, **10**, 189–196.

McEntire, D.A. (2004a) Development, disasters and vulnerability: a discussion of divergent theories and the need for their integration. *Disaster Prevention and Management*, **13**, 193–198.

McEntire, D.A. (2004b) Tenets of vulnerability: an assessment of a fundamental disaster concept. *Journal of Emergency Management*, **2**, 23–9.

McEntire, D.A. (2005) Why vulnerability matters: exploring the merit of an inclusive disaster reduction concept. *Disaster Prevention and Management*, **14**, 206–222.

Meier, G.M. and Rauch, J.E. (2000) *Leading Issues in Economic Development*, 7th edn. New York: Oxford University Press.

Mitchell, J.K. (2004) Re-Conceiving Recovery. In Norman, S. (Ed.) *New Zealand Recovery Symposium Proceedings*, 12–13 July, Napier, New Zealand. Ministry of Civil Defence and Emergency Management.

Nateghi-A, F. (2000) Disaster mitigation strategies in Teheran, Iran. *Disaster Prevention and Management*, **9**, 205–211.

Sen, A. (1998) The concept of development. In: Chenery, H. and Srinivasam, T.N. (Eds) *Handbook of Development Economics*. Volume 1. Amsterdam: Elsevier Science. pp. 12–23.

Shaluf, I.M. and Ahmadun, F. (2006) Disaster types in Malaysia: an overview. *Disaster Prevention and Management*, **15**, 286–298.

Stephenson R.S. and DuFrane, C. (2005) Disasters and development: Part 2. Understanding and exploiting disaster-development linkages. *Prehospital and Disaster Medicine*, **20**, 61–65.

Thiruppugazh V. (2007) Post-Disaster Reconstruction and the Window of Opportunity: A Review of Select Concepts, Models and Research Studies. JTCDM Working Paper Series No. 3, Jamsetji Tata Centre for Disaster Management, Tata Institute of Social Sciences, Malati and Jal A.D. Naoroji (New) Campus, India.

Thompson, D. (Ed.) (1995) *The Concise Oxford Dictionary of Current English*, 9th edn. Oxford: Oxford University Press.

United Nations (2007) *Millennium Development Goals Report*. New York: United Nations.

United Nations Development Programme (UNDP) (2004) A Global Report – Reducing Disaster Risk: A Challenge For Development. New York: Bureau for Crisis Prevention and Recovery.

United Nations Development Programme (UNDP) (2006) *Post-Conflict Reconstruction of Communities And Socio-Economic Development*. New York: Bureau for Crisis Prevention and Recovery.

United Nations Development Programme and United Nations Disaster Relief Organization (1992) *An Overview of Disaster Management*. Geneva: UNDP/UNDRO.

United Nations Economic and Social Commission for Asia and the Pacific (UN/ESCAP) (2006) *Enhancing Regional Cooperation in Infrastructure Development Including that Related to Disaster Management*. Bangkok: United Nations.

United Nations International Strategy for Disaster Reduction (UN/ISDR) (2003) Disaster reduction and sustainable development. http://www.unisdr.org/eng/riskreduction/sustainable-development/dr-and-sd-english.pdf [accessed 20/12/2007].

United Nations International Strategy for Disaster Reduction (UN/ISDR) (2004a) Terminology: basic terms of disaster risk reduction. www.unisdr.org/eng/library/lib-terminology-eng%20home.htm [accessed on 18/06/2007].

United Nations International Strategy for Disaster Reduction (UN/ISDR) (2004b) *Living With Risk: A Global Review Of Disaster Reduction Initiatives*. Geneva: ISDR, United Nations Inter-Agency Secretariat.

United Nations International Strategy for Disaster Reduction (UN/ISDR) (2009) UN/ISDR terminology on disaster risk reduction. http://www.unisdr.org/eng/library/lib-terminology-eng.htm [accessed 20/02/2009].

Warfield, C. (2008) *The Disaster Management Cycle*. www.gdrc.org/uem/disasters/1-dm_cycle.html [accessed 16/08/2007].

Weichselgartner, J. (2001) Disaster mitigation: the concept of vulnerability revisited. *Disaster Prevention and Management*, **10**, 85–94.

Willoughby, C. (2004) Infrastructure and the Millennium Development Goals. Session on Complementarity of Infrastructure for Achieving the MDGs. Berlin.

Yaoxian, Y. and Okada, N. (2002) Integrated Relief and Reconstruction Management Following a Natural Disaster. Second Annual IIASA-DPRI Meeting, Integrated Disaster Risk Management: Megacity Vulnerability and Resilience, 29–31 July, 2002, Luxembourg.

16 Disaster Risk Reduction and its Relationship with Sustainable Development

Kanchana Ginige

16.1 Introduction

During the past four decades, natural hazards have caused major loss of human life and livelihoods, the destruction of economic and social infrastructure, as well as environmental damage (UN/ISDR, 2003). Thus, there is an urgent need for disaster risk reduction.

In this context, it is important to understand how a disaster occurs in order to identify the ways of reducing it. Natural hazard events such as earthquakes, cyclones or tsunamis are not themselves disasters (Commission on Climate Change and Development, 2008). A disaster emerges as a combination of a hazard (triggering agent) and vulnerabilities (McEntire, 2001; Sahni and Ariyabandu, 2003). Incidentally, McEntire (2001) shows that vulnerability acts as the dependent component of a disaster while the triggering agent stands as the independent component. In particular, it is the vulnerabilities that could be controlled in order to reduce disaster risks. However, vulnerabilities are largely dependent on development practices that are carried out without taking the susceptibilities to natural hazards into account (UN/ISDR, 2003). Hence, controlling and minimising vulnerabilities to reduce disaster risk paves a way towards more sustained development.

Accordingly, this chapter explores the links between risk reduction for natural disasters and sustainable development, and discusses in detail the role of the built environment and construction in creating links.

Post-Disaster Reconstruction of the Built Environment: Rebuilding for Resilience, First Edition.
Edited by Dilanthi Amaratunga and Richard Haigh.
© 2011 Blackwell Publishing Ltd. Published 2011 by Blackwell Publishing Ltd.

16.2 Disasters: a result of poor development

16.2.1 Vulnerabilities: the dependent variable of a disaster

All disasters, irrespective of whether they are natural or man-made, emerge as a combination of a hazard and vulnerabilities (McEntire, 2001; Sahni and Ariyabandu, 2003). A hazard or triggering agent is defined as a potentially damaging physical event, phenomenon or human activity that may cause the loss of life or injury, property damage, social and economic disruption or environmental degradation (UN/ISDR, 2004). Further, as United Nations International Strategy for Disaster Reduction (UN/ISDR) (2004) describes, hazards can include latent conditions that may represent future threats and can have different origins: natural (geological, hydrometeorological and biological) or induced by human processes (environmental degradation and technological hazards). Earthquakes, cyclones, tsunamis, landslides, etc. are the hazards that can be transformed into disasters when they are met with vulnerabilities. On the other hand, vulnerability is known as a set of conditions that affect the ability of countries, communities and individuals to prevent, mitigate, prepare for and respond to triggering agents (Ariyabandu and Wickramasinghe, 2003). The concept that disasters emerge as a combination of a hazard and vulnerabilities suggests a significant fact in relation to disaster risk reduction. It implies that vulnerabilities or the conditions which affect the capacity of a society in responding to a triggering agent are the controllable components of a disaster and thus it is the vulnerabilities that need to reduce in order to reduce the risk of disasters.

In this context, it is important to examine disaster vulnerabilities in detail in order to understand how they can be reduced. Vulnerability has been defined in many different ways by different authors. Weichselgartner (2001) composed a list of such definitions which have been published by different scholars and institutions over two decades from 1980 to 2000 in his attempt to conceptualise vulnerability. Among them, the definition by Downing (1991) indicates some important features of vulnerability. According to Downing, vulnerability is a relative term that differentiates among socioeconomic groups or regions, a consequence rather than a cause and an adverse circumstance. According to another definition in the aforementioned list, vulnerability is a multilayered and multidimensional social space defined by the determinate, political, economic and institutional capabilities of people in specific places at specific times (Bohle *et al.*, 1994).

However, Weichselgartner (2001) indicates that there are many discrepancies over the meaning of vulnerability and these discrepancies are due to different epistemological orientations and subsequent methodological practices. Also, the way vulnerability is interpreted largely depends on the type of hazard, the scale of an event and the regions chosen for examination (Weichselgartner, 2001). According to Cutter (1996), there are three key reasons for discrepancies in the meaning of vulnerability. They can be briefly stated as follows:

- Different epistemological orientations (e.g. political ecology, human ecology, physical science, spatial analysis) and methodological practices of different studies.
- Variations in the choice of hazards and regions for examination.
- Fundamental conceptual differences – either to focus research on the likelihood of exposure (biophysical/technological risk), the likelihood of adverse consequences (social vulnerability) or some combinations of the two.

However, irrespective of the variations, Cutter (1996) believes the way vulnerability has been viewed by scholars can be categorised into three major groups, namely, vulnerability as exposure to risk/hazard, vulnerability as social response, and vulnerability of places.

In this context, a more recent definition given by UN/ISDR (2004) for disaster vulnerability can be taken as a consolidated definition for the context of disaster risk reduction. UN/ISDR (2004) defines disaster vulnerabilities as the conditions determined by physical, social, economic, and environmental factors or processes, which increase the susceptibility of a community to the impact of hazards. This definition is further elaborated by the Working Group on Climate Change and Disaster Risk Reduction of the United Nations Inter Agency Task Force for Disaster Reduction (IATF/DR-UN) (2006) by identifying four different categories of vulnerabilities similar to the aforementioned four factors. They are:

- *Physical vulnerability*: Susceptibilities of the built environment and may be described as 'exposure'.
- *Social factors of vulnerability*: Levels of literacy and education, health infrastructure, the existence of peace and security, access to basic human rights, systems of good governance, social equity, traditional values, customs and ideological beliefs and overall collective organisational systems.
- *Economic vulnerability*: Characterises people less privileged in class or caste, ethnic minorities, the very young and old, the disadvantaged, and often women who are primarily responsible for providing essential shelter and basic needs.
- *Environmental vulnerability*: The extent of natural resource degradation.

The aforementioned definition by the UN/ISDR (2004) and the categorisation of vulnerabilities by the IATF/DR-UN (2006) demonstrate that disaster vulnerabilities are closely linked with factors related to development activities of the world. This is emphasised by McEntire's (2001) classification of variables which interact to increase disaster vulnerabilities. The variables are classified under the categories of physical, social, cultural, political, economic and technological headings as listed below and most of them are clearly linked with deficiencies in the process of development.

Physical

- The proximity of people and property to triggering agents
- Improper construction of buildings
- Inadequate foresight relating to the infrastructure
- Degradation of the environment

Social

- Limited education (including insufficient knowledge about disasters)
- Inadequate routine and emergency healthcare
- Massive and unplanned migration to urban areas
- Marginalisation of specific groups and individuals

Cultural

- Public apathy towards disaster
- Defiance of safety precautions and regulations
- Loss of traditional coping measures
- Dependency and an absence of personal responsibility

Political

- Minimal support for disaster programmes among elected officials
- Inability to enforce or encourage steps for mitigation
- Over-centralisation of decision making
- Isolated or weak disaster-related institutions

Economic

- Growing divergence in the distribution of wealth
- The pursuit of profit with little regard for consequences
- Failure to purchase insurance
- Sparse resources for disaster prevention, planning and management

Technological

- Lack of structural mitigation devices
- Over-reliance upon or ineffective warning systems
- Carelessness in industrial production
- Lack of foresight regarding computer equipment/programmes

In this context, the next subsection of this chapter explains how factors related to poor development can increase the risk of disasters.

16.2.2 Disasters and development: the co-relational factors

El-Masri and Tipple (2002) state that disasters must be considered as unresolved development problems and thus they are not unpredictable, isolated or independent events; they are results of 'failures in development'. A policy briefing report by the Department for International Development (DFID) (2005) confirms this by stating that disasters to a larger extent result from failures of development which increase vulnerability to hazard events. 'Development processes may increase exposure or susceptibility to hazard more directly. Increased exposure can result from global level climate change exacerbating extreme weather events, or local level destruction of mangrove stands which protect coasts from tidal storm surges to make way for shrimp farms. Rapid urban growth may increase exposure to landslides, earthquakes or fires' (DFID, 2005, page 3).

According to Bendimerad (2003), the following are the main factors which correlate disasters and development and the reasons for the world's increasing exposure to natural hazards as a result of unsustainable development over a long period of time:

- Poor land management
- Increased population concentrations in hazard areas
- Environmental mismanagement, resulting in environmental degradation
- Lack of regulation and a lack of enforcement of regulation
- Social destitution and social injustice
- Unprepared populations and unprepared institutions
- Inappropriate use of resources

16.2.3 The difference: an overview of disaster losses in developed and developing countries

Confirming that deficiencies in development efforts have a significant positive co-relationship with increased disaster vulnerabilities, it is clearly visible that natural disasters are becoming more severe and more frequent in the case of developing countries where there is an increase in human settlements in vulnerable areas with rapid uncontrolled urbanisation and unsteady economic conditions (El-Masri and Tipple, 2002). According to DFID (2005), although only 11% of people exposed to hazards live in low human development countries, more than half of disaster deaths occur in these countries. The low developed countries suffer far greater economic losses relative to their GDP than richer countries (DFID, 2005).

According to UN/ISDR (2003), whilst no country is safe from natural hazards, limited capacity to tackle the impact of hazards increase disaster susceptibilities of developing countries. Due to inequitable access to resources, with lack of means to cope and recover from negative environmental changes, poor people in developing countries become more vulnerable than their wealthier counterparts to face such changes (UN/ISDR, 2003).

The best example for this difference is the two near identical earthquakes in December 2003 in California and in the city of Bam in Iran. On 22 December 2003, an earthquake with a magnitude of 6.5 on the Richter Scale (with a 7.6 kilometre depth) struck central California. A few days later, on 26 December, the city of Bam in Iran experienced an earthquake of similar magnitude (6.6 on the Richter Scale, with a 10 kilometre depth), making both shallow-focus earthquakes (Royal Geographical Society, 2004). However, according to a report by the Royal Geographical Society (2004), there were more than 25 000 confirmed deaths in Bam due to the immediate collapse of poorly-constructed multistorey homes with heavy roofs. In contrast, California reported only three deaths in the town of Paso Robles, where the population is around 25 000.

The key reason behind the vast difference in earthquake damage between these two countries was the disaster risk reduction measures which were in place. Developed countries are equipped with better disaster mitigation and preparedness measures including more resilient buildings and stronger infrastructure. They have sufficient funds and resources to analyse and develop necessary coping mechanisms for potential hazards. In this context, poverty emerges as a key reason for higher disaster vulnerabilities, disproportionately affecting poor countries and poor communities, since day to day living is a higher priority than disaster planning (Martin, 2002). In addition, the livelihood decisions about how to use the resources, which are motivated by poverty, migration, illness, etc., may also have an intense impact on the environment in poorer communities leading to deforestation, land degradation, etc. (UN/ISDR, 2003).

The next section highlights the obstructions to development as a result of disaster.

16.3   Disasters: a barrier for development

16.3.1   *Implications of disasters on development*

The escalation of severe disaster events triggered by natural hazards and related technological and environmental disasters is increasingly threatening both sustainable development and poverty-reduction initiatives (UN/ISDR, 2003). The impact of disasters has long-lasting implications for national development as they can shatter development efforts and drain economic resources of the community they affect through exacerbating poverty, disrupting small business and industry activities, and disabling lifelines vital for economic activity and service delivery (Boulle *et al.*, 1997). The prime goal of development is to achieve human well-being and thus human development incorporates all aspects of individuals' well- being, from their health status to their economic and political freedom (Soubbotina, 2004). UN/ISDR (2003) states that disaster response and humanitarian assistance around the world has absorbed significant amounts of resources that could have been allocated for development activities.

Incidentally, as Bendimerad (2003) illustrates, disasters delay development programmes by reducing available assets and interrupting planning. Disasters also reduce human capital as a result of the deaths, injuries and long-term trauma suffered by affected individuals (Bendimerad, 2003). In this context, the damages caused by the earthquake of 7.3 on the Richter scale on January 17, 1995 to the city of Kobe, Japan illustrate the enormous impact of disasters towards development efforts. As a result of this earthquake, the affected areas sustained heavy damage and many casualties. According to the United States Government Accountability Office (GAO) (2009), over 6000 people were killed and 40 000 injured. In addition to destroying over 400 000 homes and buildings, the earthquake caused extensive damage to roads, railroads, highways, and subway stations (GAO, 2009). Further, the port of Kobe, Japan's leading container shipping port at the time, also experienced heavy damage to almost all container berths (GAO, 2009). These damages illustrate how a disaster reverses a country's development. A country must incur huge costs for compensation and rehabilitation of devastated lives and repair and reconstruction of the destroyed built environment after this type of damage. It may take years for a society to recover, with people suffering the loss of their relatives and friends and damage to property and infrastructure. All these factors delay development activities by reducing available funds, resources and time, and changing priorities from development to rehabilitation and reconstruction.

The adverse impact of disasters on development can be further elaborated by looking into the impacts of disasters on the globally agreed development agendas. The following section gives an account of the impact of disasters on millennium development goals.

16.3.2   *Impact of disasters on millennium development goals*

In September 2000, the United Nations General Assembly adopted eight specific tasks named as Millennium Development Goals (MDGs), to be achieved by 2015. The eight goals are as follows:

- Eradicate extreme poverty and hunger
- Achieve universal primary education
- Promote gender equality and empower women
- Reduce child mortality
- Improve maternal health
- Combat HIV/AIDS, malaria and other diseases
- Ensure environmental sustainability
- Develop a global partnership for development

However, the adverse impact of disasters has been seen as an enormous barrier for achieving MDGs (UNDP, 2004; DFID, 2005). According to DFID (2005), many countries are not on course to meet MDG1, the prime goal of halving extreme poverty and hunger, and one of the key reasons for this is the effects of disasters. Disasters affect poverty reduction in several ways, bringing both direct and indirect macroeconomic impacts (DFID, 2005). The direct impact is brought through the physical damage to infrastructure, productive capital and stocks whilst the long-term, indirect impacts can be visible through inefficient productivity, slow growth and insufficient macroeconomic performance (DFID, 2005). Similarly, DFID (2006) demonstrates how disasters slow down progress towards the remaining MDGs in Table 16.1.

Table 16.1 shows the hindering effects of disasters on MDGs. According to UN/ISDR, in many earthquakes around the world, school buildings which were not built as to hazard resistant standards collapsed, causing severe setbacks to primary education. In 1999, 74% of schools were damaged by an earthquake in Pereira, Colombia, whilst 130 schools suffered extensive to complete damage in Boumerdes, Algeria in 2003 (UN/ISDR). Furthermore, an estimated US$131 billion of economic damage was caused by hurricanes Katrina and Rita in the USA in 2005 (EM-DAT). Thus, it is clear that there is a need to reduce the risk of disasters to clear the path towards global development. In confirming this, Table 16.2 illustrates how reducing the risk of disasters can accelerate the world's journey towards MDGs.

It is clear that disasters are a significant problem for development efforts around the world and it is imperative to reduce the risk of disasters. The following section details how disaster risk reduction can be achieved and how it facilitates sustainable development.

16.4   Disaster risk reduction for sustainable development and vice versa

16.4.1   Disaster risk reduction through reducing vulnerabilities

Since disasters cause large-scale damage to human life, livelihoods, economic and social infrastructure and environment (UN/ISDR, 2002), it is essential to take necessary action to reduce the risk of disasters. As it was mentioned in the earlier sections of this chapter, disaster vulnerabilities are the controllable component of a disaster. Thus disaster vulnerabilities need to be controlled and minimised in order to reduce the risk, i.e. the probability of a disaster and its negative consequences. In this regard, it is seen that all individuals and communities in the world are vulnerable to hazards to varying degrees but everyone has intrinsic capacities to reduce their vulnerability (Working Group on Climate Change and Disaster Risk Reduction of the IATF/DR 2006).

McEntire (2001) shows invulnerable development or vulnerability management as a process whereby decisions and activities are intentionally designed

Table 16.1   Disaster impacts on meeting Millenium Development Goals (MDGs) (source: DFID, 2006)

MDG	Direct impacts	Indirect impacts
1. Eradicate extreme poverty and hunger	• Damage to housing, service infrastructure, savings, productive assets and human losses reduce livelihood sustainability	• Negative macroeconomic impacts including severe short-term fiscal impacts and wider, longer-term impacts on growth, development and poverty reduction • Forced sale of productive assets by vulnerable households pushes many into long-term poverty and increases inequality
2. Achieve universal primary education	• Damage to education infrastructure • Population displacement interrupts schooling	• Increased need for child labour for household work, especially for girls • Reduced household assets make schooling less affordable, girls probably affected most
3. Promote gender equality and empower women	• As men migrate to seek alternative work, women/girls bear an increased burden of care • Women often bear the brunt of distress 'coping' strategies, e.g. by reducing food intake	• Emergency programmes may reinforce power structures which marginalise women • Domestic and sexual violence may rise in the wake of a disaster
4. Reduce child mortality	• Children are often most at risk, e.g. of drowning in floods • Damage to health, water and sanitation infrastructure • Injury and illness from disaster weakens children's immune systems	• Increased numbers of orphaned, abandoned and homeless children • Household asset depletion makes clean water, food and medicine less affordable
5. Improve maternal health	• Pregnant woman are often at high risk from death/injury in disasters • Damage to health infrastructure • Injury and illness from disaster can weaken women's health	• Increased responsibilities and workload create stress for surviving mothers • Household asset depletion makes clean water, food and medicine less affordable
6. Combat HIV/AIDS, malaria and other diseases	• Poor health and nutrition following disasters weakens immunity • Damage to health infrastructure. Increased respiratory diseases associated with damp, dust and air pollution linked to disaster	• Increased risk from communicable and vector borne diseases, e.g. malaria and diarrhoeal diseases following floods • Impoverishment and displacement following disaster can increase exposure to disease, including HIV and AIDS, and disrupt health care
7. Ensure environmental sustainability	• Damage to key environmental resources and exacerbation of soil erosion or deforestation. Damage to water management and other urban infrastructure • Slum dwellers/people in temporary settlements often heavily affected	• Disaster-induced migration to urban areas and damage to urban infrastructure increase the number of slum dwellers without access to basic services and exacerbate poverty
8. Develop a global partnership for development	• Impacts on programmes for small island developing states from tropical storms, tsunamis etc. • Impacts on commitment to good governance, development and poverty reduction – nationally and internationally	

Table 16.2 Contribution of disaster risk reduction towards meeting the Millenium Development Goals (MDGs) (source: www.unisdr.org/eng/mdgs-drr/dfid.htm)

MDG	Examples of what risk reduction can contribute
1. Eradicate extreme poverty and hunger	• Disaster risk reduction and MDG1 are interdependent. Reducing livelihood vulnerability to natural hazards is key both to eradicating income poverty and improving equity, and to improving food security and reducing hunger. Reducing disaster impacts on the macro economy will promote growth, fiscal stability and state service provision, with particular benefits for the poor • Disaster risk reduction and MDG1 share common strategies and tools: this overlap means that giving development more security from natural hazards can be very cost effective
2. Achieve universal primary education	• In hazard-prone areas, the case for building schools and encouraging attendance becomes much stronger if buildings are safe and students and teachers are trained in emergency preparedness. Promoting safer structures may encourage better maintenance even in non-disaster times • Reduced vulnerability will allow households to invest in priorities other than mere survival. Education is often a high priority. Girls (as 60% of non-attendees) may benefit disproportionately
3. Promote gender equality and empower women	• Better risk reduction will help protect women from disproportionate disaster impacts • Collective action to reduce risk by households and communities provides entry points for women (and other marginalised social groups) to organise for other purposes too, providing a catalyst for economic and social empowerment
4. Reduce child mortality	• Disaster risk reduction will help protect children from direct deaths and injuries during hazard events, and will lower mortality from diseases related to malnutrition and poor water and sanitation following disasters • Health infrastructure and personnel in hazard-prone areas will be better protected. This may also promote better maintenance of infrastructure
5. Improve maternal health	• Disaster-related illness and injury will be reduced • Improved household livelihood and food security will lower women's workloads and improve family nutrition • Health infrastructure and personnel in hazard-prone areas will be better protected. This may also promote better maintenance of infrastructure
6. Combat HIV/AIDS, malaria and other diseases	• Public health risks, e.g. from flood waters, will be reduced, and nutrition and health status improved, boosting resistance to epidemic disease • Fewer disasters will free up social sector budgets for human development • Livelihood security will reduce the need to resort to work in the sex industry • Community organisations and networks working in disaster risk reduction are a resource for family and community health promotion, and vice versa
7. Ensure environmental sustainability	• Reduced disaster-related migration into urban slums and reduced damage to urban infrastructure will improve urban environments • An emphasis on governance for risk reduction and more secure livelihoods will help curb rural and urban environmental degradation • Risk reduction partnerships that include community level actors and concerns will offer more sustainable infrastructure planning, and enable expansion of private sector contributions to reducing disasters • Housing is a key livelihood asset for the urban poor. Disaster risk reduction programmes that prioritise housing will also help preserve livelihoods
8. Develop a global partnership for development	• Creating an international governance regime to reduce risk from climate change and other disasters will help overcome disparities in national negotiating weight • Efforts to build equal global partnerships for risk reduction will have particular relevance for small island developing states and HIPCs • Disaster risk reduction initiatives could promote better public-private partnerships
All MDGs	• Reducing disaster impacts will free up resources, including ODA, to meet MDGs

and implemented to take into account and eliminate possible disaster to the fullest extent. Furthermore, it is important to control the aforementioned variables (McEntire, 2001), which interact to increase disaster vulnerabilities, to achieve disaster risk reduction. In this context, UN/ISDR (2009) defines disaster risk reduction as the concept and practice of reducing disaster risks through systematic efforts to analyse and manage the causal factors of disasters including through reduced exposure to hazards, lessened vulnerability of people and property, wise management of land and the environment, and improved preparedness for adverse events. This definition confirms that invulnerable development or vulnerability management is the prerequisite of disaster risk reduction.

Incidentally, disaster risk reduction through minimising vulnerabilities to avoid (prevention) or to limit (mitigation and preparedness) the adverse impacts of hazards has been identified as a better approach to face disasters than post-disaster responsiveness (Yokohoma Strategy of Disaster Reduction, 1994; Sahni and Ariyabandu, 2003). According to Goodyear (2003), creating a culture of prevention is essential to address the hazards and the adverse consequences of a disaster. Thus, disaster reduction is all about taking measures in advance to address risk reduction involving environmental protection, social equity and economic growth, the three cornerstones of sustainable development, to ensure that development efforts do not increase the vulnerability to hazards (UN/ISDR, 2002). Accordingly, the next section looks into the interrelationship between disaster risk reduction and sustainable development more closely.

16.4.2 *Interchangeable terms: sustainable development goals and vulnerability reduction*

As illustrated in previous sections, development activities have a significant interrelationship with disaster risk reduction since development can increase or/and decrease disaster vulnerability (McEntire, 2004). Therefore, it is essential to take disaster risk reduction into consideration in all development activities. As UNDP (2004) emphasises, although disaster risk is not inevitable, it can be managed and reduced through appropriate development policy and action. Thus, disaster risk reduction policies and measures need to be implemented to build disaster resilient societies and communities, with a twofold aim (UN/ISDR, 2003); to reduce the level of risk in societies, while ensuring, on the other hand, that development efforts do not increase the vulnerability to hazards but instead consciously reduce such vulnerability.

In this context, sustainable development, which consists of activities aimed at economic growth, social equity and environmental protection, has a major contribution to disaster risk reduction as both are aimed at vulnerability reduction (Stenchion, 1997). Sustainable development was defined in 1983 by the World Commission on Environment and Development as the development that meets the needs of the present without compromising the ability of future generations to meet their own needs (Anderson, 1995). Thus, disaster risk reduction is an important requisite for sustainable development (UN/ISDR, 2003) since disasters damage existing development, and consume resources available for future development activities. According to Soubbotina (2004), sustainable development is not only about meeting current needs without compromising the ability of future generations, but is development with equity and balance so that the well-being of people belonging to other groups or living in other parts of the world are not threatened. However, it has been emphasised that sustainable development is not possible without addressing vulnerability to hazards (UN/ISDR, 2003). According to UN/ISDR (2005), sustainable development, poverty reduction, good governance and disaster risk reduction are mutually supportive objectives, and in order to meet the challenges ahead, accelerated efforts must be made to build the necessary capacities at community and national levels to manage and reduce disaster risk.

The close relationship between disaster risk reduction and sustainable development can be drawn through the similarities that have been illustrated between the three main sustainable development goals, i.e. economic growth, social equity and environmental protection, and the aforementioned definition of disaster vulnerabilities by the UN/ISDR (2004) and categorisation of vulnerabilities by the IATF/DR-UN (2006). The three sustainable development goals directly exhibit a close link with reducing economic, social and environmental vulnerabilities, whilst reducing the physical type of vulnerabilities fall into all three sustainable development goals. In particular, although the category of physical vulnerabilities does not appear to be directly linked with any sustainable development goals, reducing these vulnerabilities through reducing susceptibilities of the built environment contributes to all three sustainable development goals. The built environment is the backbone of any development activity. The built environment comprises of everything humanly made, arranged or maintained to fulfil human purposes, i.e. needs, wants and values (Bartuska and Young, 1994). Thus, the built environment is extremely important for sustainable development in which the prime objective is to ensure human well-being and is discussed in more detail below.

16.4.3 *Integrating disaster risk reduction into the built environment and construction*

Since the built environment comprises the substantive physical framework for human society to function in its many aspects – social, economic, political and institutional (Geis, 2000) – its disaster vulnerabilities can bring serious disruption to sustainable development. Improper construction of buildings and infrastructure, or the susceptibilities of the built environment, has always been a major disaster vulnerability which has links to all other types of vulnerabilities and to the variables that increase vulnerabilities. In this regard, Duque (2005) points out that it is the characteristics of the built environment that can be managed to reduce disaster vulnerabilities since disasters occur as a result of hazards intersecting with the built environment, particularly in poorly located and constructed development. The construction industry represents a core component of the built environment.

The destruction of the built environment by disasters has serious implications – injury, death and economic damage to society. As already mentioned, most deaths and injuries in the Iran earthquake of 2003 were caused by the collapse of buildings. According to Scheuren *et al.* (2008), although the death toll was limited to 46, the damage to infrastructure by cyclone Kyrill, which affected some European countries in early 2007, was extensive. It damaged US$9 billion worth of infrastructure such as buildings, harbours, roads, and air and rail transportation infrastructure.

Therefore, it is necessary to integrate disaster risk reduction into the design–construction operation process which consists of four broad phases – preliminary, pre-construction, construction and post-completion – to reduce the vulnerabilities in the built environment (Bosher *et al.*, 2007a). In this context, Table 16.3 shows some issues and measures that need to be considered when integrating disaster risk management into the process. Disaster risk management has been identified by UN/ISDR (2009) as the systematic process of using administrative directives, organisations and operational skills and capacities in order to lessen disaster risk.

Table 16.3   Issues to be considered in integrating disaster risk management into construction (based on Bosher *et al.*, 2007a)

Process phase	Issues to be considered	Measures to be adopted
Preliminary	• What materials are to be used • Where is the development to be built • What type of building/structures are to be built • How the development will be built	• Proactive risk assessment
Pre-construction	• How the service networks (roads, railways, pipelines and cables) could be carefully planned to reduce the risk of widespread failure • Potential susceptibility of modern construction materials and processes to the climate of the future • Potential terrorist targets	• Hazard identification and mapping • Incentives for resilient building designs • Tax breaks for companies that build to hazard resistant standards • Modern standards for building regulations
Construction	• Access given to emergency services during construction	• Information exchanges and liaison between key construction and emergency management personnel
Post-completion	• Retrofitting of buildings and infrastructure at risk • Re-consideration of planning, design and engineering issues associated with the preliminary phase	• Research to appreciate the impact of natural hazards on historical buildings and infrastructure

Further, Bosher *et al.* (2007b) identifies two common types of natural hazards mitigation in the built environment as follows:

- *Structural mitigation*: Such as the strengthening of buildings and infrastructure exposed to hazards (via building codes, engineering design and construction practices, etc.).
- *Non-structural mitigation*: Includes directing new development away from known hazard locations through land use plans and regulations, relocating existing developments to safer areas and maintaining protective features of the natural environment (such as sand dunes, forests and vegetated areas that can absorb and reduce hazard impacts).

It can be seen that these two categories of mitigation address the factors which increase the susceptibilities of the built environment shown by McEntire (2001) such as: the proximity of people and property to triggering agents; improper construction of buildings; and inadequate foresight relating to the infrastructure. Furthermore, Wamsler (2006) states that the construction sector can play an important role in the structural elements of mitigation, while the planners and developers are responsible for non-structural mitigation.

16.5 Summary

Although a disaster emerges as a combination of both hazard/s and vulnerabilities, the controllable component of a disaster is the vulnerabilities. The occurrence of a hazard is almost inevitable. Thus, disaster vulnerabilities which appear mainly in the forms of economic, social, environmental and physical variables need to be minimised to reduce the risk of a disaster.

However, disaster reduction cannot be addressed without referring to the development activities of the world, since development can regulate the level of disaster vulnerabilities. The link between disaster risk and development is extremely strong as disasters can be both a cause and a product of improper development. The impact of disasters has been identified as an enormous barrier to achieve the global development goals since they reverse the development efforts of the world by destroying existing property and available resources for development. On the other hand, the characteristics of improper development such as improper land usage, inappropriate human settlements, misuse of natural resources and deficiencies in regulations increase the risk of disasters by increasing economic, social and environmental vulnerabilities to disasters.

Therefore, sustainable development that minimises the world's vulnerabilities to disasters is extremely important in achieving disaster risk reduction. Sustainable development with its three goals of economic growth, social equity and environmental protection needs to be considered as a parallel interlinked issue with disaster risk reduction since their prime objective is common, i.e. vulnerability reduction.

In the context of disaster risk reduction and sustainable development, reducing disaster vulnerabilities of the built environment becomes significant since the built environment provides the necessary settings for development. Reducing the disaster susceptibilities of these settings reduces the risks of disasters whilst achieving more sustained development through reduced physical exposure. In particular, activities within the construction industry need, as much as is possible, to reduce the risk of disasters and achieve sustainable development, since the industry is responsible for the creation and maintenance of homes and infrastructure which provide services for human well-being.

References

Anderson, M.B. (1995) Vulnerability to disaster and sustainable development: a general framework for assessing vulnerability. In: Munasinghe, M. and Clarke, C. (Eds), *Disaster Prevention for Sustainable Development: Economic and Policy Issues*. Washington, DC: International Decade for Natural Disaster Reduction (IDNDR), pp. 41–59.

Ariyabandu, M.M. and Wickramasinghe, M. (2003) *Gender Dimensions in Disaster Management – A Guide for South Asia*. Colombo: ITDG South Asia.

Bartuska, T.J. and Young, G.L. (1994) The built environment: definition and scope. In: Bartuska, T.J. and Young, G.L. (Eds), *The Built Environment: A Creative Inquiry into Design and Planning*. Menlo Park, CA: Crisp Publications, pp. 3–12.

Bendimerad, F. (2003) Disaster Risk Reduction and Sustainable Development, World Bank Seminar on The Role of Local Governments in Reducing the Risk of Disasters, Istanbul, Turkey, 28 April–2 May 2003.

Bohle, H.G., Downing, T.E. and Watts, M.J. (1994) Climate change and social vulnerability: the sociology and geography of food insecurity. *Global Environmental Change*, **4**, 37–48.

Bosher, L., Dainty, A., Carrillo, P. and Glass J. (2007a) Built-in resilience to disasters: a pre-emptive approach. *Engineering, Construction and Architectural Management*, **14**, 434–446.

Bosher, L., Dainty, A., Carrillo, P., Glass, J. and Price, A. (2007b) Integrating disaster risk into construction: a UK perspective. *Building Research and Information*, **35**, 163–177.

Boulle, P., Vrolijks, L. and Palm, E. (1997) Agency report – vulnerability reduction. *Sustainable Urban Development*, **5**, 179–188.

Commission on Climate Change and Development (2008) *Links Between Disaster Risk Reduction, Development and Climate Change*. Geneva: UN/ISDR.

Cutter, S.L. (1996) Vulnerability to environmental hazards. *Progress in Human Geography*, **20**, 529–539.

DFID (Department for International Development) (2005) *Disaster Risk Reduction: A Development Concern*. London: DFID.

DFID (Department for International Development) (2006) *Reducing the Risk of Disasters – Helping to Achieve Sustainable Poverty Reduction in a Vulnerable World: A DFID Policy Paper*. London: DFID.

Downing, T.E. (1991) Vulnerability to hunger and coping with climate change in Africa. *Global Environmental Change*, **1**, 365–380.

Duque, P.P. (2005) Disaster Management and Critical Issues on Disaster Risk Reduction in the Philippines. International Workshop on Emergency Response and Rescue, Taipei, Taiwan, 31 October–1 November 2005.

El-Masri, S. and Tipple, G. (2002) Natural disaster, mitigation and sustainability: the case of developing countries. *International Planning Studies*, **7**, 157–175.

EM-DAT (The International Disaster Database). http://www.emdat.be/Explanatory Notes/explanotes.html [accessed 12/08/2008].

GAO (United States Government Accountability Office) (2009) Disaster Recovery, Highlights of GAO-09-811, A Report to the Committee on Homeland Security and Governmental Affairs, US Senate, July 2009.

Geis, D.E. (2000) By design: the disaster resistant and quality-of-life community. *Natural Hazards Review*, **1**, 151–160.

Goodyear, E.J. (2003) Risk reduction through managing disasters and crisis. In: Sahni, P. and Ariyabandu, M.M. (Eds), *Disaster Risk Reduction in South Asia*. New Delhi: Prentice Hall of India Private Limited, pp. 44–56.

Martin, B. (2002) Are disaster management concepts relevant in the developing countries: the case of the 1999–2000 Mozambican floods. *The Australian Journal of Emergency Management*, **16**, 25–33.

McEntire, D.A. (2001) Triggering agents, vulnerabilities and disaster reduction: towards a holistic paradigm. *Disaster Prevention and Management*, **10**, 189–196.

McEntire, D.A. (2004) Development, disasters and vulnerability: a discussion of divergent theories and the need for their integration. *Disaster Prevention and Management*, **13**, 93–198.

Royal Geographical Society (2004) Twin Quakes, Geography in the News. http://www.geographyinthenews.rgs.org/news/article/default.aspx?id=251 [accessed 31/01/2009].

Sahni, P. and Ariyabandu, M.M. (2003) Introduction. In: Sahni, P. and Ariyabandu, M.M. (Eds), *Disaster Risk Reduction in South Asia*. New Delhi: Prentice Hall of India Private Limited, pp. 1–25.

Scheuren, J.M., de Waroux, O. le P., Below, R., Guha-Sapir, D. and Ponserre, S. (2008) *Annual Disaster Statistical Review – The Numbers and Trends 2007*. Brussels: Centre for Research on the Epidemiology of Disasters.

Soubbotina, T.P. (2004) *Beyond Economic Growth: An Introduction to Sustainable Development*, 2nd edn. Washington DC: The World Bank.

Stenchion, P. (1997) Development and disaster management. *Australian Journal of Emergency Management*, **12**, 40–44.

UN/ISDR (United Nations International Strategy for Disaster Reduction) (1994) Yokohama Strategy and Plan of Action for a Safer World – Guidelines for Natural Disaster Prevention, Preparedness and Mitigation, World Conference on Natural Disaster Reduction, 23–27 May 1994, Yokohama, Japan.

UN/ISDR (United Nations International Strategy for Disaster Reduction) (2002) *Gender Mainstreaming in Disaster Reduction*. Geneva: UN/ISDR.

UN/ISDR (United Nations International Strategy for Disaster Reduction) (2003) *Disaster Reduction and Sustainable Development*. Geneva: UN/ISDR.

UN/ISDR (United Nations International Strategy for Disaster Reduction) (2004) Terminology: Basic Terms of Disaster Risk Reduction. http://www.unisdr.org/eng/library/lib-terminology-eng%20home.htm [accessed 18/06/2007].

UN/ISDR (United Nations International Strategy for Disaster Reduction) (2005) Hyogo Framework for Action 2005–2015: Building the Resilience of Nations and

Communities to Disasters, Final Report of the World Conference on Disaster Reduction. Geneva: UN/ISDR.

UN/ISDR (United Nations International Strategy for Disaster Reduction) (2009) *2009 UN/ISDR Terminology on Disaster Risk Reduction*. Geneva: UN/ISDR.

UN/ISDR (United Nations International Strategy for Disaster Reduction), Millennium Development Goals and Disaster Risk Reduction. http://www.unisdr.org/eng/mdgs-drr/dfid.htm [accessed 18/12/2009].

UNDP (2004) *Reducing Disaster Risk: A Challenge for Development*. Geneva: UNDP.

Wamsler, C. (2006) Mainstreaming risk reduction in urban planning and housing: a challenge for international aid organizations. *Disasters*, **30**, 151–177.

Weichselgartner, J. (2001) Disaster mitigation: the concept of vulnerability revisited. *Disaster Prevention and Management*, **10**, 85–94.

Working Group on Climate Change and Disaster Risk Reduction of the Inter-Agency Task Force on Disaster Reduction (IATF/DR) (2006) *On Better Terms – A Glance at Key Climate Change and Disaster Reduction Concepts*. Geneva: United Nations.

17 Conclusion

Richard Haigh and Dilanthi Amaratunga

The built-environment industries are usually associated with a range of critical activities in post-disaster recovery, including temporary shelter and housing after the disaster, and the restoration of critical infrastructure such as hospitals, schools, water supply, power and communications. However, in order to achieve the challenge laid out in the opening chapter of this book – to create a more resilient built environment that can contribute to broader societal resilience – the impact of reconstruction, positively or negatively, needs to be evaluated far more carefully. Disaster planners have begun to realise the link between disaster and development – a large and well-established field relating to physical, social, natural and economic aspects of society. Although reconstruction of the built environment by itself will not eliminate the broad ranging consequences of disasters, there is increasing recognition that the reconstruction process can contribute to the development of communities beyond merely the building of their physical environment.

This potential contribution of the reconstruction process to the broader goal of a more resilient society can be viewed with the aid of the Asset-Based Community Development (ABCD) approach, developed by Kretzmann and McKnight (1993) as a methodology that seeks to uncover and highlight the strengths within communities as a means for sustainable development. The basic tenet is that a capacities-focused approach is more likely to empower a community and therefore mobilise citizens to create positive and meaningful change from within. Instead of focusing on a community's needs, deficiencies and problems, the ABCD approach helps them become stronger and more self-reliant by discovering, mapping and mobilising all their local assets. Few people realise how many assets any community has. As highlighted by many of the authors' contributions to this book, the reconstruction process has the potential to utilise and impact, positively or negatively, a community's assets. The construction and maintenance of a community's infrastructure and buildings, or physical assets, are the first obvious impact, as outlined by Keraminiyage's discussion on the links between the restoration of major infrastructure and rehabilitation of communities in Chapter 13. These physical assets

Post-Disaster Reconstruction of the Built Environment: Rebuilding for Resilience, First Edition.
Edited by Dilanthi Amaratunga and Richard Haigh.
© 2011 Blackwell Publishing Ltd. Published 2011 by Blackwell Publishing Ltd.

address material needs (infrastructure, water, housing, waste, energy, transport, work), social and educational needs (schools, play areas, meeting places), and, spiritual or cultural needs (places of worship). As noted by Palliyaguru and Amaratunga in Chapter 15, reconstruction of the physical environment is vital to secure sustainable economic development. Further, by incorporating appropriate mitigation measures, effective reconstruction of the physical environment is also an opportunity to reduce the community's vulnerability to hazards in the future, a point emphasised in Chapter 16 by Ginige's consideration of the links between disaster risk reduction and sustainable development. Several of the chapters highlighted the challenge of securing sufficient capacity or resources to deliver all this reconstruction activity. In Chapter 2, Ginige and Amaratunga provided a broad overview of the capacity gaps typically encountered in reconstruction projects following major disasters, while further specific examples from Sri Lanka and China were described by Seneviratne in Chapter 3, and by Chang, Wilkinson, Potangaroa and Seville in Chapter 4 respectively. However, the challenge posed by the scale of reconstruction can also be viewed as an opportunity: to develop livelihood competencies or human assets in construction-related trades. As Kulatunga outlines in Chapter 8, this human asset development is not just required at the trade level; project management and professional skills are also vital. This opportunity can help address a problem frequently encountered following a disaster, particularly in the type of conflict-affected environment described by Amaratunga and Seneviratne in Chapter 10: how to develop the skills of displaced peoples and ex-combatants who, for a variety of reasons, are unable to return to their original livelihoods?

Reconstruction also enables the development of economic assets within the community through opportunities to initiate market linkages in the construction supply chain. As Sutton and Haigh caution in Chapter 11, excessive reliance on external private enterprises can be counterproductive and hinder local economic development. Local intermediary and long-term income generation opportunities provided by reconstruction activity may lay an important platform for economic development in the region. Micro, small and medium enterprises are a vital component of any economy and the construction sector is largely comprised of micro and small enterprises. Reconstruction is thus an opportunity to provide market access for these local enterprises.

A community's natural assets are frequently impacted by reconstruction. On the positive side, locally sourced and contextually appropriate materials can provide an important contribution to reconstruction while also offering market opportunities for local businesses. It is, however, vital to consider the community's long-term sustainability and thereby ensure that its natural assets are not exploited to the detriment of the community. For example, in Chapter 14 Karunasena highlights the challenge of construction waste, but also the opportunities for reuse and recycling. Protection of the natural environment is an imperative to avoid making the community vulnerable to other natural hazards in the future.

Fundamental to the recovery of any disaster-affected community is the idea of connectedness. There is growing evidence that collective reconstruction contributes to social cohesion and builds social assets. Reconstruction is an opportunity for cooperation and working across diverse groups, which is particularly useful in conflict-affected environments, as noted by Amaratunga and Seneviratne in Chapter 10.

Also, as illustrated in Chapters 5 and 6 by Thurairajah and Ophiyandri respectively, engaging the community in reconstruction has the added benefit of moving them away from being passive recipients of aid, which can increase the sense of helplessness.

Finally, reconstruction can impact a community's institutional assets. In Chapter 9 Rotimi, Wilkinson and Myburgh warn that the organisation and coordination of recovery is usually complex because a wide range of activities occur simultaneously with an equally wide range of needs that have to be met, including those of the most vulnerable members of the community. Reconstruction can provide members of a community with an opportunity to influence policies, decisions and interventions that affect them, including assessment, planning, construction and monitoring. Further, the community can develop links to important stakeholders, a point emphasised by Siriwardena and Haigh in Chapter 7. In summary, the reconstruction process will have a far greater impact on the affected community than the physical buildings and infrastructure. An asset-based approach does not remove the need for outside resources. However, it will make their use more effective. It will also go a long way to creating the type of resilient society that was put forward in Chapter 1 as a goal to aspire to. Indeed many of the characteristics of resilience that were presented in the book's opening chapter can be developed through an assets-based approach. Asset-based community development starts with what is present in the community. It concentrates on the agenda-building and problem-solving capacity of the residents and stresses local determination, investment, creativity and control. Weak communities are places that fail to mobilise the skills, capacities and talents of their residents or members. As highlighted in many of the earlier chapters, ignoring a community's assets during reconstruction may inadvertently leave communities more dependent and ultimately less resilient to the threat posed by future hazards. In contrast, post-disaster reconstruction programmes where the capacities of the community are identified, valued and used, will lay the platform for a more resilient society.

Reference

Kretzmann, J. P. and McKnight, J. L. (1993) *Building Communities from the Inside Out*. Evanston, IL: Asset-Based Community Development Institute, Northwestern University.

Index

Page numbers in *italics* denote figures, those in **bold** denote tables.

Post-Disaster Reconstruction of the Built Environment: Rebuilding for Resilience, First Edition.
Edited by Dilanthi Amaratunga and Richard Haigh.
© 2011 Blackwell Publishing Ltd. Published 2011 by Blackwell Publishing Ltd.